The Interface Structure and Electrochemical Processes at the Boundary Between Two Immiscible Liquids

Edited by Vladimir E. Kazarinov

With Contributions by
L.I. Boguslavsky, T. Kakiuchi, T. Kakutani, V.E. Kazarinov,
Yu.I. Kharkats, Z. Koczorowski, J. Koryta, V.S. Krylov, M.G. Kuzmin,
A.M. Kuznetsov, V. Mareček, T. Osakai, A.N. Popov, Z. Samec,
M. Senda, A.G. Volkov, N.K. Zaitsev

With 109 Figures and 17 Tables

Springer-Verlag Berlin Heidelberg New York
London Paris Tokyo

Editor:

Professor Dr. Vladimir E. Kazarinov

A.N. Frumkin Institute of Electrochemistry
Academy of Sciences of the USSR
Leninsky Prospect, 31
SU-117071 Moscow V-71, USSR

Library of Congress Cataloging-in-Publication Data
The Interface structure and electrochemical processes at the boundary between two immiscible liquids.
Includes bibliographies and index.
1. Electrochemistry. 2. Surface chemistry. 3. Liquids. I. Kazarinov, V. E. II. Boguslavskiĭ, L. I. (Leonid
Isaakovich) III. Title: Immiscible liquids.
QD553.I48 1987 541.3′7 87-12755

ISBN-13: 978-3-642-71883-0   e-ISBN-13: 978-3-642-71881-6
DOI: 10.1007/978-3-642-71881-6

© Springer-Verlag Berlin Heidelberg 1987
Softcover reprint of the hardcover 1st edition 1987

Typesetting: Brühlsche Universitätsdruckerei, Giessen

2152/3020-543210

List of Authors

Professor Leonid I. Boguslavsky
A. N. Frumkin Institute of Electrochemistry, Academy of Sciences of the USSR
Leninsky Prospect, 31, SU-117071 Moscow V-71, USSR

Professor Takashi Kakiuchi
Department of Agricultural Chemistry and Research Center for Cell
and Tissue Culture, Faculty of Agriculture,
Kyoto University, Kyoto 606, Japan

Professor Tadaaki Kakutani
Department of Agricultural Chemistry and Research Center for Cell
and Tissue Culture, Faculty of Agriculture,
Kyoto University, Kyoto 606, Japan

Professor Vladimir E. Kazarinov
A. N. Frumkin Institute of Electrochemistry, Academy of Sciences of the USSR
Leninsky Prospect, 31, SU-117071 Moscow V-71, USSR

Professor Yurij I. Kharkats
A. N. Frumkin Institute of Electrochemistry, Academy of Sciences of the USSR
Leninsky Prospect, 31, SU-117071 Moscow V-71, USSR

Professor Zbigniew Koczorowski
Department of Chemistry, University of Warsaw
ul. Pasteura 1, PL-02-093 Warsaw, Poland

Professor Jiří Koryta
J. Heyrovský Institute of Physical Chemistry and Electrochemistry,
Czechoslovak Academy of Sciences
Opletalova 25, CS-11000 Prague 1, Czechoslovakia

Professor Valentin S. Krylov
A. N. Frumkin Institute of Electrochemistry, Academy of Sciences of the USSR
Leninsky Prospect, 31, SU-117071 Moscow V-71, USSR

Professor Michael G. Kuzmin
    Department of Chemistry, Moscow State University
    Leninskie Gory, SU-117234 Moscow, USSR

Professor Alexander M. Kuznetsov
    A. N. Frumkin Institute of Electrochemistry, Academy of Sciences of the USSR
    Leninsky Prospect, 31, SU-117071 Moscow V-71, USSR

Professor Vladimir Mareček
    J. Heyrovský Institute of Physical Chemistry and Electrochemistry,
    Czechoslovak Academy of Sciences
    CS-10200 Prague 10-Hostivar, Czechoslovakia

Professor Toshiyuki Osakai
    Department of Agricultural Chemistry and Research Center for Cell
    and Tissue Culture, Faculty of Agriculture,
    Kyoto University, Kyoto 606, Japan

Professor Alexander N. Popov
    Institute of Inorganic Chemistry, Academy of Sciences of the Latvian SSR
    ul. Miera, 34, SU-Salaspils, USSR

Professor Zdeněk Samec
    J. Heyrovský Institute of Physical Chemistry and Electrochemistry,
    Czechoslovak Academy of Sciences
    CS-10200 Prague 10-Hostivar, Czechoslovakia

Professor Mitsugi Senda
    Department of Agricultural Chemistry and Research Center for Cell
    and Tissue Culture, Faculty of Agriculture,
    Kyoto University, Kyoto 606, Japan

Professor Alexander G. Volkov
    A. N. Frumkin Institute of Electrochemistry, Academy of Sciences of the USSR
    Leninsky Prospect, 31, SU-117071 Moscow V-71, USSR

Professor Nikolay K. Zaitsev
    Department of Chemistry, Moscow State University
    Leninskie Gory, SU-117234 Moscow, USSR

Preface

Studies on the electrochemical processes at the interface between two immiscible liquids began a long time ago: they date back to the end of the last century. Such celebrated scientists as Nernst and Haber, and also young A. N. Frumkin were among those who originated this science. Later A. N. Frumkin went a long way in furthering the studies at the Institute of Electrochemistry. The theory of the appearance of potential in a system of two immiscible electrolytes was developed and experimentally verified before the beginning of the thirties. In later years the studies in this area considerably lagged behind those conducted at metal electrodes which were widely used in different industries. In the past 15 years, however, the situation has radically changed and we have witnessed a drastic increase in the number of publications on the electrochemistry of immiscible electrolytes. We are glad to note that the investigations show not only a quantitative but also a qualitative change. The theoretical works on the oil/water interface test not only the thermodynamic aspects of the interface but also recreate the molecular picture of the process. Along with the now conventional oil/water system, electrochemical studies are made on various membranes, including the finest bilayer lipid membranes, and also on microemulsion systems. A prominent place in the investigation of the oil/water interface is occupied by photoprocesses that come into play at the interface between two ionic conductors.

In short, studies on the oil/water interface are making rapid advances and we may expect new achievements both in the theory and in its various applications. I hope that these achievements will not be long in coming, especially if the cooperation between scientists in different countries continues to make good progress. The first result of this friendly cooperation is the present volume.

Moscow, June 1987   V. E. Kazarinov

Contents

**Galvani and Volta Potentials at the Interface Separating
Immiscible Electrolyte Solutions**
Z. Koczorowski 77

**Electrocapillarity and the Electric Double Layer Structure
at Oil/Water Interfaces**
M. Senda, T. Kakiuchi, T. Osakai, T. Kakutani 107

Introduction

V. E. Kazarinov

The electrochemical reaction proceeding at the interface between immiscible liquids – ion conductors – is an essential aspect of the most various technical procedures as well as of natural phenomena observed in the biological world. It is impossible to conceive of a modern industry that would not involve extraction processes. Today, ion-selective electrodes are used not only in research for control and monitoring of ionic media, but also in medical practice, chemical industry and in environmental protection. Finally, animal respiration, photoprocesses in plants providing for light energy storage, transfer of a nervous impulse and visual reaction – all are due to the existence of extensive chains of coupled chemical reactions and, last but not least, to the occurrence at the membrane/electrolyte interface of chemical processes involving charge separation.

At present it seems that the elucidation of the physico-chemical principles of processes occurring at the interface between immiscible liquids, particularly their kinetics and corresponding effects of the interface structure, is the main task of the electrochemistry of immiscible liquids.

Our insight into the main features of the processes at the interface between immiscible liquids determines the progress in those areas where these processes control complex chains of transformations. Two mutually complementing lines of research, have developed:

– Fundamental studies of interface structure and of processes occurring at the interface between immiscible liquids.
– Use of the interface for investigating important processes for chemical engineering, medicine, and biology.

This division is of course a rather arbitrary one, which is quite obvious from the contents of the present monograph. The book opens with an article by Prof. Koryta, who in a concise and sufficiently detailed form elucidates certain main stages in the development of research on the electrolysis at the interface between two immiscible electrolytes. The pioneer work of Prof. Koryta and his coworkers forms a new branch in the Czech electrochemical school, and has not only essentially contributed to the investigation of electrolysis at the interface between immiscible electrolytes but has also fostered the interest in these systems throughout the world.

The next two chapters have been written by scientists of the A.N. Frumkin Institute of Electrochemistry. Chapter 2 is concerned with the quantum theory of charge transfer across the interface between immiscible liquids. This is one of the traditional

The Interface Structure and Electrochemical Processes at the Boundary Between Two Immiscible Liquids
Editor: V. E. Kazarinov
© Springer-Verlag Berlin, Heidelberg 1987

fields of theoretical studies carried out at this Institute. Here, the concepts developed earlier for the metal/electrolyte interface have been extended in an appropriately modified form to a system of two immiscible liquids. This chapter ought to both stimulate new ideas for experiments and show suitable directions for choosing parameters that may contribute to improvement of extraction processes.

Chapter 3, written by the late V. S. Krylov, reveals the thermodynamic aspect of the charge-transfer process in oil/water systems. Mass transfer being in the majority of cases controlled by diffusion to the interface, the intensification of this process involves first of all achievement of a more intensive mass transfer. Thus, both chapters deal with the factors controlling charge – and mass transfer under kinetic and diffusion conditions.

The subsequent three chapters are devoted to the electric double-layer structure at the interface between immiscible electrolytes examined by the electrocapillary curves method (Prof. Senda and coauthors) and by measurement of the electric double-layer capacity (Dr. Samec and Dr. Mareček) as well as to the investigation of the Galvani and Volta potentials in the above-mentioned systems (Prof. Koczorowski). These chapters will be of interest to many electrochemists since the results obtained here are comparable with the thoroughly studied metal/electrolyte solution interface. An insignificant potential shift in the compact layer at the interface between immiscible electrolytes in the absence of specific ion adsorption – this is the main conclusion arrived at by the authors of Chaps. 4 and 5. Chapter 6 deals with the scale of potentials in a system of immiscible electrolytes and the thermodynamic relation between the distribution coefficients and the Volta potentials.

The last four chapters cover various special cases in which the interface between immiscible liquids is used.

In Chap. 7, L. I. Boguslavsky and A. G. Volkov consider various redox reactions occurring at the interface between immiscible liquids. Here, the octane/water interface is taken as a biological membrane model. With its use it is possible to observe redox processes and their coupling with other reactions, involving not only the simplest ingredients but also the respiratory chain enzymes.

Chapter 8, written by M. G. Kuz'min and N. K. Zaitsev, is concerned with the specific case of the photoseparation of charges in micellar systems. The particular importance of this specific case is due to the fact that these systems have an enormous interface but a small total volume.

In Chap. 9. A. N. Popov discusses the most important conditions at the oil/water interface affecting the extraction characteristics of the system. The author has proved convincingly the existence of a relationship between the surface activity of extractants during adsorption from the nonaqueous phase at the oil/water interface and the extraction characteristics of the system.

In conclusion, I would like to note that as a matter of course the book, devoted to the structure of the interface between immiscible liquids and the electrochemical processes occurring at this interface, could not cover all the studies on the problem. I hope, however, that at the present moment it will be helpful and promote progress in this new field of electrochemistry.

Electrolysis at the Interface
Between Two Immiscible Electrolyte Solutions

J. Koryta

The search for models of biological membranes started at the end of the 19th century. This general interest in membranes was expressed by the famous German physico-chemist Wilhelm Ostwald who wrote: "Not only mysterious phenomena of electric fish but also processes occurring in muscles and nerves will be, in the future, explained by semipermeable membranes" [1].

The first important investigation of a liquid membrane and, at the same time, of the interface of two immiscible electrolyte solutions (ITIES) was carried out by Nernst and Riesenfeld [2] at the beginning of this century. They measured not only the electrical potential difference between both the phases but also the effect of current flow across ITIES. Their ITIES was represented by a boundary between an aqueous solution and a solution in an organic solvent. Their main interest was not, however, in the current-potential characteristics but mainly in the proportion of cations and anions carrying the charge across ITIES. On the basis of their theory they could measure experimentally the transport numbers in the organic phase.

Theory

The further work on this subject connected with the names of Beutner [3], Bonhoeffer, Kahlweit and Strehlow [4] and Karpfen and Randles [5], was mainly devoted to equilibrium potentials at ITIES, which can be described by the Nernst equation, derived in the following way. At the water/organic solvent (immiscible with water) phase boundary (w/o), an equilibrium exists between a particular univalent *cationic* species I in both the phases, described by equality of electrochemical potentials,

$$\tilde{\mu}_i(w) = \tilde{\mu}_i(o), \tag{1}$$

$$\varphi(w) - \varphi(o) = \Delta_o^w \varphi = (\mu_i^0(o) - \mu_i^0(w))/F$$
$$+ RT/F \ln(a_i(o)/a_i(w)), \tag{2}$$

where the φ's are the inner electrical potentials, the μ_i^0's the standard chemical potentials and the a_i's the activities of the ionic species I in both the phases.

As pointed out by Gibbs already in 1875 it is not possible to determine a difference of electrical potentials of two chemically different phases by a purely thermodynamic procedure. However, several authors suggested extrathermodynamic approaches in

The Interface Structure and Electrochemical Processes
at the Boundary Between Two Immiscible Liquids
Editor: V. E. Kazarinov
© Springer-Verlag Berlin, Heidelberg 1987

order to overcome this difficulty. Particularly important is the suggestion put forward by Parker [6] which will now be discussed in more detail.

For a distribution coefficient for a substance x between the phases o and w we have

$$k_x^{w,o} = a_x(o)/a_x(w) = \exp[(\mu_x^0(w) - \mu_x^0(o))/RT]$$
$$= \exp(\Delta G_{tr,x}^{0;o \to w}/RT) \tag{3}$$

where $\Delta G_{tr,x}^{0;o \to w}$ is the standard Gibbs energy of transfer of x from o to w. The distribution coefficient of the univalent electrolyte BA is given by the equation

$$k_{BA}^{w,o} = a_\pm(o)/a_\pm(w) = \left(\frac{a_+(o)a_-(o)}{a_+(w)a_-(w)}\right)^{1/2}$$
$$= \exp(\Delta G_{tr,BA}^{0;o \to w}/RT). \tag{4}$$

So far we have used a purely thermodynamic approach. The present problem is how to split the value of the standard transfer Gibbs energy of the salt BA as a whole into two contributions, belonging to the cation B^+ and to the anion A^-, which are defined by means of the ionic distribution coefficients

$$k_{B^+}^{w,o} = a_+(o)/a_+(w) = \exp[(\mu_{B^+}^0(w) - \mu_{B^+}^0(o))/RT]$$
$$= \exp(\Delta G_{tr,B^+}^{0;o \to w}/RT)$$
$$k_{A^-}^{w,o} = a_-(o)/a_-(w) = \exp[(\mu_{A^-}^0(w) - \mu_{A^-}^0(o))/RT]$$
$$= \exp(\Delta G_{tr,A^-}^{0;o \to w}/RT) \tag{5}$$

Thus,

$$2\Delta G_{tr,BA}^{0,o \to w} = \Delta G_{tr,B^+}^{0,o \to w} + \Delta G_{tr,A^-}^{0,o \to w}. \tag{6}$$

According to Parker [6] the standard Gibbs energy for transfer of the tetraphenylarsonium ion (TPAs$^+$) and of the tetraphenylborate ion (TPB$^-$) are equal for any pair of solvents. The standard Gibbs energy for transfer of the TPAs$^+$ and TPB$^-$ ions can be determined from the distribution coefficients between any pair of immiscible solvents. If the distribution coefficient for the TPAsA salt is found for any arbitrary ion A^-, then its standard Gibbs transfer energy is

$$\Delta G_{tr,A}^{0,o \to w} = 2\Delta G_{tr,TPAsA}^{0,o \to w} - \Delta G_{tr,TPAs}^{0,o \to w} \tag{7}$$

In electrical units

$$\Delta G_{tr,i}^{0,o \to w} = -z_i F \Delta_o^w \varphi_i^0 \tag{8}$$

so that the Nernst equation acquires the general form

$$\Delta_o^w \varphi = \Delta_o^w \varphi_i^0 + RT/z_i F \ln(a_i(o)/a_i(w)) \tag{9}$$

By means of this approach we can define the electrical potential difference between the phases w and o, in spite of the fact that they are chemically different. We can also list the standard Gibbs transfer energies of individual ions as well as the standard electrical potential differences between the phases concerned for these ions in the same way as, for example, the hydrogen scale for standard electrode potential is composed.

The standard transfer Gibbs energies help us to rationalise the concepts of hydrophobicity and hydrophility of ions. First, these concepts are not only connected with the ions, but, at the same time, with the pair of solvents concerned (i.e. water and the given organic solvent, as for example, nitrobenzene, 1,2-dichlorethane, dichloromethane, o-nitrophenyloctylether etc., which have been mainly used in investigations of electrolysis at ITIES). Second, these properties are directly linked to the position of the standard Gibbs energy for a given ion on the scale.

A remarkable situation is observed in the case where a single salt is in a distribution equilibrium between the solvents w and o. With respect to Eqs. (2) and (5) we have for both the cation and the anion

$$\Delta_o^w \varphi = -\Delta G_{tr,B^+}^{0;o \to w}/F + RT/F \ln[a_+(o)/a_+(w)]$$

$$= \Delta G_{tr,A^-}^{0;o \to w}/F - RT/F \ln[a_-(o)/a_-(w)]. \tag{10}$$

For a dilute solution, $a_+(o) \approx a_-(o)$ and $a_+(w) \approx a_-(w)$ giving the result for the distribution potential [5]

$$\Delta_o^w \varphi_{distr} = (\Delta G_{tr,A^-}^{0;o \to w} - \Delta G_{tr,B^+}^{0;o \to w})/F. \tag{11}$$

Obviously, if both the cation and the anion have the same hydrophobicity (or hydrophility) $\Delta_o^w \varphi_{distr} = 0$. If the cation is more hydrophobic than the anion ($\Delta G_{tr,A^-}^{0;o \to w} > \Delta G_{tr,B^+}^{0;o \to w}$), the distribution potential is negative and vice versa. This finding is directly connected with the charge in the electric double-layer formed at ITIES. When the cation is more hydrophobic that the anion, for example, it has the tendency to get across ITIES from the aqueous phase to the organic phase while the anion, being more hydrophilic, stays preferentially in the aqueous phase. They cannot, of course, be transferred into the bulk of these phases as this would violate the electroneutrality condition but they can remain at ITIES forming thus the charged envelopes of the electric double-layer.

The rule for the electric double-layer at the metal/electrolyte interface is also valid for ITIES, that the surface charges in the aqueous side and in the non-aqueous side of the electric double-layer must be equal in absolute values and opposite in signs.

The double-layer structure at ITIES shows different features than that formed at the metal/electrolyte interface. The charge distribution in the both phases preserves its diffuse property even in rather concentrated solutions [7]. The potential difference in the compact double-layer is much smaller than the potential differences in the adjacent diffuse layers. Thus, practically, the overall potential difference only consists of the potential difference in the diffuse layers $\phi_2(w)$ and $\phi(o)$

$$\Delta_o^w \varphi = \varphi_2(w) + \varphi_2(o). \tag{12}$$

This behaviour was already predicted by Verwey and Niessen [7] as early as in 1939. They described theproperties of the electric double-layer at ITIES by the simple Gouy-Chapman model. For potential differences $|\Delta_o^w \varphi| \gg RT/F$ the potential difference in the diffuse layer in the aqueous phase

$$\varphi_2(w) = \frac{1}{2}\Delta_o^w \varphi \mp RT/2F \ln \frac{\varepsilon(w)c(w)}{\varepsilon(o)c(o)} \tag{13}$$

and in the organic phase

$$\varphi_2(\text{o}) = \tfrac{1}{2}\Delta_\text{o}^\text{w}\varphi \pm \frac{RT}{2F}\left[\ln\frac{\varepsilon(\text{w})c(\text{w})}{\varepsilon(\text{o})c(\text{o})}\right] \tag{14}$$

where the ε's are permittivities and the c's univalent electrolyte concentrations in each phase.

If the basic laws of electrocapillarity should be valid for ITIES it should show the basic phenomena like interfacial tension γ, surface charge q and differential capacity being linked together by the Lippmann equation

$$\frac{\partial\gamma}{\partial\Delta_\text{o}^\text{w}\varphi} = -q(\text{w}) = q(\text{o}). \tag{15}$$

This has been proved by Senda and his coworkers [8] who have shown that the dependence of $\Delta_\text{o}^\text{w}\varphi$ on $q(w)$ is the same when the surface charge is determined by differentiation of the interfacial tension or by integration of the differential capacity.

The kinetics of ion transfer across ITIES is closely linked to the double-layer structure. The concentration of the ion transferred in each phase close to the plane dividing the both phases is related to the concentration outside the space charge region by the Boltzmann formula, for example, for the aqueous phase (cf. [9])

$$c_i(\text{w},0) = c_i(\text{w})\exp[-z_i F\varphi_2(\text{w})/RT] \tag{16}$$

where $c_i(w)$ is the concentration of the species I outside the diffuse layer (this quantity may be influenced by transport processes in the bulk of the phase). Further, $\varphi_2(w)$ is the potential difference between a point outside the space charge region and the interface which can be calculated using the simple Gouy-Chapman theory.

Let us use as an intuitive approach [10] to the problem, the Butler-Volmer-Frumkin equation for the current density j

$$j = z_i F[k_i^{\text{w}\to\text{o}}c_i(\text{w},0) - k_i^{\text{o}\to\text{w}}c_i(\text{o},0)] \tag{17}$$

with

$$\begin{aligned}
&k_i^{\text{w}\to\text{o}}k_i^0\exp\{z_i F[\Delta_\text{o}^\text{w}\phi - \varphi_2(\text{w}) - \Delta_\text{o}^\text{w}\varphi_i^0]/RT\}, \\
&k_i^{\text{o}\to\text{w}} = k_i^0\exp\{-(1-\alpha)z_i F[\Delta_\text{o}^\text{w}\varphi - \varphi_2(\text{o}) - \Delta_\text{o}^\text{w}\varphi_i^0]/RT\}
\end{aligned} \tag{18}$$

where α is the true charge transfer coefficient and $\varphi_2(w)$ and $\varphi_2(o)$ are obtained from the Gouy-Chapman theory.

For $|\Delta_\text{o}^\text{w}\varphi| \gg RT/F$ we obtain

$$\begin{aligned}
j = {}& z_i F\left(\frac{\varepsilon(\text{o})c(\text{o})}{\varepsilon(\text{w})c(\text{w})}\right)^{\pm 1/2}\exp[(\tfrac{1}{2}-\alpha)(z_i F\Delta_\text{o}^\text{w}\varphi_i^0/RT)] \\
&\times \{c_i(\text{w})\exp[z_i F/2RT(\Delta_\text{o}^\text{w}\varphi - \Delta_\text{o}^\text{w}\varphi_i^0)] \\
&- c_i(\text{o})\exp[-z_i F/2RT(\Delta_\text{o}^\text{w}\varphi - \Delta_\text{o}^\text{w}\varphi_i^0)]\}.
\end{aligned} \tag{19}$$

The apparent charge transfer coefficient 1/2 comes from the Gouy-Chapman theory and has nothing to do with the symmetry of the potential barrier for ion transfer. The true transfer coefficient α appears in the equation for the transfer rate constants corrected for the potential difference in the diffuse double-layer

$$k_{i,\text{true}}^{w\to o} = k_i^0 \exp(-\alpha z_i F \Delta_o^w \varphi_i^0 / RT)$$
$$= k_i^0 \exp(\alpha \Delta G_{\text{tr}}^{0,\,o\to w} / RT). \tag{20}$$

Experimental results show that, assuming k_i^0 is constant, the true transfer coefficient seems to remain constant with the approximate value $\alpha = 0.5$ [11].

The membranologists denote as *facilitated transport* [12] a definite sort of mass transfer across the membrane which is enhanced by a chemical reaction in the phase towards which this transport is directed. Facilitated ion transfer is also observed among processes taking place at ITIES [13].

In cases so far investigated the chemical reaction takes place in the organic phase and is mainly represented by complex formation. The ions transferred are usually alkali or alkaline earth metals. The complex formers are ionophores, i.e. substances that are capable of assisting these ions in traversing biological membranes and their models. Practically all of these species have also found application in the field of ion-selective electrodes [14]. They include either neutral carriers (valinomycin, macrotetrolides, crown polyethers, synthetic acyclic substances introduced by Simon and his coworkers [15] to ISEs) or acid anions like monensin or nigericin. A typical ionophore comprises in its structure an internal cavity equipped with polar groups which helps to accommodate a metal ion with suitable size and charge inside the cavity. The outside of the ionophore is hydrophobic by which the solubilisation of the complex in a rather low polarity medium is achieved. As ionophores certain acyclic substances are also considered which are able, in the course of complex formation, to form the internal cavity of identical properties as in the preceding case.

The effect of transfer facilitation can be shown using the Nernst equation (9). Let the univalent metal ion I participate in the reaction of complex formation in the organic phase

$$I + X \rightleftarrows IX \tag{21}$$

characterised by the stability constant

$$K_{ix} = \frac{a_{ix}(o)}{a_i(o)a_x(o)} \tag{22}$$

Equation (9) is then transformed to

$$\Delta_o^w \varphi = \Delta_o^w \varphi_i^0 - \frac{RT}{F} \ln[K_{ix}a_x(o)]$$
$$+ \frac{RT}{F} \ln[a_{ix}(o)/a_i(w)] \tag{23}$$

Certain hydrophobic anions function as proton carriers like, for example, dipicrylaminate [17]. The facilitation effect is then described by Eq. (23) as well.

The antibiotic monensin functions simultaneously as a sodium ion carrier and as a proton carrier [18]. In this case Eq. (23) is modified to the form

$$\Delta_o^w \varphi = \text{const} + \frac{RT}{F} \ln[k_1/a_i(w) + k_2/a_H(w)] \tag{24}$$

where $a_i(w)$ is the sodium activity in the aqueous phase, $a_H(w)$ the hydrogen ion activity in the aqueous phase and

$$k_1 = \exp(F\Delta_o^w \varphi_i^0/RT)K_1^{-1},$$
$$k_2 = \exp(F\Delta_o^w \varphi_H^0/RT)K_2^{-1} \qquad\qquad (25)$$

with K_1 and K_2 denoting equilibrium constants of formation of the complex, $NaHX^+$, in the organic phase

$$K_1 = \frac{a(NaHX^+)_o}{a_{Na^+}(o)a_{HX}(o)},$$

$$K_2 = \frac{a(NaHX^+)}{a_{H^+}(o)a_{NaX}(o)} \qquad\qquad (26)$$

where HX denotes the undissociated form of monensin.

Experimental Procedures

At present mainly potential-sweep voltammetry at ITIES is used for investigation of electrolysis phenomena [19]. Given a proper composition of both the phases, ITIES behaves similarly to a metal-electrode/electrolyte solution interface [13]. When the organic phase contains a strongly hydrophobic indifferent ("base") electrolyte and the aqueous phase a strongly hydrophilic base electrolyte then, under polarization, a potential window is formed in the current-potential curve where the main contribution to the current is due to charging of the interface (a non-faradaic current). Suppose that an ion whose concentration is considerably lower than that of the base electrolytes and which has a suitable value of the standard Gibbs energy of transfer between both the phases, is present in one of the phases. Under polarization in a triangular potential-pulse mode both positive and negative current peaks are observed on the polarization curves. The rates of the ion transfer are rapid so for low polarization rates the concentration of the ion concerned, close to the ITIES, follows the Nernst potential equation. In this "reversible" case the current-potential dependence is exactly described by the Randles-Ševčík equation, which was originally deduced for a mercury-electrode/electrolyte-solution interface [20, 21].

At higher scan rates the kinetics of charge transfer across ITIES could be detected which is governed by Eqs. (17) to (20). The basic component of the experimental arrangement for voltammetry at ITIES is a four electrode potentiostat [22]. It makes it possible to impose a programmed potential difference at two Luggin capillaries with tips placed close to ITIES. This experimental approach is identical with the voltage-clamp method introduced earlier to electrophysiology [23].

A number of other methods (chronopotentiometry [24], polarography with dropping electrolyte electrode [25], faradaic impedance measurement [26], current scan voltammetry [27] etc.) were also applied to the study of electrolysis at ITIES.

Results and Applications

The simple ion transfer across ITIES was observed with alkali metal ions, tetraalkyl-ammonium cations, choline, acetylcholine, picrate, perchlorate, iodide, thiocyanate, nitrate, dodecylsulphate, cationic forms of various tetracycline derivatives, etc. The facilitated ion transfer was mainly studied with alkali and alkaline earth metal ions the transfer of which was mediated by ionophores already mentioned in the first section of this lecture (for reviews see [13, 18]).

Besides determination of data of theoretical interest such as standard Gibbs transfer energies, distribution coefficients and stability constants, voltammetry at ITIES has found the following applications:

i. determination of selectivity coefficients of liquid-membrane ion-selective electrodes [29, 30].

ii. determination of ions which are complexed in the organic phase by ionophores [30],

iii. determination of ionophores [31].

For determination of a low concentration of a given ion present in the aqueous phase a higher concentration of the complex-former in the organic phase is used. Under these conditions the current peak of the determinand is shifted to more negative potentials so that it can be accurately measured. An enhancement of the sensitivity of this method is achieved when using a hanging organic electrolyte drop [32].

In the case of ionophore determination a rather high concentration of the ion to be complexed in the aqueous phase is used while the ionophore is present at a low concentration in the organic phase. Under increasing potential scan the current peak is controlled by diffusion of the ionphore to ITIES and by diffusion of the complex formed from ITIES into the bulk of the organic phase while after scan reversal opposite processes take place. The peak currents are proportional to ionophore concentration. This method has been applied to the determination of monensin in cultures of *Streptomyces cinnamonensis* [33].

The electrochemical phenomena included under the heading of electrolysis at ITIES can be also characterised as a voltammetric version of electrochemistry of ion-selective electrodes. It helps to elucidate the processes taking place at liquid-membrane ISEs and has some interesting analytical applications.

The subject of electrolysis at ITIES has been dealt with in the author's review where the detailed references are also listed [13, 28].

References

1. Ostwald, W.: Z. phys. Chem. *6*, 71 (1891)
2. Nernst, W., Riesenfeld, E.H.: Ann. Physik *8*, 600 (1902)
3. Beutner, R.: Z. Elektrochem. *19*, 319 (1913)
4. Bonhoeffer, K.F., Kahlweit, M., Strehlow, H.: ibid. *57*, 614 (1953)
5. Karpfen, F.M., Randles, J.E.B.: Trans. Faraday Soc. *49*, 823 (1953)
6. Parker, A.J: Electrochim. Acta *21*, 671 (1976)
7. Verwey, E.J.W., Niessen, K.F.: Phil. Mag. *28*, 435 (1939)
8. Kakiuchi, T., Senda, M.: Bull. Chem. Soc. Japan *56*, 1322 (1983)
9. D'Epenoux, B., Seta, P., Amblard, G., Gavach, C.: J. Electroanal. Chem. *94*, 77 (1979)

10. Koryta, J.: Anal. Chim. Acta *139*, 1 (1982)
11. Samec, Z., Mareček, V.: J. Electroanal. Chem. *200*, 17 (1986)
12. Ward, W.J.: AIChE J. *16*, 805 (1970)
13. Koryta, J.: Electrochim. Acta *24*, 293 (1979)
14. Koryta, J., Štuli'k, K.: Ion-Selective Electrodes, Cambridge University Press, Cambridge, 1983
15. Ammann, D., Güggi, M., Pretsch, E., Simon, W.: Anal. Lett. *8*, 709 (1975)
16. Pressman, B.C.: Proc. Nat. Acad. Sci. *53*, 1076 (1965)
17. Samec, Z.: private communication
18. Guo Du, Koryta, J., Ruth, W., Vanýsek, P.: J. Electroanal. Chem. *159*, 413 (1983)
19. Samec, Z., Mareček, V., Koryta, J., Khalil, W.: ibid. *83*, 393 (1977)
20. Randles, J.E.B.: Trans. Faraday Soc. *44*, 327 (1948)
21. Ševčik, A.: Coll. Czech. Chem. Comm. *13*, 349 (1948)
22. Samec, Z., Mareček, V., Weber, J.: J. Electroanal. Chem. *100*, 841 (1979)
23. Hodgkin, A.L., Huxley, A.F., Katz, B.: Arch. Sci. Physiol. *3*, 129 (1949)
24. Gavach, C., Henry, F.: J. Electroanal. Chem. *54*, 361 (1974)
25. Koryta, J., Vanýsek, P., Březina, M.: ibid. *75*, 211 (1977)
26. Osakai, T., Kakutani, T., Senda, M.: Rev. Polarogr., Kyoto *27*, 51 (1981)
27. Kihara, S., Yoshida, Z., Fujinaga, T.: Bunseki Kagaku *31*, E297 (1982)
28. Koryta, J.: Electrochim. Acta *29*, 445 (1984)
29. Koryta, J.: Hung. Sci. Instr. *49*, 25 (1980)
30. Samec, Z., Homolka, D., Mareček, V.: J. Electroanal. Chem. *135*, 265 (1982)
31. Koryta, J., Kozlov, Yu.N., Hofmanová, A., Hung, Le Q., Khalil, W., Guo Du, Ruth, W., Vanýsek, P.: Antibiotiki, Moscow, 810 (1983)
32. Mareček, V., Samec, Z.: Anal. Lett. B *15*, 1241 (1981)
33. Kozlov, Yu.N., Koryta, J.: Anal. Lett. *16*B, 255 (1983)

Problems of a Quantum Theory of Charge Transfer Reactions at the Interface Between Two Immiscible Liquids

A. M. Kuznetsov and Yu. I. Kharkats

The theory of charge transfer processes in solutions is more than two decades old and continues to develop intensively both in respect to solving some theoretical problems of principal importance and from the viewpoint of its application to various systems and processes. The phenomena and processes of the given type, though being quite different in nautre, have some common features which allow us to consider them from a single point of view. One of the important factors inherent to all these processes is the fact that these are strongly influenced by the solvent. A great role of the polar solvent in the formation of the Franck-Condon barrier in electron transfer reactions was supposed for the first time by Libby [1]. This idea was found to be fruitful and became one of the essential components of modern theories of chemical kinetics in solutions. Within the framework of the classical non-equilibrium thermodynamics method this concept was implemented by Marcus [2]. Others use the ideas of a quantum-mechanical theory of non-radiative transitions [3]. The results obtained within the framework of this concept have been presented in a most complete way in Refs. [4–6]. The studies have resulted not only in establishing the physical mechanism of elementary chemical transformations, but also in the methods for their calculation with due account of the peculiarities of specific systems. This paper will briefly outline the basic ideas underlying the physical mechanism of an elementary act of charge transfer processes in polar media, which is common to various classes of processes, and it will also consider some peculiarities related to the specificity of the interface between two immiscible liquids. For the sake of certainty, the electron transfer reactions will be mainly considered. Our consideration, however, will also take into account the possible reorganization of the intramolecular structure of reactants which bears a direct relation to the transfer of heavy particles, and the transfer of ions through the interface will also be briefly discussed.

1. The Franck-Condon Principle and the Physical Mechanism of the Transition

Prior to discussing the role of the medium in the kinetics of electron transfer reactions between the donor A and acceptor B, which are situated at some fixed distance R from each other, we shall first consider the isolated molecular system. The stable molecule, being in the ground electronic state, possesses some specific structure, i.e. it is characterized by specific equilibrium positions of nuclei r_{0i}. The change of the

The Interface Structure and Electrochemical Processes at the Boundary Between Two Immiscible Liquids
Editor: V. E. Kazarinov
© Springer-Verlag Berlin, Heidelberg 1987

electronic state of a molecule (the electronic excitation, ionization, etc.) results generally in the change of its nuclear configuration, i.e. in the change of equilibrium positions r_{0f} of nuclei. This displacement of equilibrium positions of nuclei, $\Delta r = r_{0f} - r_{0i}$, is the greater, the stronger the interaction of intramolecular vibrations with a transferred electron and the greater the variation of its interaction with the change of an electronic state.

To separate the motions of strongly interacting electrons and nuclei, the Born-Oppenheimer approximation is used, which assumes that the state of an electron, i.e. its energy ε_i and the wave function ψ_i, follow adiabatically the variation of nuclei positions, i.e. it depends on the instantaneous values of the nuclear coordinates $r(t)$. The dependence of the electron energy ε_γ on nuclear coordinates differs for different electronic states, and for some values of nuclear coordinates r^* the electronic energies for various electronic states may be close to each other: $\varepsilon_i(r^*) \simeq \varepsilon_f(r^*)$. The Born-Oppenheimer approximation works well everywhere except in the vicinity of these points r^*. At points r^* this approximation is violated, and the change of the electronic state without the change of the system's energy is possible. Far from these points in the absence of external influences on the system, the probability of change of an electronic state is negligible. This is the essense of the Franck-Condon principle.

Thus, for changing an electronic state the nuclear configuration r^* is necessary for which $\varepsilon_i(r^*) \simeq \varepsilon_f(r^*)$. This configuration is called transitional. In a system, which is in the thermal equilibrium state, the probability of existence of some nuclear configuration is determined by the respective quantum-mechanical distribution function

$$\Phi(r) = \sum_n \exp(-E_n^i/kT) |\chi_n^i(r)|^2 \Big/ \sum_n \exp(-E_n^i/kT), \tag{1}$$

where E_n^i and $\chi_n^i(r)$ are the energies and wave functions of nuclei at the initial state.

The wave functions are determined by the shape of the potential energy surface which defines the motion of nuclei. The field in which the nuclei move is determined not only by their direct mutual interaction, but also by the average field produced by an electron in the given quantum state. The transitional configurations r^* correspond to the intersection points between the potential energy surfaces for the initial and final electronic states, $U_i(r^*) = U_f(r^*)$. The greater the variation of the electron-nuclei interaction, the larger the displacement of equilibrium positions of nuclei with the change of an electronic state and the larger the separation between the point r^* of intersection of potential energy surfaces $U_i(r)$ and $U_f(r)$ and the initial equilibrium state r_{0i} and, respectively, the larger the Franck-Condon barrier which must be surmounted by the system.

Equation (1) allows us to analyse qualitatively the character of the system's behaviour in the vicinity of a transitional configuration. If the greatest contribution to the sum in n is made by the term with $n = 0$, that corresponds to an unexcited vibrational state of nuclei, then the behaviour of nuclei is quantum one, and the transition into the transitional configuration is accomplished by the penetration into a classically inaccessible region.

If, however, the main contribution to the sum in n is made by a large group of terms corresponding to highly excited vibrational states, then the motion of nuclei is close to the classical one, and the probability of finding such a nucleus configuration is governed by the Boltzmann equation. The implementation of one or the other limiting case depends on the type of energy spectrum E_n, on the shape of the potential barrier and on the temperature. At room temperature a classical behaviour is found when the barriers are wide. For narrow barriers the behaviour of nuclei is the quantum type and is described, in the respective limit, by the wave function of an unexcited vibrational state. When the shape of potential energy surfaces is parabolic, the form of an energy spectrum and the shape of a barrier are interrelated and the approximate criteria for classical and quantum behaviour are, respectively, as follows:

$$\hbar\omega_{\text{class.}} \ll kT, \qquad \hbar\omega_{\text{quant.}} \gg kT. \tag{2}$$

Note that when the second inequality of Eq. (2) is met, the behaviour of nuclei is quantum because the probability of the system existing at an excited vibrational state is very low. Equation (1) takes into account all possibilities of implementation of the given nuclear configuration including those associated with the system existing at various excited vibrational states. Merely, within the limit under consideration the contribution of the ground vibrational state to the probability of implementation of the given nuclear configuration is much larger than the contribution made by excited vibrational states.

The occurrence of the transitional configuration r^* is the necessary condition for the electronic state variation. However, the probability of the occurrence of such an electronic state variation depends on the time of existence of a transitional configuration. If the time τ_* of a system existing in the vicinity of a transitional configuration is short, then the mentioned probability is low and the process is non-adiabatic. If τ_* is large enough, the probability of transformation of an electronic state equals unity and the process is adiabatic.

For the system to be transferred into the final state the rearrangement only of an electronic wave function in the transitional configuration is unsufficient. It is also necessary for the system to pass this transitional configuration and enter the region of classical motion in the potential well corresponding to the final electronic state. This is manifested in the fact that the transition probability is determined not by the distribution function over nuclear coordinates, described by Eq. (1), but by the Franck-Condon factor averaged over the thermal distribution, which represents the averaged square of overlap of nuclear wave functions for initial and final vibrational states. The Franck-Condon factor is exponentially small for weakly-excited vibrational states and grows in passing to higher excited states.

The averaged Franck-Condon factor is the smaller, the greater the value of a barrier formed by the intersection of potential energy surfaces for the initial and final states, i.e. $U_i(r)$ and $U_f(r)$. The transition of a system from the initial state to the final one may be treated as the motion of a point, which represents the system in the configurational space of nuclei, on potential energy surfaces U_i and U_f. In this case the trajectories bringing the system from the initial equilibrium state into the final one may have in a general case both sections of classical motion and sections corresponding to the sub-barrier passage. In a case where motion of nuclei in the course of a process is fully classical, the most probable trajectories pass near the "saddle" point at the intersection of potential energy surfaces. Thus, knowing the potential energy surfaces for nuclei, electronic wave functions for the initial and final states, and the interaction leading to transition, the probability of this transition may be calculated.

However, as shown in Refs. [7, 8], to calculate the transition probability it is sufficient to have in some cases less detailed information than that provided by potential energy surfaces. In cases where the normal coordinates of classical and quantum degrees of freedom of nuclei are not intermixed during the process, it is sufficient for calculating transition probabilities to know the free energy surfaces for the system in the initial and final states, for fixed vibrational states of quantum degrees of freedom, as functions of coordinates of reactive modes, i.e. as functions of values of configurational parameters of the system, the change of which leads to its transition from the region near the initial equilibrium configuration into the final configuration region [8]. Physically, this is due to the fact that for finding the averaged probability of a system's transition from the initial to the final state it is sufficient to know the probabilities of finding the configurations of coordinates of reactive modes averaged over configurations of the remaining, non-reactive modes. These probabilities are determined by the configurational part of the system's free energy in the corresponding electronic and quantum vibrational states. Warshel [9] came to the same conclusion independently.

Thus, in interpreting the transition, one may use both the potential energy surfaces and the free energy surfaces; however, for complicated systems the latter require less detailed information on the system and, hence, are more accessible in some cases.

2. The Role of a Polar Medium and the Solvent Model

For reactions proceeding in a condensed medium, especially in a polar one, there exists, as a rule, a strong interaction of reactants with a medium which changes with changing the charge state of reactants. This implies that such an interaction must have, generally, a strong influence on the electron transfer process. The mechanism of such an interaction is basically similar to that caused by the interaction between electrons and nuclei in isolated molecular systems. The presence of a medium, however, introduces some features which have to be discussed.

First of all one should note that in separating the motions of nuclei and transferrable electrons one has used the Born-Oppenheimer approximation which means that the electron is more rapid with respect to all remaining degrees of freedom. As far as the interaction of an electron in a donor or in an acceptor with the medium is concerned, one should bear in mind that there are various types of motions in a medium with their various characteristic times. Therefore, one should distinguish the motions which are more rapid with respect to a transferrable electron, and the motions which are slower with respect to the electron's motion.

Accordingly, the medium polarization as a whole is subdivided into the inertialess and inertial polarizations. The inertialess polarization characterizes the screening of a high-frequency electrical field in a medium. It follows adiabatically the variation of the electronic state of reactants and, while neglecting some fine effects, it does not form the Franck-Condon barrier for electron transfer.

The role of inertial polarization is similar to the role of intramolecular vibrational degrees of freedom. Since this polarization is slow with respect to a transferrable electron, it "feels" the average field produced by an electron in the given state. This average field induces in a medium the average polarization near the donor and acceptor, which depends on the charge states of the latter. The stronger the variation of an equilibrium polarization in the electronic transfer, the larger the value of the Franck-Condon barrier produced by this polarization.

One should emphasize that the coordinates describing the vibrations of an inertial polarization should be considered as the reactive coordinates equivalent to intramolecular coordinates, and in the absence of the latter, the vibrations of an inertial polarization of the medium are the only factor which provides matching of the electronic energy levels of a donor and of an acceptor. However, as shown in Ref. [10], the effect of a medium in electron transfer reactions is not reduced to matching of the electronic energies only. There are some additional effects caused by the dynamical behaviour of a medium in the electronic transfer process. One of them consists in the fact that the vibration of polarization near an acceptor produces the electric field which is the interaction, additional with respect to the direct interaction between an electron and an acceptor, leading to electron transfer to an acceptor. In some cases this fluctuational interaction exceeds the direct interaction with an acceptor.

Besides, the fluctuations of polarization give rise to the modulation of electronic wave functions of a donor and of an acceptor. This results in two effects: the distortion of the shape of free energy surfaces and the change in overlapping of electronic wave functions of a donor and an acceptor. All these effects influence the kinetic parameters of a transfer, but their values differ for different systems. The role of each of these effects is most significant in the case where weakly bound electrons are transferred over long distances.

The dynamical influence of a polar medium on the electron transfer process kinetics can be taken into account quantitatively only within the framework of specific models for describing the solvent. As shown in Ref. [4], to calculate transition probabilities for slow reactions, for which the distribution over vibrational states can be considered to be equilibrium, the inertial polarization of the medium can be represented as a set of effective oscillators

$$H_s = \tfrac{1}{2} \sum_k \hbar \omega_k (q_k^2 - \partial^2/\partial q_k^2), \tag{3}$$

where ω_k and q_k are frequencies and dimensionless normal coordinates of oscillators.

The parameters of the Hamiltonian (3), i.e. the frequencies and the number of oscillators (more specifically, the strength of oscillators) are determined by the imaginary part of a complex dielectric function $\varepsilon(k, \omega)$ which characterizes the dielectric losses for polarization fluctuations in a medium. This model is, strictly speaking, applicable to homogeneous isotropic media in which the spatial correlations of polarization fluctuations $\langle \delta P(r) \delta P(r') \rangle$, which determine the dependence of $\varepsilon(k, \omega)$ on the wave vector k, depend on the difference of coordinates $r - r'$ only.

In the presence of an interface, as in particular, for the two immiscible liquids, the correlation between polarization fluctuations depend on each of coordinates r

and r' separately and the formulation of a model is greatly complicated. If, however, the correlation radius is small enough ($\Lambda \to 0$), then the properties of both liquids can be considered to be approximately equal to the properties of the corresponding bulk phases up to the interface, and the model (3) can be used assuming that the Hamiltonian consists of the sum of two Hamiltonians $H_s = H_{s1} + H_{s2}$ each of which is determined by dielectric properties of the respective liquid: $\varepsilon_1(\omega)$ and $\varepsilon_2(\omega)$.

The interaction of an electron, which is in a state with wave function $\varphi_\gamma(x)$, with the inertial polarization of a medium $P(r)$ has the form

$$V_{es} = -\int P(r) E^V(r) d^3 r, \tag{4}$$

where $E^V(r)$ is the electric field (in vacuum) produced by the charge distribution $\varrho_e(x) = -e|\varphi_\gamma(x)|^2$, where e is the positive charge equal to the electron charge in magnitude.

This interaction results in a shift of the equilibrium coordinates of effective oscillators of the medium. As a result, the effective Hamiltonians for the initial and final states have the form

$$H_i = H_e^i + \frac{1}{2} \sum_k \hbar \omega_k \left[(q_k - q_{k0i})^2 - \frac{\partial^2}{\partial q_k^2} \right], \tag{5}$$

$$H_f = H_e^f + \frac{1}{2} \sum_k \hbar \omega_k \left[(q_k - q_{k0f})^2 - \frac{\partial^2}{\partial q_k^2} \right], \tag{6}$$

where H_e^i and H_e^f are Hamiltonians for an electron in a donor and in an acceptor at equilibrium values of normal coordinates of oscillators q_{k0i} and q_{k0f}.

For the sake of simplicity we shall confine ourselves to transfer processes for tightly bound electrons and neglect the effects of distortion of electronic wave functions by polarization fluctuations. Furthermore, we shall suppose that the whole inertial polarization can be described in a classical manner (i.e. $\hbar \omega_k \ll kT$). In the case of water the quantum "tail" of polarization fluctuations is small and, as shown in Ref. [11], can be formally taken into account by introducing a factor (< 1) into the reorganization energy (see below). For the other liquids, used in the experiments with immiscible liquids, the detailed measurements of $\varepsilon(\omega)$ are not available as a rule.

The probability of electron transfer per unit time between two ions, situated at some given distance from each other, has the following form within the framework of the above approximations [4]:

$$W = \frac{\omega_{\text{eff}}}{2\pi} \varkappa \exp(-F_a/kT), \tag{7}$$

where

$$F_a = [E_s + \Delta F]^2 / 4E_s. \tag{8}$$

The quantity $\Delta F(R)$ in Eq. (8) is the local free energy of a transfer, i.e. the difference between free energies of a system at the final and initial states for the given position of ions. It contains, along with the difference $\Delta \varepsilon$ of electronic energies in an acceptor and in a donor, also the difference $\Delta U(R)$ between free energies of donor-acceptor interaction at various charge states. The relation between quantity $\Delta F(R)$ and the reaction free energy ΔF_0, corresponding to the separation of the reactants for a large distance, depends on the phases in which the reactants are located and will be considered below.

The quantity E_s is the reorganization energy of the solvent determined by the relation

$$E_s = U_i(q_{0f}) - U_i(q_{0i}) = \frac{1}{2} \sum_k \hbar\omega_k (q_{k0f} - q_{k0i})^2 . \tag{9}$$

The exponential factor in Eq. (7) characterizes the probability of finding the transitional configuration. As shown in Ref. [7], the quantity F_a is the difference between configurational parts of the system's free energy corresponding to transitional and initial equilibrium configurations. Thus, the correspondence is established between the physical meanings of activation parameters in expressions for transition probability obtained by the methods of a quantum theory of non-radiative transitions [4] and those obtained by the methods of classical non-equilibrium thermodynamics [2].

The quantity $\varkappa$ in Eq. (7) is the transmission coefficient which characterizes the probability of rearrangement of the electronic state when the system passes the transitional configuration. If the time of a system's existence in the vicinity of a transitional configuration is large enough, then $\varkappa = 1$ and the reaction is adiabatic. For the opposite limit $\varkappa \ll 1$ and the reaction is non-adiabatic. The condition for the transfer to be non-adiabatic is as follows:

$$\varkappa = \frac{2\pi}{\hbar\omega_{\mathrm{eff}}} V_{if}^2 (\pi/kTE_s)^{1/2} \ll 1 , \tag{10}$$

where V_{if} is the electronic resonance integral

$$V_{if} = \int \varphi_i(x) V(x) \varphi_f(x) \mathrm{d}^3x \tag{11}$$

whose value is determined by the interaction $V(x)$ leading to an electron transfer as well as by overlapping of electronic wave functions for the initial and final states φ_i and φ_f.

3. The Reorganization Energy of the Medium

As seen from Eqs. (7) and (8), the free activation energy is determined to a considerable degree by the value of the medium reorganization energy E_s. This value is an important physical parameter of the theory. Using the relation between Hamiltonian parameters and the dielectric properties of the medium, the reorganization energy may be related to the properties of a solvent. Formulae of such a type were first obtained for a homogeneous, structureless dielectric medium in Ref. [12] for electronic processes in polar crystals and in Ref. [2] – for electron transfer reactions in solutions. For the charge transfer processes in an inhomogeneous medium for E_s we have [7]

$$E_s = \frac{1}{8\pi} \int c(r) [\mathscr{D}_i(r) - \mathscr{D}_f(r)]^2 \mathrm{d}^3r , \tag{12}$$

where $\mathscr{D}_i$ and $\mathscr{D}_f$ are electrostatic inductions produced in a medium at the initial and final states,

$$c(r) = \frac{1}{\varepsilon_0(r)} - \frac{1}{\varepsilon_s(r)}, \tag{13}$$

where ε_0 and ε_s are the high-frequency and static dielectric constants of a medium. Integration in Eq. (12) is carried out over the two liquids allowing for the fact that the dielectric constants are different in both liquids.

As shown in Ref. [7], the quantity E_s corresponds to a free energy. It is the work consumed for changing the medium inertial polarization from the initial equilibrium value P_{0i} to the final equilibrium value P_{0f} with the fixed charge state of reactants (Fig. 1). The medium polarization states corresponding to points P_{0i} and P_{0f} on the free energy surface are shown schematically in Fig. 2 [13]. The solid arrows show the inertial polarization of a medium and the dotted arrows – the inertialess one. For simplicity this figure shows the case where both a donor and an acceptor are neutral in the absence of an electron. Two arrows of the same type but in the opposite directions imply the absence of the respective polarization component. As Fig. 2 shows, the state corresponding to point P_{0f} differs from the initial equilibrium state by the value of the inertial polarization component only. The inertialess polarization remains unchanged in transition from P_{0i} to P_{0f}. The work for changing the inertial polarization from P_{0i} to P_{0f} with the fixed charge state of reactants may be calculated as the work consumed in the following reversible process (Fig. 3). First, the charges of a donor and of an acceptor are slowly changed to $-e \rightarrow 0$ and $0 \rightarrow -e$, respectively. In this case the induction changes from $\mathscr{D}_i$ to $\mathscr{D}_f$ and the system passes to configuration f (Fig. 3). Then, the charges are rapidly changed to values $0 \rightarrow -e'$ and $-e \rightarrow e''$, so that the induction changes from $\mathscr{D}_f$ to $\mathscr{D}_i$. In this case the inertialess polarization returns to its initial state and the inertial polarization retains its value, i.e. the system is in the desired configuration (Fig. 3).

The total work equals the sum of the work at the first and second stages of the process

$$E_s = -\frac{1}{4\pi} \int d^3r \left[\int_{\substack{0 \\ (\text{slow})}}^{\Delta\mathscr{D}} E\delta\mathscr{D} + \int_{\substack{\Delta\mathscr{D} \\ (\text{rapid})}}^{0} E\delta\mathscr{D} \right], \tag{14}$$

where $\Delta\mathscr{D} = \mathscr{D}_f - \mathscr{D}_i$.

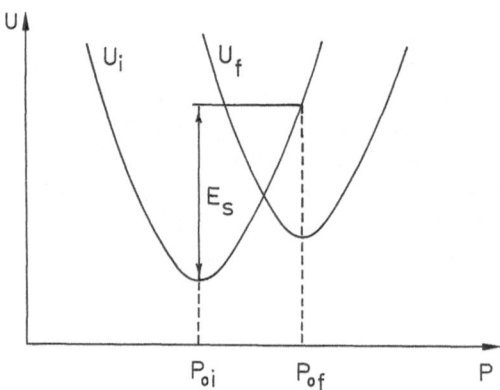

Fig. 1. Scheme of free energy surfaces for the initial and final states as functions of medium polarization P

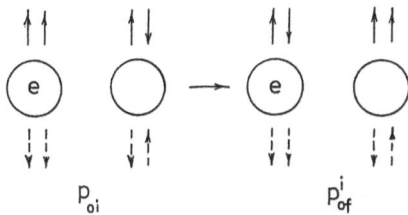

Fig. 2. Scheme of distribution of inertial (solid arrows) and inertialess (dotted arrows) polarization near a donor and an acceptor at the initial equilibrium state (P_{0i}) and at the state (P_{0f}^i) corresponding to the final equilibrium value of inertial polarization (P_{0f}) for a fixed charge state of reactants (the electron is on the donor)

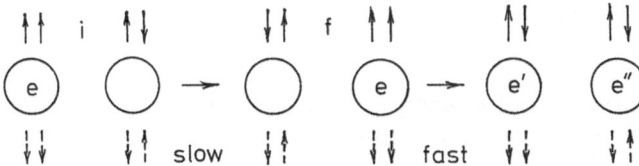

Fig. 3. Scheme of the process for calculating the medium reorganization energy

It is essential that the same value $\Delta\mathcal{D} = \mathcal{D}_f - \mathcal{D}_i$ is present in both terms, where $\mathcal{D}_i$ and $\mathcal{D}_f$ are the electrostatic inductions produced by the initial and final distributions of charge, respectively, in the medium with dielectric constant ε_s. Equation (14) gives rise to Eq. (12). As seen from Eq. (14), in order for Eq. (12) for the reorganization energy to be valid, the dielectric must give a linear response when the field changes by the value $\Delta\mathcal{D}$.

Note that the inductions, produced by a donor and by an acceptor in the absence of an electron, are mutually cancelled in $\Delta\mathcal{D}$, and $\Delta\mathcal{D}$ is equal to the difference of inductions produced by an electron situated, respectively, on an acceptor and on a donor

$$\Delta\mathcal{D}(r) = e\left[\frac{\vec{r}}{r^3} - \frac{(\vec{r} - \vec{R})}{|\vec{r} - \vec{R}|^3}\right]. \tag{15}$$

Thus, quantity $\Delta\mathcal{D}(r)$ formally coincides with the electrostatic induction $\mathcal{D}_d$ produced by a dipole with the dipole moment $d = eR$, and Eq. (12) in the case of a homogeneous medium differs from the expression describing the solvation of dipole d in a liquid

$$F_d = -\frac{1}{8\pi}\left(1 - \frac{1}{\varepsilon_s}\right)\int \mathcal{D}_d^2 d^3r, \tag{16}$$

only by a multiplier. It is clearly seen that in this case the reorganization energy is formally equal to [14]

$$E_s = F_d(\varepsilon_0) - F_d(\varepsilon_s), \tag{17}$$

where $F_d(\varepsilon_0)$ and $F_d(\varepsilon_s)$ are free energies of solvation of dipole d in media with dielectric constants ε_0 and ε_s.

An alternative method for calculating E_s was proposed in Ref. [15] where the reorganization energy was considered as a difference between the free energy of the electron transfer process from a donor to an acceptor in the medium with the

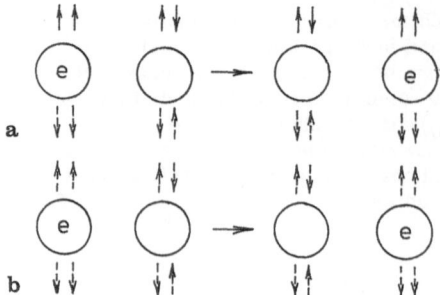

Fig. 4. Scheme of the processes for calculating the medium reorganization energy according to Ref. [13]

dielectric constant ε_s (Fig. 4a) and the free energy of the electron transfer process from a donor to an acceptor in the medium with dielectric constant ε_0 (Fig. 4b). This value is equal to [13]

$$E_s = -\frac{1}{4\pi} \int d^3r \left[\int_0^{\Delta\mathscr{D}(\varepsilon_s)} E\delta\mathscr{D} - \int_0^{\Delta\mathscr{D}(\varepsilon_0)} E\delta\mathscr{D} \right], \tag{18}$$

where $\Delta\mathscr{D}(\varepsilon_s)$ and $\Delta\mathscr{D}(\varepsilon_0)$ are the variations of electrostatic induction in the course of the mentioned processes.

Equations (17) and (18) may be shown to yield the correct values of reorganization energy, coinciding with Eq. (12), in some specific cases only [13], namely: 1) the homogeneous medium, 2) the medium adjacent to metal bodies, 3) the inhomogeneous medium in the case of a problem with the spherical symmetry and in some other cases. In other words, this situation occurs only if $\Delta\mathscr{D}(\varepsilon_s) = \Delta\mathscr{D}(\varepsilon_0)$.

In all other situations, when $\Delta\mathscr{D}(\varepsilon_s) \neq \Delta\mathscr{D}(\varepsilon_0)$, Eqs. (17)–(18) give a wrong result. Indeed, in this case, taking into account the relation between $\mathscr{D}$ and E for slow ($\mathscr{D} = \varepsilon_s E$) and rapid ($\mathscr{D} = \varepsilon_0 E$) processes, we have from Eq. (18), for example:

$$\tilde{E}_s = \frac{1}{8\pi} \int \frac{1}{\varepsilon_0} [\Delta\mathscr{D}(\varepsilon_0, r)]^2 d^3r - \frac{1}{8\pi} \int \frac{1}{\varepsilon_s} [\Delta\mathscr{D}(\varepsilon_s, r)]^2 d^3r. \tag{19}$$

Note that the distinction between Eq. (19) and Eq. (12) lies in the term containing $1/\varepsilon_0$, i.e. in the term which makes a maximum contribution to the reorganization energy. The difference between the two approaches considered is most essential in the cases where there exist inhomogeneities in a dielectric medium, in particular, for processes at the interface between the two immiscible liquids, as well as when the finite sizes of ions are taken into account in the absence of a spherical symmetry. In these cases the second approach may give rise to considerable errors, whereas Eqs. (12)–(13) remain valid here as well.

At the same time, for some situations the distinction between the mentioned approaches is insignificant. This is true, for example, for the electronic transfer in the bulk of a solution, and near the surface of a metal electrode (electrochemical and photoelectrochemical processes) when neglecting the finite sizes of ions, as well as for the photoelectron emission from spherical ions in the bulk of a solution. In the latter case the finite size of ions can be taken into account. The mentioned coincidence of results is due here to the fact that in these cases the induction

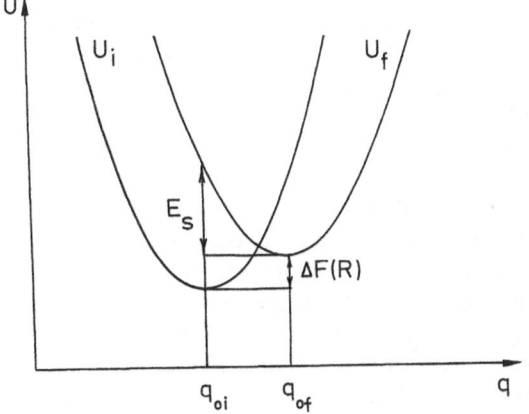

Fig. 5. Scheme of free energy surfaces for the initial and final states of a system as functions of normal coordinates q

coincides with the field in vacuum which does not depend on dielectric properties of a medium.

In the cases where the results do not coincide in the two approaches the distinction is due to the difference in the electrostatic inductions and, as the comparison of Eqs. (12) and (13) shows, due to the difference in the state of inertialess polarization in the media with dielectric constants ε_0 and ε_s (Figs. 4a and b). Thus, though the second approach yields correct results in some situations, the first approach is more reasonably used for giving the physical picture of a process.

And, finally, one further method for calculating the reorganization energy has been proposed in Ref. [9]. As seen from Fig. 5, if the shape of free energy surfaces is the same at the initial and final states, then the reorganization energy is also equal to

$$E_s = U_f(q_{0i}) - U_i(q_{0i}) - \Delta F(R) = U_i(q_{0f}) - U_f(q_{0f}) + \Delta F(R). \tag{20}$$

Such a form of presentation is convenient when the equilibrium values of normal coordinates q_k at the initial or final states and the form of interaction energy between the electron and the vibrational subsystem at initial and final states $V_{es}^{(i)}(q)$ and $V_{es}^{(f)}(q)$ are known. Indeed, since the free energy surfaces U_i and U_f can be written as

$$U_i(q) = U_s(q) + V_{es}^i(q),$$
$$U_f(q) = U_s(q) + V_{es}^f(q), \tag{21}$$

where $U_s(q)$ is the free energy surface in the absence of interaction with an electron, Eq. (20) takes the form

$$E_s = V_{es}^f(q_{0i}) - V_{es}^i(q_{0i}) - \Delta F(R). \tag{22}$$

For symmetric reactions $\Delta F(R) = 0$, and the reorganization energy equals the difference between the energies of interaction of an electron with a medium at initial (or final) equilibrium values of coordinates of effective oscillators. As seen from Eq. (22), in this case it is sufficient to know only the type of interaction between an electron and a vibrational subsystem at the initial and final states, and it may seem at a glance that no additional data on the properties of a vibrational subsystem itself are required.

However, in this approach the main calculational problem is merely transferred to calculating equilibrium coordinates q_{0i} or q_{0f}, since this calculation requires a knowledge of the response function of a vibrational subsystem [in the case of a polar dielectric this function is $\varepsilon(k, \omega)$] to the external field.

Equations of type (20) and (22) are more suitable, however, in the cases where there are no data on force constants or vibration frequencies for a vibrational subsystem, which are necessary for calculating E_s by Eq. (9), and the equilibrium values of q_{k0f} or q_{k0i} coordinates are experimentally known. Such an approach may be used in the most effective way for calculating the contribution of intramolecular degrees of freedom into the reorganization energy (see below). In this case, however, one must be convinced that the experimental methods for determining equilibrium positions of atoms (the X-ray method, for example) yield the same values of coordinates as in Eqs. (20) and (22).

The fact is that the interaction energies V_{es}^i and V_{es}^f depend usually on some number of coordinates of specific groups of atoms, rather than on coordinates of all atoms, and represent the values averaged over the positions of remaining atoms. Various experiments may involve various types of averaging, and the values of $V_{es}^i(q_{k0i})$ and $V_{es}^f(q_{k0i})$ calculated at points q_{k0i}, which are found by minimizing the free energy surface $U_i(q_k)$, may not coincide, generally speaking, with the values of the same quantities calculated at points q_{k0i}^{ex} which are found from experimental measurements of other type. Such a situation is most probable in the cases where the difference between free energy surfaces and potential energy surfaces is significant. If this difference is insignificant, both approaches are equivalent.

4. The Role of Intramolecular Vibrations and Quantum Degrees of Freedom

If the intramolecular structure of reactants changes as a result of electron transfer, then this factor also influences the transition probability in accordance with the discussion given in Sect. 1. The transitional configuration r^* over the coordinates of intra-molecular degrees of freedom r is reached in a classical or quantum way depending on the shape of potential energy surfaces at the initial and final states. The motion along the intramolecular coordinates from the initial equilibrium configuration r_{0i} into the final one r_{0f} is one of the components of the system's motion in a configurational space including the motion over all coordinates forming the Franck-Condon barrier. It is essential, that in electron transfer reactions the role of a medium is not reduced to solvation of initial and transitional states over intramolecular coordinates. Coordinates q_k, describing the solvent's polarization state, appear as independent dynamic variables of a system which compose a part of its configurational space. This is for two reasons: First, the frequencies of intramolecular vibrations are larger or of the same order as the frequencies characterizing the vibrations of the most part of the medium polarization. As a result, the medium has no time to follow adiabatically the variation of intramolecular structure of the reactants. Second, the transitional configurations r_f^* and $r_i^*(r_i^* = r_f^*)$ for intramolecular degrees of freedom at the initial and final states differ by the charge distribution in a system and, hence, for achieving the transitional configuration over all degrees of freedom the medium polarization should be shifted from the equilibrium state with respect to configuration r_i^* into some intermediate position P^*. Thus, in this case the medium also plays a dynamic role in the charge transfer process.

The expressions for transition probability in the general case, when the frequencies of intramolecular vibrations and the systems of normal coordinates may change during the transition, are rather cumbersome and are given in Refs. [4, 5]. In the systems where the intramolecular vibrations can be subdivided into two groups – classical and quantum ones – and there is no entangling of normal coordinates and no variation of vibration frequencies, the expression for transition probability is simplified and takes the form

$$W = \frac{V_{if}^2}{\hbar} \left(\frac{\pi}{kTE_r^t} \right)^{1/2} e^{-\sigma} e^{-(E_f^\ddagger + \Delta F)^2/4E_r^\ddagger kT} , \tag{23}$$

where

$$E_r^t = E_s + E_r^{in} \tag{24}$$

is the total reorganization energy of the system which includes, along with the medium reorganization energy E_s, also the reorganization energy E_r^{in} of classical intramolecular vibrations ($\hbar\omega \ll kT$).

Quantity σ is the tunnelling factor which characterizes the probability of rearrangement of the state of quantum vibrational degrees of freedom. In the harmonic approximation, σ is expressed as

$$\sigma = \tfrac{1}{2} \sum_n (q_{n0i} - q_{n0f})^2 . \tag{25}$$

As noted above, water contains some small part of polarization which is characterized by quantum frequencies of vibrations. The change of the state of these effective oscillators during the transition process also makes a contribution to the tunnelling and though the spectrum of dielectric losses in water is continuous and the polarization cannot be strictly separated into the quantum and classical parts, nevertheless, the quadratic dependence (23) for a free activation energy remains approximately valid on replacing E_s by $0.8E_s$ [11].

5. General Regularities in Charge Transfer Processes at the Interface Between Immiscible Liquids

There exist two types of charge transfer process at the interface between the two immiscible electrolytes 1 and 2. These are the transfer of ion X^z with charge z from phase 1 into phase 2 and back

$$X^z(1) \rightleftarrows X^z(2) \tag{26}$$

and secondly the transfer of an electron from the reactant in phase 1 to the reactant in phase 2; this transfer can be written as

$$Red_1(1) + Ox_2(2) \rightleftarrows Ox_1(1) + Red_2(2) . \tag{27}$$

In the general case there is a difference of Galvani potentials at the interface: $\Delta\varphi = \varphi(2) - \varphi(1)$. In equilibrium the sum of electrochemical potentials of reactants equals the sum of electrochemical potentials of reaction products. Introducing the quantity $\mu^\theta = \mu^0 + RT\ln\gamma$, where μ^0 is the standard chemical potential and γ is the activity coefficient, we obtain for the process of the first type:

$$\Delta\varphi = \Delta\varphi^\theta + \frac{RT}{zF} \ln \frac{c(1)}{c(2)}. \tag{28}$$

Here the difference of Galvani potentials for $c(1)/c(2) = 1$ is associated with the Gibbs free energy of ion transfer $\Delta G_{tr}^{\theta, 1\to 2}$ from phase 1 into phase 2:

$$\Delta\varphi^\theta = -(\Delta G_{tr}^{\theta, 1\to 2}/zF) = -[\mu^\theta(2) - \mu^\theta(1)]/zF. \tag{29}$$

In a similar way, we have in the case of electron transfer [23]

$$\Delta\varphi = \Delta\varphi^\theta + \frac{RT}{nF} \ln \frac{C_{Ox_1}(1)C_{Red_2}(2)}{C_{Red_1}(1)C_{Ox_2}(2)}, \tag{30}$$

where n is the number of transferable electrons and $\Delta\varphi^\theta$ is associated with the Gibbs energy ΔG^θ by

$$\Delta\varphi^\theta = \Delta G^\theta/nF = [\mu_{Ox_1}^\theta(1) + \mu_{Red_2}^\theta(2) - \mu_{Red_1}^\theta(1) - \mu_{Ox_2}^\theta(2)]/nF. \tag{31}$$

When the system is not in equilibrium, a predominant charge transfer between phases 1 and 2 in one of the directions occurs. The kinetics of ion transfer at the water/nitrobenzene interface has been studied in recent years in some work by galvanostatic [16–19] and potentiostatic [20–22] methods. The results of these studies have shown that the charge transfer processes at the interface between the two immiscible electrolytes have the same character as those at the electrode/solution interface. These processes include in a general case the stage of charge transfer through the interface and the diffusion-migration transport of reactants and products towards and from the interface. The theoretical analysis of the mentioned processes requires the derivation of the equation for the dependence of the interphase charge transfer rate on the difference of Galvani potentials in phases 1 and 2. This derivation is based on an analogy with electron transfer processes in electrode reactions [18, 21, 23, 24] and on the basic results of a quantum theory of chemical reactions in polar media, which have been presented above. The derivation further uses the idea of formation at the interface of phases of a compact layer composed of oriented molecules of both solvents [25, 26], which separates two diffuse double layers (see Fig. 6).

Within the framework of the above approach the expression for the rate of ion transition through the interface is as follows [23, 24]:

$$v_{\text{Ion}}(1\to 2) = K^{1\to 2}c(1) - K^{2\to 1}c(2), \tag{32}$$

where $K^{1\to 2}$ and $K^{2\to 1}$ depend on Galvani potential difference $\Delta\varphi$:

$$K^{1\to 2} = K^\theta \exp[-\alpha ze(\Delta\varphi - \Delta\varphi^\theta)/kT], \tag{33}$$

$$K^{2\to 1} = K^\theta \exp[+(1-\alpha)ze(\Delta\varphi - \Delta\varphi^\theta)/kT], \tag{34}$$

$$K^\theta = Bd \exp(-E_a^0/kT) \exp\{-ze[(1-\alpha)\varphi_2(1) + \alpha\varphi_2(2)]/kT\}. \tag{35}$$

Here B is a constant which depends on quantum characteristics of the system, d is the distance of an ion jump at the interface, E_a^0 is the activation energy for a symmetric

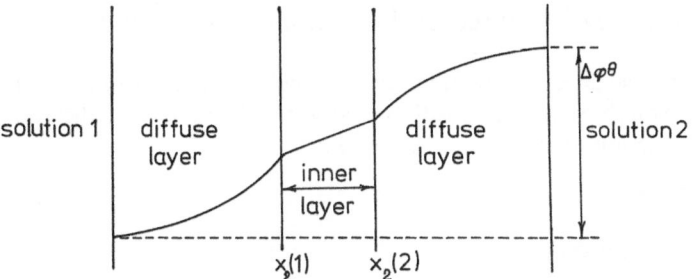

Fig. 6. Distributions of the electric potential at the interface of the two immiscible electrolyte solitions

transition (see Sect. 7). The distribution of the electrical potential over the interface is shown in Fig. 6.

The transfer coefficient α in Eqs. (33)–(35) depends in general on the overpotential, whereas the potential barrier can be described approximately by the barrier formed by the intersection of two potential surfaces of parabolic shape, then α is given by the relation

$$\alpha = \frac{1}{2} + \frac{ze}{2E_r} [\Delta\varphi - \Delta\varphi^\theta + \varphi_2(2) - \varphi_2(1)], \tag{36}$$

where E_r is the reorganization energy.

In a similar way, for the electron transfer rate at the interface one obtains [23]:

$$v_{el}(1 \rightarrow 2) = K^{1 \rightarrow 2} C_{Red_1}(1) C_{Ox_2}(2) - K^{2 \rightarrow 1} C_{Ox_1}(1) C_{Red_2}(2), \tag{37}$$

where

$$K^{1 \rightarrow 2} = K^\theta \exp[\alpha ne(\Delta\varphi - \Delta\varphi^\theta)/kT], \tag{38}$$

$$K^{2 \rightarrow 1} = K^\theta \exp[-(1-\alpha)ne(\Delta\varphi - \Delta\varphi^\theta)/kT], \tag{39}$$

$$K^\theta = BV_m d \exp(-E_s/4kT) \exp\{-e[(z_{Red_1}+\alpha n)\varphi_2(1) + (z_{Ox_2}-\alpha n)\varphi_2(2)]/kT\}, \tag{40}$$

$$\alpha = \frac{1}{2} - \frac{ne}{2E_s}(\Delta\varphi - \Delta\varphi^\theta + \varphi_2(2) - \varphi_2(1)), \tag{41}$$

where V_m is the reaction volume and the other notations are the same as for Eqs. (33)–(35).

The equations given in this section were obtained under the assumption that the system contains the supporting electrolyte which does not participate in a reaction, and the concentration of this electrolyte in each of phases is higher than the concentration of species participating in the charge transfer.

6. Electron Transfer at the Interface of Two Immiscible Liquids

In this section we shall discuss the question of the dependence of the electron transfer rate in a two-phase system on the reactants' position relative to the interface. In the case of the electron transfer reaction in the one-phase system

$$A^{z_1} + B^{z_2} \rightarrow A^{z_1+n} + B^{z_2-n} \tag{42}$$

the process rate can be expressed as

$$K = kc_A c_B, \tag{43}$$

where c_A and c_B are concentrations of reactants in the bulk of a solution and k is the rate constant of a bimolecular electron transfer reaction

$$k = k_0 \exp\left\{ -\frac{V_i}{kT} - \frac{[E_s + \Delta F_0 + V_f - V_i]^2}{4E_s kT} \right\}. \tag{44}$$

Here k_0 is the pre-exponential factor proportional to the effective frequency ω_0 of a vibrational subsystem and to the transmission coefficient of the reaction $\varkappa$, ΔF_0 is the reaction free energy, V_i is the free energy of approaching of reactants from the infinity to the reaction configuration corresponding to some effective distance R between the reactants, V_f is the free energy of separating the reaction products and E_s is determined by Eq. (12).

The approximate calculation of the solvent reorganization energy for the charge transfer between spherical reactants was carried out by Marcus [2]. Under the assumption that the distance between the centres of reactants R is much larger than their radii a and b and that the reactants can be described by non-polarizable spheres with the charges strictly and uniformly distributed over their surfaces, the expression for E_s obtained in Ref. [2] has the form

$$E_s = e^2 n^2 \left(\frac{1}{\varepsilon_0} - \frac{1}{\varepsilon_s} \right) \left(\frac{1}{2a} + \frac{1}{2b} - \frac{1}{R} \right). \tag{45}$$

In some papers [27–29] an improved expression for E_s was obtained, which takes into account the effect of mutual polarizability of reactants and the effect of a finite size of reactants in calculating the integral in Eq. (12). Both the above effects give rise to additional terms of the order of a^3/R^4 and b^3/R^4 in the geometric factor in Eq. (45). Energies V_i and V_f in Eq. (44) can be expressed as

$$V_i = \frac{1}{8\pi} \int_{\infty - V_a - V_b} E_i \mathscr{D}_i \, d^3 r - \frac{z_1^2 e^2}{2a\varepsilon_s} - \frac{z_2^2 e^2}{2b\varepsilon_s}, \tag{46}$$

$$V_f = \frac{1}{8\pi} \int_{\infty - V_a - V_b} E_f \mathscr{D}_f \, d^3 r - \frac{(z_1+n)^2 e^2}{2a\varepsilon_s} - \frac{(z_2-n)^2 e^2}{2b\varepsilon_s}, \tag{47}$$

where the integration is carried out over the solvent volume exclusive of volumes of reactants, the second and third terms represent the Born solvation energies in the spherical model of ions. Calculations of V_i and V_f under the same approximations

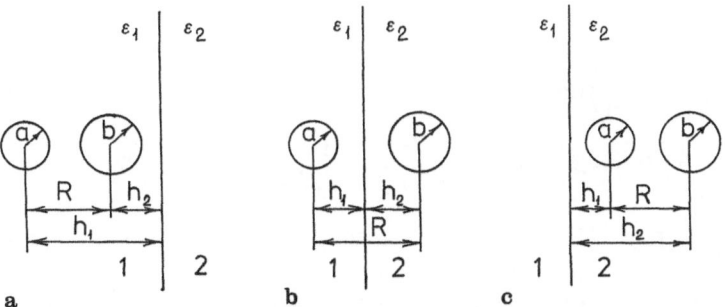

Fig. 7. Scheme of mutual position of reactants relative to the interface between the phases with dielectric constants ε_1 and ε_2: **a** case I – both reactants are in phase 1; **b** case II – one reactant is in phase 1 and the second one in phase 2; **c** case III – both reactants are in phase 2

as in deriving Eq. (45) result in simple Coulomb expressions

$$V_i = \frac{z_1 z_2 e^2}{\varepsilon_s R}; \quad V_f = \frac{(z_1 + n)(z_2 - n)e^2}{\varepsilon_s R}. \tag{48}$$

Now we shall consider the charge transfer process near the interface between the two immiscible liquids which have static dielectric constants ε_1 and ε_2 and optical dielectric constants ε_{01} and ε_{02}. We shall distinguish three basic cases of the reactants' positioning relative to the interface: I. Both reactants are placed in the medium with static dielectric constant ε_1 (Fig. 7a). II. One of the reactants (A^{z_1} for certainty) is placed in the medium with dielectric constant ε_1, and the second one in the medium with dielectric constant ε_2 (Fig. 7b). III. Both reactants are placed in the medium with dielectric constant ε_2 (Fig. 7c). We shall suppose for simplicity that both reactants are situated on the straight line which is perpendicular to the interface between the two dielectric media, reactant A^{z_1} being at distance h_1 and reactant B^{z_2} at distance h_2 from the interface plane. Then, in cases I, II, and III parameters a, b, R, h_1, and h_2 satisfy the relations (see Fig. 7):

$$\left.\begin{array}{l} R \geqq a+b \\ h_2 \geqq b \\ h_1 = h_2 + R \end{array}\right\} \text{ case I} \tag{49}$$

$$\left.\begin{array}{l} h_1 \geqq a \\ h_2 \geqq b \\ R = h_1 + h_2 \end{array}\right\} \text{ case II} \tag{50}$$

$$\left.\begin{array}{l} R \geqq a+b \\ h_1 \geqq a \\ h_2 = h_1 + R \end{array}\right\} \text{ case III} \tag{51}$$

When the charge transfer reaction proceeds near the interface, a reorganization (re-polarization) of each of two dielectric media occurs. In this case Eq. (12) can be written as [30]:

$$E_s = \frac{1}{8\pi} \int d^3r \left(\frac{1}{\varepsilon_0(r)} - \frac{1}{\varepsilon_s(r)} \right) (\mathscr{D}_i - \mathscr{D}_f)^2$$

$$= \frac{1}{8\pi} \left(\frac{1}{\varepsilon_{01}} - \frac{1}{\varepsilon_1} \right) \int_1 (\mathscr{D}_i - \mathscr{D}_f)^2 d^3r + \frac{1}{8\pi} \left(\frac{1}{\varepsilon_{02}} - \frac{1}{\varepsilon_2} \right) \int_2 (\mathscr{D}_i - \mathscr{D}_f)^2 d^3r$$

$$= \frac{1}{8\pi} c_1 I_1 + \frac{1}{8\pi} c_2 I_2. \tag{52}$$

Here $c_1 = 1/\varepsilon_{01} - 1/\varepsilon_1$; $c_2 = 1/\varepsilon_{02} - 1/\varepsilon_2$; 1 and 2 are the regions of integration which represent semispaces of media with static dielectric constants ε_1 and ε_2 exclusive of the volumes of reactants which are contained in these media; I_1 and I_2 are the integrals over regions 1 and 2.

Regions 1 and 2 and their respective distributions of inductions $\mathscr{D}_i$ and $\mathscr{D}_f$ at the initial and final states exist for each of three cases of the reactants' position relative to the interface which are shown in Fig. 7.

The corresponding generalizations should also be done for free energies V_i and V_f for the three cases mentioned above. In case I, V_i describes the work for approaching of the reactants, which initially were placed at an infinite distance in the medium with dielectric constant ε_1, into the reactive situation where these reactants are situated at distance R from each other in the same medium near the interface. V_f describes the work for separation of reaction products in the same medium. Thus, we have for case I

$$V_i^{\mathrm{I}} = \frac{1}{8\pi} \int E_i \mathscr{D}_i d^3r - \frac{z_1^2 e^2}{2a\varepsilon_1} - \frac{z_2^2 e^2}{2b\varepsilon_1}, \tag{53}$$

$$V_f^{\mathrm{I}} = \frac{1}{8\pi} \int E_f \mathscr{D}_f d^3r - \frac{(z_1+n)^2 e^2}{2a\varepsilon_1} - \frac{(z_2-n)^2 e^2}{2b\varepsilon_1}. \tag{54}$$

Here electric fields $E_{i,f}$ and inductions $D_{i,f}$ should be found from the solution of the electrostatic problem with due account of polarization of the interface between the two media.

Similar equations for V_i, V_f in case III may be obtained from Eqs. (53) and (54) by replacing $\varepsilon_1 \to \varepsilon_2$ and, besides, by replacing $\varepsilon_1 \rightleftarrows \varepsilon_2$ in the solution of the electrostatic problem for $E_{i,f}$ and $D_{i,f}$ for case I.

In case II, V_i describes the work for approaching of the reactants, which initially were placed at an infinite distance from each other in different phases, into the reactive situation where these reactants are situated at distance R from each other on different sides of the interface. V_f describes the work for separating the reactants, where the latter are brought from the reactive situation to an infinite distance from each other each retaining its own phase. V_i and V_f may be written for this case in the form:

$$V_i^{\mathrm{II}} = \frac{1}{8\pi} \int E_i \mathscr{D}_i d^3r - \frac{z_1^2 e^2}{2a\varepsilon_1} - \frac{z_2^2 e^2}{2b\varepsilon_2}, \tag{55}$$

$$V_f^{\mathrm{II}} = \frac{1}{8\pi} \int E_f \mathscr{D}_f d^3r - \frac{(z_1+n)^2 e^2}{2a\varepsilon_1} - \frac{(z_2-n)^2 e^2}{2b\varepsilon_2}. \tag{56}$$

To study the character of behaviour of the charge transfer reaction rate near the interface we shall relate the rates in cases II and III to the rate in case I, where the reactants are in the bulk of a phase with dielectric constant ε_1. In the given case we shall consider, for simplicity, the situation where the potential drop in diffuse layers can be neglected, and by the bulk of phases 1 and 2 are meant the distances of the order of tens of Ångstroems when the image forces can also be neglected. Furthermore, we shall confine ourselves to considering the simplest model of the interface, where the influence of a compact layer is not taken into account in calculating the ion interaction with the interface. The influence of a sublayer corresponding to the compact layer in the charge transfer kinetics was considered in Ref. [32, 33]. Taking into account the resolution of one of the reactants in case II and that of both reactants in case III, which results in the respective change of their concentration as compared to case I, we write the expression for the charge transfer reaction rate as:

$$K_{\mathrm{I}} = k_0 c_A^{\mathrm{I}} c_B^{\mathrm{I}} \exp \left\{ -\frac{V_i^{\mathrm{I}}}{kT} - \frac{(E_s^{\mathrm{I}} + \Delta \tilde{F}_0 + J_{fs}^{\mathrm{I}} - J_{is}^{\mathrm{I}} + V_f^{\mathrm{I}} - V_i^{\mathrm{I}})^2}{4 E_s^{\mathrm{I}} kT} \right\}, \tag{57}$$

$$K_{\mathrm{II}} = k_0 c_A^{\mathrm{I}} c_B^{\mathrm{I}} \exp \left\{ -\frac{J_{is}^{\mathrm{II}} - J_{is}^{\mathrm{I}}}{kT} - \frac{V_i^{\mathrm{II}}}{kT} - \frac{(E_s^{\mathrm{II}} + \Delta \tilde{F}_0 + J_{fs}^{\mathrm{II}} - J_{is}^{\mathrm{II}} + V_f^{\mathrm{II}} - V_i^{\mathrm{II}})^2}{4 E_s^{\mathrm{II}} kT} \right\}, \tag{58}$$

$$K_{\mathrm{III}} = k_0 c_A^{\mathrm{I}} c_B^{\mathrm{I}} \exp \left\{ -\frac{J_{is}^{\mathrm{III}} - J_{is}^{\mathrm{I}}}{kT} - \frac{V_i^{\mathrm{III}}}{kT} - \frac{(E_s^{\mathrm{III}} + \Delta \tilde{F}_0 + J_{fs}^{\mathrm{III}} - J_{is}^{\mathrm{III}} + V_f^{\mathrm{III}} - V_i^{\mathrm{III}})^2}{4 E_s^{\mathrm{III}} kT} \right\} \tag{59}$$

Here

$$J_{is}^{\mathrm{I}} = \frac{z_1^2 e^2}{2 a \varepsilon_1} + \frac{z_2^2 e^2}{2 b \varepsilon_1}, \tag{60}$$

$$J_{is}^{\mathrm{II}} = \frac{z_1^2 e^2}{2 a \varepsilon_1} + \frac{z_2^2 e^2}{2 b \varepsilon_2}, \tag{61}$$

$$J_{is}^{\mathrm{III}} = \frac{z_1^2 e^2}{2 a \varepsilon_2} + \frac{z_2^2 e^2}{2 b \varepsilon_2}. \tag{62}$$

The quantities $J_{fs}^{\mathrm{I, II, III}}$ are obtained from $J_{is}^{\mathrm{I, II, III}}$ by the substitutions $z_1 \to z_1 + n$, $z_2 \to z_2 - n$. In Eqs. (57)–(59) for cases I, II, III the contribution $\Delta \tilde{F}_0$, independent of the dielectric functions ε_1 and ε_2, is separated from the reaction free energy ΔF_0, i.e.:

$$\Delta F_0^{\mathrm{I, II, III}} = \Delta \tilde{F}_0 + J_{fs}^{\mathrm{I, II, III}} - J_{is}^{\mathrm{I, II, III}}. \tag{63}$$

Expressions for K_{I}, K_{II}, and K_{III} have been written under the assumption that the quantity k_0 remains unchanged for the three cases under consideration. For non-adiabatic processes this condition implies neglecting the variation of the overlapping integral for electronic wave functions and also neglecting the variation of the reorganization energy in the pre-exponential factor for the three cases of reactions near the interface under consideration.

To calculate reorganization energies $E_s^{\mathrm{I, II, III}}$ and interaction energies $V_{i,f}^{\mathrm{I, II, III}}$ it is convenient to express the potential in the form of a sum of contributions of

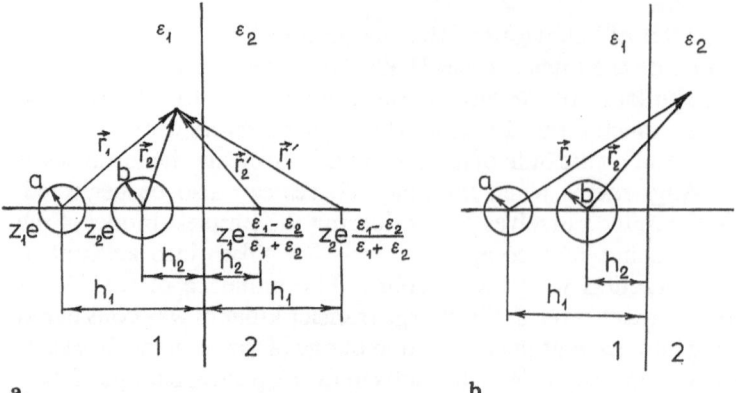

Fig. 8. Scheme of position of reactant charges and image charges corresponding to the expression of potential φ_i^I in Eq. (64): **a** φ_i^I in region I; **b** φ_i^I in region 2

reactant potentials and potentials produced by image charges [30]. So, in considering case I, potential ϕ_i may be written in the form (see Fig. 8):

$$\phi_i^I = \begin{cases} \dfrac{z_1 e}{\varepsilon_1 r_1} + \dfrac{z_1 e}{\varepsilon_1 r_1'}\left(\dfrac{\varepsilon_1-\varepsilon_2}{\varepsilon_1+\varepsilon_2}\right) + \dfrac{z_2 e}{\varepsilon_1 r_2} + \dfrac{z_2 e}{\varepsilon_1 r_2'}\left(\dfrac{\varepsilon_1-\varepsilon_2}{\varepsilon_1+\varepsilon_2}\right) & \text{in region 1} \\[4mm] \dfrac{z_1 e}{\varepsilon_2 r_1}\left(\dfrac{2\varepsilon_2}{\varepsilon_1+\varepsilon_2}\right) + \dfrac{z_2 e}{\varepsilon_2 r_2}\left(\dfrac{2\varepsilon_2}{\varepsilon_1+\varepsilon_2}\right) & \text{in region 2}. \end{cases} \tag{64}$$

Values ϕ_f^I are obtained from ϕ_i^I by substituting $z_1 \rightarrow z_1 + n$, $z_2 \rightarrow z_2 - n$. Using the relation $D_{i,f}^I = -\varepsilon \nabla \varphi_{i,f}^I$, the vector $\mathcal{D}_i^I - \mathcal{D}_f^I$ included in the expression for E_s can be represented as

$$\mathcal{D}_i^I - \mathcal{D}_f^I = n\nabla \begin{cases} \dfrac{e}{r_1} + \dfrac{e}{r_1'}\left(\dfrac{\varepsilon_1-\varepsilon_2}{\varepsilon_1+\varepsilon_2}\right) - \dfrac{e}{r_2} - \dfrac{e}{r_2'}\left(\dfrac{\varepsilon_1-\varepsilon_2}{\varepsilon_1+\varepsilon_2}\right) & \text{in region 1} \\[4mm] \dfrac{e}{r_1}\left(\dfrac{2\varepsilon_1}{\varepsilon_1+\varepsilon_2}\right) - \dfrac{e}{r_2}\left(\dfrac{2\varepsilon_2}{\varepsilon_1+\varepsilon_2}\right) & \text{in region 2}. \end{cases} \tag{65}$$

Then, passing in calculating integrals I_1 and I_2 in Eq. (52) for E_s from integration over a volume to integration over the surfaces which confine regions 1 and 2 (for region 1 these surfaces are the interface plane and surfaces of spheres a and b, and for region 2, the interface plane), we obtain to an accuracy of terms of the order of a^3/R^4 and b^3/R^4:

$$E_s^I = e^2 n^2 c_1 \left\{ \frac{1}{2a} + \frac{1}{2b} - \frac{1}{R} + \left[\left(\frac{2\varepsilon_1}{\varepsilon_1+\varepsilon_2}\right)^2 - 2\right]\frac{1}{8}\left(\frac{1}{h_1} + \frac{1}{h_2} - \frac{4}{h_1+h_2}\right)\right\}$$

$$+ \frac{e^2 n^2}{8} c_2 \left\{\frac{1}{h_1} + \frac{1}{h_2} - \frac{4}{h_1+h_2}\right\}\left(\frac{2\varepsilon_2}{\varepsilon_1+\varepsilon_2}\right)^2. \tag{66}$$

In a similar way one may calculate E_s^{III}

$$E_s^{III} = e^2 n^2 c_2 \left\{ \frac{1}{2a} + \frac{1}{2b} - \frac{1}{R} + \left[\left(\frac{2\varepsilon_2}{\varepsilon_1 + \varepsilon_2} \right)^2 - 2 \right] \frac{1}{8} \left(\frac{1}{h_1} + \frac{1}{h_2} - \frac{4}{h_1 + h_2} \right) \right\}$$
$$+ \frac{e^2 n^2 c_1}{8} \left\{ \frac{1}{h_1} + \frac{1}{h_2} - \frac{4}{h_1 + h_2} \right\} \left(\frac{2\varepsilon_1}{\varepsilon_1 + \varepsilon_2} \right)^2. \tag{67}$$

The expression for E_s^{II} was calculated in Ref. [30], and is determined, with an accuracy of the order of a^3/R^4, b^3/R^4, by the formula:

$$E_s^{II} = e^2 n^2 c_1 \left\{ \frac{1}{2a} - \frac{1}{4h_1} + \left(\frac{2\varepsilon_1}{\varepsilon_1 + \varepsilon_2} \right)^2 \frac{1}{8} \left(\frac{1}{h_1} + \frac{1}{h_2} - \frac{4}{h_1 + h_2} \right) \right\}$$
$$+ e^2 n^2 c_2 \left\{ \frac{1}{2b} - \frac{1}{4h_2} + \left(\frac{2\varepsilon_2}{\varepsilon_1 + \varepsilon_2} \right)^2 \frac{1}{8} \left(\frac{1}{h_1} + \frac{1}{h_2} - \frac{4}{h_1 + h_2} \right) \right\}. \tag{68}$$

The calculation of $V_{i,f}^I$ may also be carried out by passing in integrals over regions 1 and 2 in Eqs. (53)–(56) from volume to surface integration, which yields the following result:

$$V_i^I = \frac{z_1 z_2 e^2}{\varepsilon_1 R} + \frac{e^2(\varepsilon_1 - \varepsilon_2)}{4\varepsilon_1(\varepsilon_1 + \varepsilon_2)} \left(\frac{z_1^2}{h_1} + \frac{z_2^2}{h_2} + \frac{4z_1 z_2}{h_1 + h_2} \right), \tag{69}$$

$$V_f^I = \frac{(z_1 + n)(z_2 - n)e^2}{\varepsilon_1 R} + \frac{e^2(\varepsilon_1 - \varepsilon_2)}{4\varepsilon_1(\varepsilon_1 + \varepsilon_2)} \left(\frac{(z_1 + n)^2}{h_1} + \frac{(z_2 - n)^2}{h_2} \right.$$
$$\left. + \frac{4(z_1 + n)(z_2 - n)}{h_1 + h_2} \right). \tag{70}$$

The corresponding equations for V_i^{III} and V_f^{III} are obtained from the expressions for V_i^I and V_f^I by replacing $\varepsilon_1 \rightleftarrows \varepsilon_2$.

The equations for V_i^{II} and V_f^{II} were obtained in Ref. [30] and have the form:

$$V_i^{II} = \frac{2z_1 z_2 e^2}{(\varepsilon_1 + \varepsilon_2)(h_1 + h_2)} + \frac{z_1^2 e^2(\varepsilon_1 - \varepsilon_2)}{4\varepsilon_1(\varepsilon_1 + \varepsilon_2)h_1} + \frac{z_2^2 e^2(\varepsilon_2 - \varepsilon_1)}{4\varepsilon_2(\varepsilon_1 + \varepsilon_2)h_2}, \tag{71}$$

$$V_f^{II} = \frac{2(z_1 + n)(z_2 - n)e^2}{(\varepsilon_1 + \varepsilon_2)(h_1 + h_2)} + \frac{(z_1 + n)^2 e^2(\varepsilon_1 - \varepsilon_2)}{4\varepsilon_1(\varepsilon_1 + \varepsilon_2)h_1} + \frac{(z_2 - n)^2 e^2(\varepsilon_1 - \varepsilon_2)}{4\varepsilon_2(\varepsilon_1 + \varepsilon_2)h_2}. \tag{72}$$

Equations (57)–(59) together with Eqs. (60)–(62) and (66)–(72) determine the character of variation of the charge transfer reaction rate near the interface.

Now we shall discuss the character of behaviour of the medium reorganization energy $E_s^{I, II, III}$ with the changing position of reactants relative to the interface. As follows from Eqs. (66)–(68), E_s is proportional to the square of the charge transferred during the reaction, and includes two contributions proportional to c_1 and c_2, respectively, from the reorganization of a medium with dielectric constant ε_1 and a medium with dielectric constant ε_2. We shall consider the case where dielectric constants ε_1 and ε_2 differ considerably from each other, for example, at the contact of an aqueous solution with an organic liquid ($\varepsilon_1 \gg \varepsilon_2$). In this case usually $c_1 > c_2$. We shall suppose for simplicity that the reactants have equal radii and are in a mutual contact position: $a=b$, $R=2a$. Assuming $\varepsilon_2 \ll \varepsilon_1$ one may neglect the term proportional to c_2 in Eq. (66), since it includes the factor $(\varepsilon_2/\varepsilon_1)^2$. Then E_s^{I} can be written as:

$$E_s^{\mathrm{I}} \simeq \frac{e^2 n^2 c_1}{2a} \left\{ 1 + \frac{1}{2}\left(\frac{1}{\varrho_1} + \frac{1}{\varrho_1 - 2} - \frac{2}{\varrho_1 - 1} \right) \right\}, \tag{73}$$

where $\varrho_1 = h_1/a$. In case I, parameter ϱ_1 may vary within the limits $3 \leq \varrho_1 < \infty$. As follows from Eq. (73), when the reactants are in a water phase, the reorganization energy is mainly determined by the reorganization of a solvent in this phase only and grows when approaching the interface, reaching 7/6 of the reorganization energy value in the bulk of a water phase $E_s^{\mathrm{I}}(\infty) = e^2 n^2 c_1/2a$.

When the reactants are in phase 2, we obtain from Eq. (67) under the same assumptions:

$$E_s^{\mathrm{III}} \simeq \frac{e^2 n^2 c_2}{2a} \left\{ 1 + \left(\frac{c_1}{c_2} - \frac{1}{2} \right)\left(\frac{1}{\varrho_2} - \frac{1}{\varrho_2 - 2} - \frac{2}{\varrho_2 - 1} \right) \right\}, \tag{74}$$

where $\varrho_2 = h_2/a$. In case III, parameter ϱ_2 varies between the limits $3 \leq \varrho_2 < \infty$. As follows from Eq. (74), when the reactants are situated in the medium with the low dielectric constant, the medium reorganization energy E_s^{III} is determined by the contributions from both region 2 and region 1 provided that the reactants are situated not too far from the interface. E_s^{III} grows on approaching the interface, quantity c_1/c_2 being the greater, the larger the value of a contribution from the reorganization of an aqueous medium and the larger the quantity E_s^{III}.

And, finally, in case II, when one of the reactants is in phase I and the other in phase 2, we obtain from Eq. (68):

$$E_s^{\mathrm{II}} = \frac{e^2 n^2}{4a} (c_1 + c_2) = \tfrac{1}{2}[E_s^{\mathrm{I}}(\infty) + E_s^{\mathrm{III}}(\infty)]. \tag{75}$$

The medium reorganization energy is equal, in this case, to the arithmetic mean value of the reorganization energies at the bulk of phases 1 and 2 and may be lower than the reorganization energy values for charge transfer in phase 1 and in phase 2 just near the interface:

$$E_s^{\mathrm{II}} < E_s^{\mathrm{I}}(\varrho_1 = 3), \quad E_s^{\mathrm{III}}(\varrho_2 = 3). \tag{76}$$

The medium reorganization energy is an important parameter whose value determines to a considerable degree the activation energy of an elementary act of charge transfer. In the general case one may state that the lower the value of c_1 and c_2 parameters in the expression for E_s, the lower the reorganization energy and, respectively, the lower the charge transfer process activation energy. The charge transfer process in the bulk of phase 2 with low c_2 and low dielectric constant ε_2 is most favourable from the viewpoint of a minimum value of reorganization energy.

However, due to the decrease in the energy of solvation when the reactants pass from phase 1 with high dielectric constant into phase 2 with low dielectric constant, their concentration in phase 2 is considerably lower than that in phase 1. As a result of a competition between the concentration factor and the factor related to decreasing reorganization energy in phase 2, the resulting charge transfer reaction rate in the bulk of phase 2 may be either lower or comparable to the similar rate in the bulk of phase 1.

When the reactants approach the interface, the rate of a transfer process decreases as a rule. In some cases it may behave non-monotonically, passing a maximum at the distances of reactants from the interface comparable with their radii. Finally, in case II, when one of the reactants is in phase 2 and the second one is in phase 1, the reaction may be considerably accelerated as compared to reaction rates in the bulk of phases 1 and 2 (Figs. 9, 10, 11). In this case the interface plays the role of a catalyst. As an example, Figs. 9, 10, and 11 show the results of calculating the reorganization energy $E_s^{I,II,III}$, the activation factor E_a,

$$E_a = \begin{cases} V_i + \dfrac{(E_s^{I} + \Delta \tilde{F}_0 + J_{fs}^{I} - J_{is}^{I} + V_f^{I} - V_i^{I})^2}{4E_s^{I}} & \text{in case I} \\[2ex] J_{is}^{II} - J_{is}^{I} + V_i^{II} + \dfrac{(E_s^{II} + \Delta \tilde{F}_0 + J_{fs}^{II} - J_{is}^{II} + V_f^{II} - V_i^{II})^2}{4E_s^{II}} & \text{in case II} \\[2ex] J_{is}^{III} - J_{is}^{I} + V_i^{III} + \dfrac{(E_s^{II} + \Delta \tilde{F}_0 + J_{fs}^{III} - J_{is}^{III} + V_f^{III} - V_i^{III})^2}{4E_s^{III}} & \text{in case III} \end{cases} \tag{77}$$

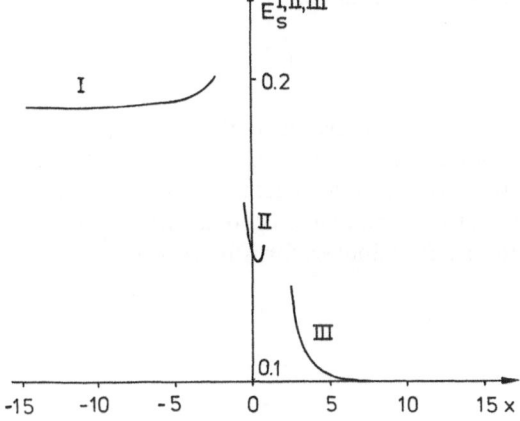

Fig. 9. Dependence of the dimensionless medium reorganization energy $E_s^{I,II,III}/(e^2/\text{Å})$ on the position of reactants relative to the interface (78), $a=b=2\text{Å}$, $R=6\text{Å}$, $z_1=-1$, $z_2=0$, $n=1$, $\Delta \tilde{F}_0=0$, $\varepsilon_{01}=1.78$, $\varepsilon_1=78$, $\varepsilon_{02}=2.5$, $\varepsilon_2=10$

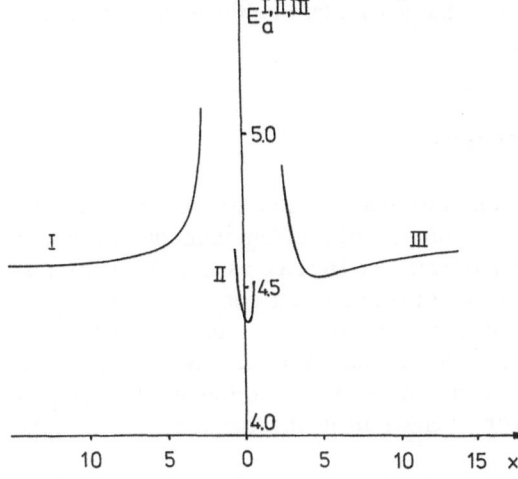

Fig. 10. Dependence of the dimensionless activation factor $E_a^{I,II,III}/(e^2/1\text{Å})$ (77) on the position of reactants relative to the interface. The values of parameters are the same as in Fig. 9

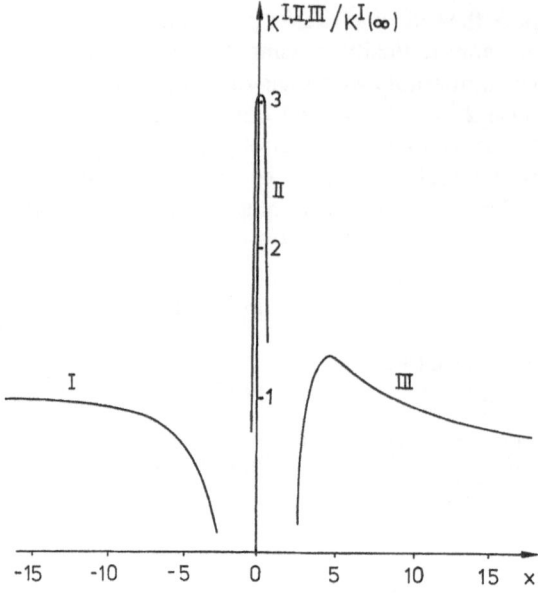

Fig. 11. Dependence of the relative charge transfer rate $K^{I, II, III}/K^I(\infty)$ on the position of reactants relative to the interface, $T = 300$ K, the other parameters are the same as in Figs. 9, 10

and the charge transfer rate near the interface related to the rate in the bulk of phase 1, $K^{I, II, III}/K^I(\infty)$, as functions of the position of reactants relative to the interface. It was assumed in these calculations that the distance between the reactants is fixed ($R = 6$ Å, $a = 2$ Å, $b = 2$ Å), and that only the distance of the center of mass of reactants relative to the interface varies. This distance is given in dimensionless form by relations:

$$x = \begin{cases} -(h_1 + h_2)/2a & \text{in case I} \\ (-h_1 + h_2)/2a & \text{in case II} \\ (h_1 + h_2)/2a & \text{in case III}. \end{cases} \tag{78}$$

As seen from Fig. 9, the medium reorganization energy in case II has an intermediate value as compared to E_s for cases I and III. The value of E_a in region II is lower than E_a^I and E_a^{III} and, accordingly, the charge transfer rate through the interface K^{II} increases as compared to K^I and K^{III}.

7. Ion Transfer Through the Interface

The calculation of the transfer rate of ions through the interface between the two immiscible liquids is generally a much more complicated problem than the calculation of the probability of an elementary electron transfer act. Here a great variety of physical situations is possible depending on the type of process under consideration.

Chemical reactions with a transfer of light ions (H^+, Li^+, etc.) between heavy molecular fragments are closest to electron transfer processes in their physical mechanism. In this case the behaviour of a transferable ion is quantum mechanical and is similar, in some respects, to an electron's behaviour in electron transfer reactions

[4, 34]. Due to the sub-barrier character of a light ion transfer the reactions are either non-adiabatic or partially adiabatic, and their probabilities are governed by equations of the type [4, 34]:

$$W = z^{-1} \frac{1}{\hbar} |V_{if}|^2 (\pi/kTE_s)^{1/2} \sum_{m,n} \exp(-E_m^i/kT)$$

$$\times |\langle \chi_n^f | \chi_m^i \rangle|^2 \exp \left\{ -\frac{(E_s + \Delta F + E_n^f - E_m^i)^2}{4E_s kT} \right\}, \tag{79}$$

where V_{if} is the electron resonance integral, z the vibrational statistical sum of an ion in the initial state, E_m^i and E_n^f and χ_m^i and χ_n^f are vibrational excitation energies and ion wave functions for the initial and final states, respectively. E_s is the medium reorganization energy, ΔF the free energy of a transfer. Summation is performed over all vibrational states of a transferable ion m and n for the initial and final electron states, which determine the bonding of an ion to an initial molecule and to an acceptor molecule.

The difference from electron transfer reactions consists in this case in the fact that the Franck-Condon factor for a transferable ion $\langle \chi_n^f | \chi_m^i \rangle^2$ is present in Eq. (79), and summation over the vibrational states of the latter is carried out. The medium reorganization energy E_s must be calculated, as in the case of electron transfer, with due account of the fact that the reactants are in different phases (see Sect. 6).

By using expressions of the type of Eq. (79) one may also describe the processes of tunnel hopping of a light ion from one cage formed by solvent molecules into the neighbouring cage [35, 36]. In this case the electron matrix element in Eq. (79) should be replaced by some energy V which plays the role of perturbation for an ion transfer between the states of zeroth-order approximation in each cage. This value can be estimated as the difference between the true potential in which an ion moves, and the potential of zeroth-order approximation which describes the motion of an ion in a separate cage [35, 36]. The processes of such a type are most probable in well structurized solvents and solid matrices, especially at low temperatures.

In the diffusion transfer of heavier ions their motion is presumably classical, and the transfer is accomplished through classical overcoming the potential barrier by a system. A transition of this type is similar in some senses to the adiabatic electron transfer reaction but its calculation is more difficult. If the potential barrier separating initial and final states of a system is sufficiently narrow and high, then the process rate is low and does not violate the equilibrium distribution in coordinates and in velocities at the initial state. Such a situation can be expected in transfer of ions which are not too heavy (with the mass of the order or less than the solvent mass) in well structurized solvents at room temperature.

The complexity of the problem follows from the fact that it is essentially many-body due to an ion interaction with medium molecules, the motions of various components of a system not being separable in the general case. Two basic approaches to describing the processes of the given type have been developed so far. These approaches correspond to two limiting cases one of which consists in a dynamical description of the motion of an ion and of a portion of the solvent molecules (including the medium polarization) which directly interact with an ion. This situation corresponds to the limiting case where the time between "collisions", leading to relaxation processes in a dynamical subsystem, is long compared to the characteristic transition time for high and narrow potential barriers [36]. In the second approach, the motion of an ion is described as a random walk in some potential due to random forces from ion interactions with fluctuating solvent molecules. This approach corresponds to the limiting case where an ion undergoes short rapid collisions and the time of correlation of the random forces is short compared to the characteristic transition times. Such a consideration is justified in describing the motion of macroscopic particles in liquids and can be extrapolated to some extent for an approximate description of large molecular fragments in solvents of low molecular weight. The discussion of the relation between the dynamical and stochastic approaches in the theory of charge transfer processes is given in Refs. [37, 38]. The difference between the results obtained within the framework of these approaches is revealed in considering the adiabatic transitions only. For non-adiabatic reactions both approaches yield the same result.

The calculation within the framework of the dynamical approach gives results which essentially coincide with those obtained using the transitional state method. The general expression for transition probability is in this case as follows [4]:

$$W = \frac{\omega_{\text{eff}}}{2\pi} \exp(-E_a/kT), \tag{80}$$

where ω_{eff} is the effective frequency and E_a the activation energy.

The activation energy is determined by a saddle point on the potential energy surface, and the effective frequency can be calculated as the ratio of vibrational statistical sums in the transitional configuration z^+ and at the initial state z_i [39]

$$\omega_{\text{eff}} = (z^+/z_i) \frac{kT}{\hbar}. \tag{81}$$

Specific expressions for E_a and ω_{eff} depend on the shape of the potential energy surface U_{ad}. If U_{ad} is approximated by the expression:

$$U_{ad} = \tfrac{1}{2}\{U_i + U_f - \sqrt{(U_i - U_f)^2 + 4V^2}\}, \tag{82}$$

where

$$U_i = \tfrac{1}{2} \sum_n \hbar\omega_n (q_n - q_{n0}^i)^2 + J_i$$

$$U_f = \tfrac{1}{2} \sum_n \hbar\omega_n (q_n - q_{n0}^f)^2 + J_f \tag{83}$$

then for sufficiently low V the expressions for ω_{eff} have the form [4, 5]:

$$\omega_{\text{eff}}^2 = \sum_n \omega_n^2 E_{rn} \Big/ \sum_n E_{rn}, \tag{84}$$

$$E_a = [E_r + \Delta J]^2 / 4E_r - 2V\sqrt{\alpha(1-\alpha)}, \tag{85}$$

where

$$E_r = \tfrac{1}{2} \sum_n \hbar\omega_n(q_{n0}^i - q_{n0}^f)^2; \quad \alpha = \tfrac{1}{2}(1 + \Delta J/E_r). \tag{86}$$

Note that summation in Eq. (84) is carried out over all degrees of freedom participating in the transition. These include both degrees of freedom Q_k, describing the motion of an ion and of closest molecules of a medium, and the degrees of freedom q_k describing the solvent polarization. For the latter, parameters $\Delta q_{k0} = q_{k0i} - q_{k0f}$, describing the displacement of equilibrium values of normal coordinates of effective oscillators in the transition, can be related to the complex dielectric function of a medium $\varepsilon(\omega)$ [4]. As a result, we obtain for ω_{eff} [4, 5]:

$$\omega_{\text{eff}}^2 = \frac{1}{E_r}\left[\sum_k \tfrac{1}{2}\hbar\omega_k^3(Q_{k0i} - Q_{k0f})^2 + \frac{2}{\pi}\int_0^\infty d\omega\,\omega^2\,\frac{\text{Im}\,\varepsilon(\omega)}{\omega|\varepsilon(\omega)|^2}\int d^3r(\mathcal{D}_i - \mathcal{D}_f)^2\right]. \tag{87}$$

Expressions for ω_{eff} at high V are obtained in Ref. [39]. Note that since $\varepsilon(\omega)$ depends on temperature in general, ω_{eff} also depends on temperature, and for some specific type of function $\varepsilon(\omega)$ it may even have an activation character, which may result in an additional contribution to the total activation energy.

Expressions for ω_{eff} and E_a may also be obtained for potentials of a more complicated form [4, 41]. In this sense the main difficulty in the given approach consists in determining the potential energy surface. For qualitative estimations one may also use model potentials of the same type as in Eqs. (82)–(83) or of a more complicated form, which take into account the discrete character of ion interaction with closest solvent molecules (see, e.g., Refs. [35, 36]).

The stochastic approach was first proposed in the classical work of Kramer [42] and has been further developed in recent years [43–46]. The stochastic description of the dynamics of heavy particles (which obey the equations of classical mechanics) is based on the equation proposed by Langevin:

$$du/dt = -\Gamma u - \frac{1}{m}\frac{\partial V}{\partial r} + A(t, r). \tag{88}$$

Here u is the microscopic velocity of a particle, m is its mass, $-\partial V/\partial r$ the force corresponding to the regular potential $V(r)$, $mA(t, r)$ the random force caused by fluctuations of a condensed medium in which the particle is situated, and Γ is the phenomenological coefficient of friction, which describes dissipative processes resulting from particle interactions with the medium.

In the general case, for solving Eq. (58) the field of random forces $A(t, r)$ must be specified. The problem, however, is considerably simplified if the characteristic time for fluctuations of these forces τ_A is short compared to the characteristic time of system evolution τ_s (in particular, $\tau_A \ll \Gamma^{-1}$). In this case (see, e.g., [47]):

$$\langle A(t)A(t')\rangle = 2\Gamma\,\frac{kT}{m}\,\delta(t - t'). \tag{89}$$

Here angular brackets denote statistical averaging, $\delta(t - t')$ is the delta-function which reflects mathematically the absence of correlations in the action of random forces over time scales of the order of τ_s (provided that $\tau_s \gg \tau_A$). The coefficient at $\delta(t-t')$ is related to Γ by virtue of the conditions imposed on the system at the thermodynamical equilibrium state (when $\langle u^2 \rangle = 3kT/2m$).

Due to the presence of the field of random forces, the behaviour of heavy particles is governed by the distribution function $F(t, r, u)$. Proceeding from Eqs. (88)–(89), one may show [47] that F satisfies the Fokker-Planck equation:

$$\left[\frac{\partial}{\partial t} + u \frac{\partial}{\partial r} - \frac{1}{m} \frac{\partial V}{\partial r} \frac{\partial}{\partial u} \right] F = - \frac{\partial}{\partial u} j_{FP}, \tag{90}$$

$$j_{FP} = - \Gamma \left(\frac{kT}{m} \frac{\partial F}{\partial u} + uF \right). \tag{91}$$

Let the equilibrium in velocities be established much more rapidly than that in coordinates. Then, considering the space-time evolution of the system, one may suppose that the Maxwell distribution in velocities occurs at all times, so that:

$$F(t, u, r) = \left(\frac{m}{2\pi kT} \right)^{3/2} \exp \left(- \frac{mu^2}{2kT} \right) f(t, r). \tag{92}$$

An equation of the type of Fick's diffusion equation follows from Eqs. (90)–(91) for the distribution function $f(t, r)$:

$$\partial f / \partial t = - \frac{\partial}{\partial r} j, \tag{93}$$

$$j = - \mathscr{D}_r [\partial f / \partial r + (f/kT) \partial V / \partial r], \tag{94}$$

where $\mathscr{D}_r = kT/\Gamma m$. The quasi-equilibrium condition used in the space of velocities is expressed by the inequality $(\mathscr{D}_r \Gamma^{-1})^{1/2} |d \ln V/dr| \gg 1$, where $|(d \ln V/dr)|$ characterizes the space scale of potential variation.

The ion transfer through the interface is associated with overcoming the multi-dimensional energy barrier of rather complicated shape, which generally fluctuates with time. Separating the average (regular) potential and including the fluctuating component into term $A(t)$ in Eq. (88), we use the formalism, described above, for calculating the ion current through the interface [46].

The ion moves in a complicated non-one-dimensional potential; however, by introducing the reaction coordinate (as has been usually done) we shall consider this motion to be one-dimensional. Under steady-state conditions we obtain from Eq. (94):

$$j = - \mathscr{D}_r \left(\frac{\partial f}{\partial x} + f \frac{\partial \psi}{\partial x} \right). \tag{95}$$

Here x is the reaction coordinate, $\psi = V/kT$ the dimensionless potential, and $j = \text{const}$ is the flow along the reaction coordinate to be determined. Using the expression for an equilibrium distribution function $f_0 = c \exp(-\psi)$, where c is a normalization constant, Eq. (95) is re-written in the form:

$$j = - \mathscr{D}_r f_0 \frac{d}{dx} \left(\frac{f}{f_0} \right). \tag{96}$$

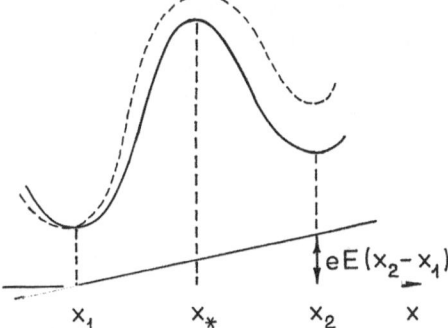

Fig. 12. The potential profile along the reaction coordinate

Integrating Eq. (96) along the reaction coordinate between points x_1 and x_2 which lie on different sides of the interface in phase 1 and in phase 2 (see Fig. 12), we obtain:

$$j \cdot \int_{x_1}^{x_2} \frac{dx}{\mathscr{D}_r f_0} = \frac{f(x_1)}{f_0(x_1)} - \frac{f(x_2)}{f_0(x_2)}. \tag{97}$$

Equation (97) is the key for calculating ion currents through the interface within the approach under consideration. For a singly charged positive ion, the current is equal to $i = ej$ where e is the positive charge equal to the electron charge in magnitude. At the equilibrium state the system under consideration is characterized by the exchange current $i_0 = ej_0$. The value of the equilibrium flow j_0 may be found from Eq. (97) under the assumption that at point x_1 the distribution function is equilibrium $f = f_0$, and at point x_2, $f = 0$, so that the current flows in one direction from x_1 to x_2. For the flow $j(x_1 \rightarrow x_2)$ we find from Eq. (97):

$$\vec{j}_0 = \left[\int_{x_1}^{x_2} \frac{dx}{\mathscr{D}_r f_0} \right]^{-1}. \tag{98}$$

Similarly, for the flow from x_2 to x_1, considering that $f(x_2) = f_0$ and $f(x_1) = 0$, we obtain:

$$\overset{\leftarrow}{j}_0 = - \left[\int_{x_1}^{x_2} \frac{dx}{\mathscr{D}_r f_0} \right]^{-1}. \tag{99}$$

Therefore,

$$|\vec{j}_0| = |\overset{\leftarrow}{j}_0| = j_0. \tag{100}$$

Calculating in the expression for $f_0 = c \exp(-\psi)$ the normalization constant by integrating near the bottom of the potential well in the vicinity of point x_1, leads to:

$$j_0 = c_{1s}^0 \omega_1 (m_1/2\pi kT)^{1/2} \left[\int_{x_1}^{x_2} \frac{1}{\mathscr{D}_r} e^{V(x)/kT} \right]^{-1}. \tag{101}$$

Here ω_1 is the frequency of vibrations of an ion in the potential well, and c_{1s}^0 is the equilibrium surface concentration of ions in phase 1.

If the height of the potential barrier at the interface ($x = x_*$) is large enough, so that $V_*/kT \gg 1$ (the potential at point x_1 is taken to be zero), then the main contribution to the integral in Eq. (101) is made by a narrow region near the top of the barrier: $x = x_*$. Approximating the potential near the maximum by the parabola:

$$V(x) = V_* - \frac{m\omega_*^2}{2}(x - x_*)^2 \tag{102}$$

leads to:

$$i_0 = \frac{ec_{1s}^0 \omega_1 \omega_*}{2\pi \Gamma} e^{-V_*/kT}. \tag{103}$$

The pre-exponential factor in Eq. (103) differs noticeably from a similar factor in the expression:

$$i_0 = \frac{ec_{1s}^0 \omega_1}{2\pi} e^{-V_*/kT} \tag{104}$$

which is obtained in the simplest version of the absolute reaction rate theory [48]. This is due to the difference in mechanisms for overcoming the surface barrier by an ion in the case $\Gamma > \omega_*$ and in the case $\Gamma < \omega_*$. Here the value of the current given by Eq. (103) is less than the value determined by Eq. (104).

We calculate now currents $\vec{i}$ and $\overleftarrow{i}$ in the presence of the external electric field E. In this case $|\vec{i}| \neq |\overleftarrow{i}|$ so that the resulting current

$$i = \vec{i} + \overleftarrow{i} \tag{105}$$

differs from zero. Near the barrier top, the potential $V(x)$ can be written, in the same approximation as before, as:

$$V(x) = V_* - \frac{\omega_*^2 m}{2}(x - x_*)^2 - eE(x - x_1). \tag{106}$$

Integrating Eq. (101) with due account of the form of $V(x)$ determined by Eq. (106), leads to:

$$\vec{i} = i_0 \exp\left\{-\left[eE(x_* - x_1) + \frac{e^2 E^2}{2m}\left(\frac{1}{\omega_*^2} + \frac{1}{\omega_1^2}\right)\right]\Big/kT\right\}, \tag{107}$$

$$\overleftarrow{i} = i_0 \exp\left\{-\left[eE(x_* - x_2) + \frac{e^2 E^2}{2m}\left(\frac{1}{\omega_*^2} + \frac{1}{\omega_2^2}\right)\right]\Big/kT\right\}. \tag{108}$$

If we introduce formally the overpotential η according to the relation which takes into account the shift of the potential minimum in the electric field:

$$e\eta = eE(x_2 - x_1) + \frac{e^2 E^2}{2m}\left(\frac{1}{\omega_1^2} - \frac{1}{\omega_2^2}\right) \tag{109}$$

and the transfer coefficient α:

$$\alpha = \frac{x_* - x_1 + \dfrac{eE}{2m}\left(\dfrac{1}{\omega_*^2} + \dfrac{1}{\omega_1^2}\right)}{x_2 - x_1 + \dfrac{eE}{2m}\left(\dfrac{1}{\omega_1^2} - \dfrac{1}{\omega_2^2}\right)} \tag{110}$$

then Eqs. (107) and (108) can be combined and written in the standard form:

$$i = i_0(e^{-\alpha e\eta/kT} - e^{(1-\alpha)e\eta/kT}). \tag{111}$$

For a case where the potential $V(x)$ near the barrier top $x = x_*$ differs noticeably from the parabolic form and can be described using an additional term, cubic in $x - x_*$:

$$V(y) = V_* - \frac{\omega_*^2 m}{2}\left(y^2 + \frac{y^3}{d}\right) - eEy, \tag{112}$$

where $y = x - x_*$ and $d \gtrless 0$ is the anharmonicity parameter, the integration in Eq. (101) yields [46]:

$$i = \tilde{i}_0(e^{-\tilde{\alpha} e\eta/kT} - e^{(1-\tilde{\alpha})e\eta/kT}), \tag{113}$$

where $\tilde{i}_0 = i_0 \sqrt[4]{1-t}$

$$\tilde{\alpha} = \frac{x_* - x_1 + \dfrac{eE}{2m}\left(\dfrac{1}{\omega_*^2} g(t) + \dfrac{1}{\omega_1^2}\right)}{x_2 - x_1 + \dfrac{eE}{2m}\left(\dfrac{1}{\omega_1^2} - \dfrac{1}{\omega_2^2}\right)}, \tag{114}$$

$$g(t) = \frac{4}{3t^2}[2(1 - \sqrt{1-t})(t-1) + t], \qquad t = 6eE/m\omega_*^2 d. \tag{115}$$

As follows from Eqs. (113)–(114), the existence of an "anharmonicity" near the barrier top is manifested only when the electric field is applied, which "reveals" its existence. A comparison of Eqs. (114)–(115) and (110) also shows that the anharmonicity introduces some additional asymmetry which is related physically to the fact that the field E may either enhance the barrier asymmetry (when the signs of E and d coincide), or weaken it; the sign of parameter t is determined by the sign of the ratio E/d.

This mechanism for the process results in a modification of the pre-exponential factor in the expression for the exchange current as compared to that given in the absolute reaction rate theory. The symmetry factor α depends, in general, on the overpotential, this dependence being determined by properties of a regular part of the potential curve near the bottom of the potential well and near the barrier top.

Recently new methods for the dynamical description of elementary processes of charge transfer in a condensed phase have been developed, which take into account the interaction of a dynamical subsystem with a thermostat (the role of which is played by a part of the medium). The method of classical trajectories is one of these. Below are given the results of calculations using this method for a simple model which allows us to

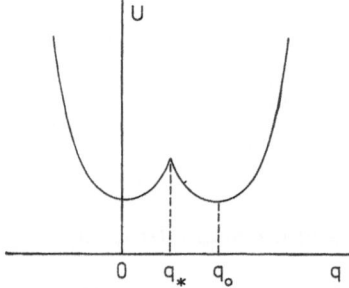

Fig. 13. Two-minimum potential well

obtain both limiting cases of the system's behaviour considered above, within the framework of a single approach [49].

We shall consider the motion of a particle in a symmetrical two-minima potential consisting of two potential wells of parabolic shape (Fig. 13). The Hamiltonian of the system, which describes the particle's states in the left potential well, is as follows

$$H = \hbar\omega a^+ a + \sum_m \hbar\Omega_m b_m^+ b_m + \sum_m \hbar\lambda_m(a^+ b_m + b_m^+ a_m), \tag{116}$$

where a^+ and a are the creation and annihilation operators for a particle, b_m^+ and b_m are the creation and annihilation operators for oscillators modelling the thermostat.

The last term in Eq. (116) describes the particle's interaction with a thermostat. We introduce the eigenfunctions and eigenvalues of creation and annihilation operators (coherent states) according to Ref. [50]:

$$\begin{aligned} a|\alpha\rangle &= \alpha|\alpha\rangle, & \langle\alpha|a^+ &= \alpha^*\langle\alpha| \\ b_m|\beta_m\rangle &= \beta_m|\beta_m\rangle, & \langle\beta_m|b_m^+ &= \beta_m^*\langle\beta_m|. \end{aligned} \tag{117}$$

Equations for the eigenvalues $\alpha(t)$ and $\beta_m(t)$ have the form:

$$\begin{aligned} \dot{\alpha}(t) &= -i\omega\alpha(t) - i\sum_m \lambda_m\beta_m(t) \\ \dot{\beta}_m(t) &= -i\Omega_m\beta_m(t) - i\lambda_m\alpha(t), \end{aligned} \tag{118}$$

where a dot above the function denotes differentiation with respect to time.

The real and imaginary parts of the eigenvalues are related to the expectation values of the respective operators for position and velocity. In particular,

$$q(t) = \sqrt{2}\,\mathrm{Re}\{\alpha(t)\}. \tag{119}$$

The initial conditions for Eq. (118) are

$$\alpha(0) = \alpha; \qquad \beta_m(0) = \beta_m. \tag{120}$$

In the classical limit, for the given initial conditions the probability $p(t)$ of finding a particle at time t in the final state, i.e. in the region of the right hand potential well ($q > q^* = q_0/2$) is obviously equal to:

$$p(t) = \theta(\mathrm{Re}\{\alpha(t)\} - q^*/\sqrt{2}) = \begin{cases} 1, & \mathrm{Re}\{\alpha(t)\} > q^*/\sqrt{2} \\ 0, & \mathrm{Re}\{\alpha(t)\} < q^*/\sqrt{2}. \end{cases} \tag{121}$$

The calculation of the average probability $P(t)$ is reduced to integration of $p(t)$ over all possible initial values of α and β_m with the respective distribution function $f(\alpha, \beta_m)$ the form of which depends on the preparation procedure of the system:

$$P(t) = \int p(t) f(\alpha, \beta_m) d^2\alpha \prod_m d^2\beta_m. \tag{122}$$

If the initial state of a system is at equilibrium in all degrees of freedom, then the normalized distribution function $f(\alpha, \beta_m)$ has the form:

$$f(\alpha, \beta_m) = \prod_m \left(\frac{1}{\pi\langle n\rangle}\right)\left(\frac{1}{\pi\langle n_m\rangle}\right) \exp\left[-\frac{|\beta_m|^2}{\langle n_m\rangle} - \frac{|\alpha|^2}{\langle n\rangle}\right], \tag{123}$$

where $\langle n_m\rangle$ and $\langle n\rangle$ are the average occupation numbers for the vibrational modes at the given temperature

$$\langle n_m\rangle = [\exp(\hbar\Omega_m/kT) - 1]^{-1} \simeq kT/\hbar\Omega_m$$
$$\langle n\rangle = [\exp(\hbar\omega/kT) - 1]^{-1} \simeq kT/\hbar\omega. \tag{124}$$

The probability of transition per unit time W is determined by differentiating $P(t)$:

$$W = dP(t)/dt. \tag{125}$$

The exact solution of Eqs. (118) with respect to $\alpha(t)$ can be written formally for any form of frequency spectrum in a thermostat. Below we shall accept the often used approximation according to which the spectrum of thermostat oscillators near the frequency ω is considered to be dense enough that the solution for $\alpha(t)$ is as follows:

$$\alpha(t) = \alpha\exp(-i\bar{\omega}t) + \sum_m \{\lambda_m\beta_m[\exp(-i\bar{\omega}t) - \exp(-i\Omega_m t)]/(\bar{\omega} - \Omega_m)\}, \tag{126}$$

where

$$\bar{\omega} = \omega + \delta\omega - i\varkappa. \tag{127}$$

The damping constant $\varkappa$ (the friction) and the frequency shift $\delta\omega$ are expressed in terms of coupling constants λ_m of an oscillator which describes the transferable particle (the q-oscillator) with thermostat oscillators (the Q-oscillator) as well as in terms of frequency characteristics of the latter [50].

The integration in Eq. (122) with due account of Eqs. (123) and (126) yields the expression for $P(t)$, which for short times ($t < \varkappa^{-1}, \omega^{-1}$) has the form [43]:

$$P(t) = Wt = (\omega_{\text{eff}} t/2\pi)\exp(-\hbar\omega q^{*2}/2kT), \tag{128}$$

where

$$\omega_{\text{eff}}^2 = \omega^2 + \sum_m \lambda_m^2\langle n_m\rangle/\langle n\rangle. \tag{129}$$

Thus, in this limit the concept of transition probability per unit time W may be introduced, which takes a form similar to that in the theory of adiabatic electron transfer processes. Equations (128)–(129) can also be generalized to the case of a multi-dimensional oscillator interacting with a thermostat. In this case the concept of transition probability per unit time, which does not depend on t, is preserved over long time intervals as well [51]. If the system is prepared in such a manner that in the initial state there exists an equilibrium in all degrees of freedom with due account of particle-thermostat interaction, then an expression for $P(t)$ is obtained, which differs from Eq. (128) only in that the energy in the exponent, corresponding to the transitional configuration q^*, should be substituted by free energy in this configuration.

Consider now the situation where the distribution function for thermostat oscillators corresponds to the thermal equilibrium distribution in coordinates and velocities, and where, at the initial time $t=0$, the reactive q-oscillator is at the coordinate origin with zero velocity, i.e.:

$$f(\alpha, \beta_m) = \delta^{(2)}(\alpha) \prod_m \left(\frac{1}{\pi \langle n_m \rangle} \right) \exp \left[-\frac{|\beta_m|^2}{\langle n_m \rangle} \right], \tag{130}$$

where $\delta^{(2)}(\alpha)$ is the two-dimensional δ-function:

$$\delta^{(2)}(\alpha) = \delta(\mathrm{Re}\,\alpha)\delta(\mathrm{Im}\,\alpha).$$

The integration in Eq. (122) using the solution of Eq. (126) yields the lower bound for $P(t)$. The quantity $W = dP/dt$ is found to be a complicated function of time [49]. It is small at low t and grows with growing t reaching a maximum at some point t_0 and then decreases. W is approximately constant near the maximum, and a transition probability per unit time may introduced as an approximation. The calculation gives for W in the vicinity of a maximum the expression [49]:

$$W|_{t=t_0} \sim (2\tau)^{-1} \exp(-E_a/kT), \tag{131}$$

where

$$\tau^{-1} = 2\varkappa; \quad E_a = \hbar\omega q^{*2}/2. \tag{132}$$

Thus, unlike the previous case, where the transition probability per unit time exists even at small times and is determined by the frequency characteristics of a reaction oscillator, in the given case the concept of transition probability per unit time arises at sufficiently long times only, and the frequency factor (the pre-exponent) is determined by an inverse relaxation time $\tau^{-1} = 2\varkappa$.

The difference between the limiting cases mentioned above follows from the physical mechanism of a transition which consists in the following. In the system there exist a number of reactive degrees of freedom (one degree in the simplest case) which interact with the other vibrational degrees of freedom composing a thermostat. The potential energy surface along the reactive degree of freedom has two minima corresponding to the equilibrium initial and final states which are separated by some potential barrier. The transition from region 1 near the first minimum into region 2 near the second minimum leading to the reaction is accomplished by means of classically overcoming the potential barrier by motion along the reactive mode.

As follows from the calculation carried out above [49], the possibility of introducing the concept of transition probability per unit time and the form of W depend on the properties of a reactive sub-system, on the time of observation, and on the interaction of the reactive sub-system with a thermostate. So, in the case of one reactive mode not interacting with the thermostat, the system oscillates between two potential wells and the reaction in the usual sense does not occur. If the reactive oscillator interacts with the thermostat, then due to exchange of their energies the motion of the oscillator is complicated. If the thermostat has temperature $T = 0$ K, then the motion of the reactive oscillator has the character of damped oscillations. If the temperature of a thermostat is high enough, then the time dependence of a reactive oscillator coordinate is more complicated and has a random character on the average while having a general tendency to damping. This implies in the dynamical language that there exist trajectories which after intersecting the potential barrier describe damped vibrations of an oscillator about the new equilibrium position, and the transition through a barrier implies for these trajectories that the reaction does actually occur. However, there are also trajectories for which the system returns to region 1 even after intersecting the barrier.

If the initial distribution in coordinates and velocities is equilibrium, then at small times the damping has no influence and the system behaves as a dynamical one. In this case the probability of reaching the top of a barrier $P(t)$ prior to time t is mainly determined by the "concentration" of systems near the barrier top. At large t the damping processes become significant.

If at the initial time the reactive oscillator was near the initial state minimum $q = 0$ with virtually zero velocity $v \simeq 0$, then at small times $P(t)$ is about zero, since time is required for the system to reach the barrier top. In this case the reactive oscillator reaches the barrier top at the cost of thermostat energy. And at large times its motion to the transitional configuration has a complicated (and on average, stochastic) character.

8. Conclusion

It follows from the above discussion that the physical mechanism of charge transfer processes at the interface between two immiscible liquids does not differ in principle from the physical mechanism of other charge transfer reactions in polar media. General concepts and methods of the quantum-mechanical theory of charge transfer processes in polar media, which has been developed for more than 20 years and continues to develop intensively at present, are also applicable for describing processes in the systems of interest here. However, in a specific analysis one must take into account the specificity stipulated by the contact of two liquids. This specificity reveals itself, first of all, in the possibility of transition of the reactants from one liquid to another and in the influence of the interface between the two dielectric media on the values of kinetic parameters. The specificity of the interface under consideration results in complication of methods for calculating the kinetics of charge transfer processes. However, it is for systems of this type that the differences in the various theoretical approaches are revealed in the most clear manner, allowing us to establish the limits of their applicability. Thus the study of these systems is of great significance for further development of the quantum-mechanical theory of chemical kinetics in solutions.

References

1. Libby, W.: J. Phys. Chem. *56*, 863 (1952)
2. Marcus, R.A.: ibid. *24*, 966 (1956), *26*, 867 (1957)
3. Levich, V.G., Dogonadze, R.R.: Dokl. AN SSSR *124*, 123 (1959) (in Russian)

4. Dogonadze, R.R., Kuznetsov, A.M.: in "Itogi Nauki i Tekhniki", ser. Physical Chemistry. Kinetics, v. 2, M.: VINITI, 1973 (in Russian)
5. Dogonadze, R.R., Kuznetsov, A.M.: in "Itogi Nauki i Tekhniki", ser. Kinetics and Catalysis, v. 5, M.: VINITI, 1978 (in Russian)
6. Ulstrup, J.: Charge Transfer Processes in Condensed Media, Springer-Verlag, Berlin-Heidelberg-New York 1979
7. Kuznetsov, A.M.: Elektrokhimiya 17, 84 (1981)
8. Kuznetsov, A.M.: Nouv. J. de Chimie 5, 427 (1981)
9. Warshel, A.: J. Phys. Chem. 86, 2218 (1982)
10. Kuznetsov, A.M.: J. Electroanal. Chem. 159, 241 (1983)
11. Vorotyntsev, M.A., Dogonadze, R.R., Kuznetsov, A.M.: Dokl. AN SSSR 195, 1135 (1970)
12. Pekar, S.I.: Untersuchungen über die Elektronentheorie der Kristalle, Berlin, Akademie-Verlag 1954
13. Kuznetsov, A.M., Ulstrup, E.: Elektrokhimiya 21, 632 (1985)
14. Cannon, R.D.: Chem. Phys. Lett. 49, 299 (1977)
15. Tembe, B.L., Friedman, H.L., Newton, M.D.: J. Chem. Phys. 76, 1490 (1982)
16. Gavach, C., Henry, F.: J. Electroanal. Chem. 54, 361 (1974)
17. Gavach, C.D., Epenoux, B.D.: ibid. 55, 59 (1974)
18. Gavach, C., D'Epenoux, B.D., Henry, F.: ibid. 64, 107 (1975)
19. Guastalla, J.: Nature 227, 485 (1970)
20. Koryta, J., Vanysek, P., Brezina, M.: J. Electroanal. Chem. 67, 263 (1976)
21. Koryta, J., Vanysek, P., Brezina, M.: ibid. 75, 211 (1977)
22. Samec, Z., Marecek, V., Koryta, J., Khalil, M.W.: ibid. 83, 393 (1977)
23. Samec, Z.: ibid. 99, 197 (1979)
24. Koryta, J.: Hungarian Sci. Instruments 49, 25 (1980)
25. Gavach, C., Seta, P., D'Epenoux, B.: J. Electroanal. Chem. 83, 225 (1977)
26. Ti Tien, H.: Bilayer Lipid Membranes (BLM), Theory and Practice, Marcel Dekker, New York, 1974, p. 181
27. Kharkats, Yu.I.: Elektrokhimiya 3, 881 (1973)
28. Kharkats, Yu.I.: ibid. 12, 592 (1976)
29. Kharkats, Yu.I., Chudin, N.I.: ibid. 20, 892 (1984)
30. Kharkats, Yu.I.: ibid. 12, 1370 (1976)
31. Kharkats, Yu.I.: ibid. 15, 246 (1979)
32. Kharkats, Yu.I., Nielsen, H., Ulstrup, J.: J. Electroanal. Chem. 196, 47 (1984)
33. Nielsen, H., Ulstrup, J., Kharkats, Yu.I.: Elektrokhimiya 21, 1541 (1985)
34. Dogondaze, R.R., Kuznetsov, A.M.: Comprehensive Treatise of Electrochemistry, v. 7, p. 1-40, Plenum Publishing Corp., New York, 1983
35. Schmidt, P.P.: J. Chem. Soc., Faraday Trans. 2 80, 157 (1984)
36. Schmidt, P.P.: ibid. 2 80, 181 (1984)
37. Kuznetsov, A.M.: Elektrokhimiya 20, 1069 (1984)
38. Kuznetsov, A.M.: ibid. 20, 1226 (1984)
39. Levich, V.G., Dogonadze, R.R.: Coll. Czech. Chem. Comm. 26, 193 (1961)
40. Dakhnovsky, Yu.I., Ovchinnikov, A.A.: Dokl. AN SSSR 270, 119 (1983)
41. Dogonadze, R.R., Urushadze, Z.D.: J. Electroanal. Chem. 32, 235 (1971)
42. Kramers, H.A.: Physica 7, 284 (1940)
43. Visscher, P.B.: Phys. Rev. B 14, 347 (1976)
44. Weiner, J.H., Korman, R.E.: ibid. B 10, 315 (1974)
45. Zusman, L.D.: Chem. Phys. 80, 29 (1983)
46. Gurevich, Yu.Ya., Kharkats, Yu.I.: J. Electroanal. Chem. 200, 3 (1986)
47. Klimontovich, Yu.L.: Statistical physics, M., Nauka, 1982 (in Russian)
48. Glasston, S., Laidler, K., Eyring, H.: The theory of Rate Processes, New York, London and Tokyo, McGraw-Hill 1941
49. Kuznetsov, A.M.: Elektrokhimiya 20, 1233 (1984)
50. Glauber, R.: in "Coherent states in a quantum theory", M., Mir, 1972 (in Russian)
51. Dogonadze, R.R., Kuznetsov, A.M.: Elektrokhimiya 3, 1324 (1967)

Hydrodynamics and Mass Exchange at the Phase Boundaries with Regular Dissipative Structures

V. S. Krylov

1. Introduction

Searching for ways to intensify the interfacial heat and mass exchange processes is one of the most significant problems of modern chemical and electrochemical technology. The extremal behaviour of the interfacial exchange (extremal with respect to the thermodynamic balance) is the most important part of this investigation [1, 2]. For the solution of technological problems the utilization of regular dissipative structures is of great interest, firstly because of the very high efficiencies of power consumption for these processes, and secondly due to the principal possibility of obtaining a quantitative kinetic description of the interfacial heat and mass exchange without requiring a model for the case of dissipative structures. A further aim is to develop reliable methods for increasing the efficiency of the equipment by using this description.

Because of the variable and complex character of physical mechanisms controlling dissipative structure formation and behaviour, we can observe only the very beginning of a theory which mathematically describes the transfer processes and the interfacial exchange in systems with this kind of structure. The present review includes consideration of current basic theoretical descriptions of regularities of hydrodynamics and mass exchange at the mobile phase boundaries with a heterogeneous distribution of surface tension. Methods of criteria definition are described for the disturbance of the interfacial hydrodynamic stability, and the conditions of rising regular circulational convective flows are discussed. A theoretical analysis is given for the kinetics of mass transfer through the interface with circulational cell hydrodynamic structure. In view the theoretical results we formulate possible ways of organization of high intensity mass exchange processes in liquid-liquid systems.

2. General Theoretical Description of Dissipative Structures

Disruption of hydrodynamical stability may take place either as a result of local fluctuations in volume properties of a continuous medium (viscosity or density) or due to disturbances introduced into the boundary conditions. A general equation system for the differential balance of forces can be written as:

$$\frac{\partial}{\partial t}(\varrho v_i) + \frac{\partial}{\partial x_k} \Pi_{ik} = 0, \tag{1}$$

$$\frac{\partial \varrho}{\partial t} + \frac{\partial(\sigma v_k)}{\partial x_k} = 0 \tag{2}$$

The Interface Structure and Electrochemical Processes at the Boundary Between Two Immiscible Liquids
Editor: V. E. Kazarinov
© Springer-Verlag Berlin, Heidelberg 1987

where Π_{ik} is the tensor of the momentum flow density including inertial, hydrodynamic, osmotic and external force contributions (taking into account gravity and electromagnetic factors); $i, k = 1, 2, 3$ are the components of hydrodynamic velocities; ϱ is a liquid density. In the presence of an external electrical field the expression for Π_{ik} is [3]:

$$\Pi_{ik} = \varrho v_i v_k + \delta_{ik} \left\{ p + kT \sum_a (c_a - c_{a0}) \right\} - \tau_{ik}^{(l)} - \tau_{ik}^{(\mu)} \tag{3}$$

where c_a and c_{a0} are the concentrations of the "a"-kind component in arbitrary point of the given phase and in the area of the uniform distribution; $\tau_{ik}^{(l)}$ and $\tau_{ik}^{(\mu)}$ are tensors of viscosity and electrical field strength*, respectively. For these tensors the following equations apply:

$$\tau_{ik}^{(l)} = \mu \left(\frac{\partial v_i}{\partial x_k} + \frac{\partial v_k}{\partial x_i} \right), \tag{4}$$

$$\tau_{ik}^{(\mu)} = \frac{\varepsilon}{4\pi} E_i E_k - \frac{\delta_{ik}}{8\pi} E^2 \left[\varepsilon - \varrho \left(\frac{\partial \varepsilon}{\partial \varrho} \right) T \right]. \tag{5}$$

In these expressions μ is the dynamic viscosity of a liquid, ε is a dielectric permeability, E_i and E_k are the electric field intensity components. These components in principle cannot be obtained from Eqs. (1)–(2); to determine them it is necessary to use the Maxwell equation:

$$\frac{\partial}{\partial x_k} (E_k \varepsilon) = 4\pi_q \tag{6}$$

where q is the density of the free electric charge in a liquid. The item in Eq. (3) which is proportional to the concentration difference for a substance distributed nonuniformly in each phase, is responsible for the osmotic pressure contribution to the resulting balance of forces. As follows from Ref. [4], this contribution is exactly equal to one of the Maxwell tensors $\tau_{ik}^{(\mu)}$ when calculating the surface tension of electrolyte solutions.

To complete Eqs. (1), (2), and (6) it is necessary to add the state equation:

$$p = p(\varrho, T) \tag{7}$$

which connects pressure, density, and temperature, as much as the convective thermoconduction equation:

$$\frac{\partial T}{\partial t} + v_k \frac{\partial T}{\partial x_k} = \frac{\partial}{\partial x_k} \left(\varkappa \frac{\partial T}{\partial x_k} \right) \tag{8}$$

where $\varkappa$ is a thermoconduction factor. Further, when the density ϱ and the viscosity μ are dependent on the concentration of the impurity dissolved in the liquid, the above considered system of equations is joined with the equation of convective diffusion:

$$\frac{\partial c_j}{\partial t} + v_k \frac{\partial c_j}{\partial x_k} = \frac{\partial}{\partial x_k} \left(D_j \frac{\partial c_j}{\partial x_k} \right) \tag{9}$$

---

* In a more general case the right hand side of Eq. (3) is to be completed with one more item such as the magnetic strength tensor. However, in the present review the aspects of magnetohydrodynamical phenomena influence upon the dissipative structures, will not be considered.

where the j integer is a number of dissolved components and where the concentrations c_j influence the viscosity and density, and D_j is the diffusion coefficient for numbered components.

For instability appearance analysis it is essential not only to develop the main equations of transfer but to formulate boundary conditions, because dissipative structures appear more frequently near phase boundaries and may be caused by fluctuations of boundary conditions. It is of great practical interest to investigate the role of capillary [5] and electrosurface [6] forces in inducing the instability of mobile phase boundaries. That is the reason for determining boundary conditions for the transfer equation system characterizing the interface of two mobile media (liquid-gas or liquid-liquid). Let $r^{(s)}(t)$ be the radius-vector drawn from any fixed space point to another on the phase boundary and n a unit vector normal to the interface and directed towards the interior of phase 1 (the second phase would be written as phase 2). The tangential unit vector at the interfacial surface would be τ. Then conjugation conditions for velocities, temperatures, concentrations and electric fields at the interface would be:

$$v_k^{(1)}\tau_k = v_k^{(2)}\tau_k, \tag{10}$$

$$v_k^{(1)}n_k = v_k^{(2)}n_k = \frac{dx_k^{(s)}}{dt}n_k, \tag{11}$$

$$p_2 - p_1 + kT\sum_a (c_a^{(2)} - c_{a0}^{(2)} - c_a^{(1)} + c_{a0}^{(1)}) = (\sigma_{ik}^{(2)} - \sigma_{ik}^{(1)})n_in_k - \tau\left(\frac{1}{R_1} + \frac{1}{R_2}\right), \tag{12}$$

$$\Gamma M\tau_k\left\{\frac{\partial v_k^{(s)}}{\partial t} + v_i^{(s)}\frac{\partial v_k^{(s)}}{\partial x_i}\right\} = \tau_k\frac{\partial \sigma}{\partial x_k} - (\sigma_{ik}^{(2)} - \sigma_{ik}^{(1)})n_i\tau_k, \tag{13}$$

$$D_1n_k\frac{\partial c^{(1)}}{\partial x_k} - D_2n_k\frac{\partial c^{(2)}}{\partial x_k} = \frac{\partial \Gamma}{\partial t} + \tau_k\frac{\partial}{\partial x_k}\left(v_k^{(s)}\Gamma - D_s\frac{\partial \Gamma}{\partial x_k}\right) + R_d - R_a + R_c, \tag{14}$$

$$c^{(1)} = f(c^{(2)}), \tag{15}$$

$$\lambda_1 n_k\frac{\partial T^{(1)}}{\partial x_k} - \lambda_2 n_k\frac{\partial T^{(2)}}{\partial x_k} = -D_1 n_1\frac{\partial c^{(1)}}{\partial x_k}Q_1(T^{(s)}), \tag{16}$$

$$T^{(1)} = T^{(2)}. \tag{17}$$

Equations (10) and (11) describe a continuous velocity field at the phase boundary; Eqs. (12) and (13) formulate the normal and tangential force balance (where σ is the surface tension, R_1 and R_2 the principal surface curvature radii, $\sigma_{ik} = \tau_{ik}^{(l)} + \tau_{ik}^{(\mu)}$ the surface tension tensor, Γ the Gibbs surface surplus of surface-active substance taken through the interface, M is its molecular mass, D_i and λ_i the diffusion and thermoconductivity factors in proper phases; R_d, R_a, and R_c are the desorption, adsorption, and surface chemical reaction velocities. Equation (15) supposes the existence at the phase boundary of local interfacial thermodynamic balance which can be described by the known equation of the adsorption isotherm. Equation (16) takes into account the surface thermic resistance caused by phase changing limited by the velocity of diffusive removal of the component into the interior of phase 1 (in case of phase 2 evaporation, the quantity Q_1 is the latent heat of evaporation).

The occurrence of unstable current conditions is connected with dynamical development features of initial slow disturbances of the velocity profile. The stability analysis is aimed at determining the conditions which promote nonlimited disturbance rising in connection with space and time coordinates. The theoretically based [7, 8] proposal is applied to systems with induced or spontaneous convection where two-dimensional disturbances are responsible for stability disruption (such disturbances have a maximum velocity at the initial stage at least). These two-dimensional disturbances can be interpreted as periodic functions of coordinates and then it is possible to decompose the given disturbance into a Fourier series. For example, in the case of an initial spontaneous convective current along the flat vertical plate $y = 0$ the disturbed current function $\psi(x, y, t)$ is written as [9]:

$$\psi(x, y, t) = \bar{\psi}(x, y) + \psi'(x, y, t) \tag{18}$$

where the disturbing function $\psi'(x, y, t)$ is given by:

$$\psi'(x, y, t) = 4\nu \sqrt[4]{c_r r_x/4}\, \varphi(\eta) \exp(i\alpha x + \beta t) \tag{19}$$

Here x is a vertical coordinate along the surface beginning from the front edge $x = 0$; $\eta = (y/x)\sqrt[4]{c_r r_x/4}$ is an automodeled variable, $c_r r\alpha$ is the local Grasgoff number, and ν the dynamic viscosity of the medium. Introducing Eq. (18) into the initial momentum transfer equations of Eqs. (1)–(2) and then linearizing these equations for disturbed parameters we can obtain the equations for the function $\varphi(\eta)$ and for the α and β parameters. The character of the disturbance evolution over time is determined by the β parameter. In complex form it can be written as:

$$\beta = \beta_1 + i\beta_2$$

and it is clear that if $\beta_1 > 0$, then the medium movement near a vertical plate will be unstable, if $\beta_1 < 0$, it will be stable and if $\beta_1 = 0$, it becomes neutral-stable. Depending on the quantity β_2 the neutral-stable movement may be realized in two forms, either in form of circular cells (when $\beta_2 = 0$) or in form of oscillating currents (when $\beta_2 \neq 0$).

The approach of a two-dimensional velocity disturbing field, with as much linearization as possible of initial transfer equations in respect to the introduced disturbance amplitude, becomes incorrect when increasing disturbances develop down the stream. More perfect analysis of the induced convective flow stability in a nonlinear theory and with three-dimensional disturbance geometry, impel us to draw the conclusion that the disturbances which were initially two-dimensional transform into three-dimensional turbulences [10]. While studying nonlinear effects in terms of spontaneous convection [11] the possibility of an averaged secondary flowing appearance with deformation of the main current was found. In particular, it was proved that there exists a system including a number of pairs of longitudinal turbulences near the heated vertical surface with one turbulence inside, just near the surface and the other being external, penetrating into the boundary layer.

3. Capillary Instability Due to the Marangoni Effect

The special kinds of hydrodynamic instability occur when external disturbances are introduced into the boundary conditions of Eqs. (10)–(17). One of these is connected with local fluctuations of interphase surface tension. To explain the physics of this phenomenon (usually called the Marangoni instability) let us consider the condition where tangential forces balance (13) at the interface. From this condition it follows that in case of nonuniform distribution at surface tension σ there would be a macroscopic movement in at least one of the phases, and due to this movement a friction occurs on the interface which is able to equilibrate the surface tension gradient. This is just the Marangoni instability. If for some reason this gradient of surface tension becomes a function of time it will cause an inevitable temporary evolution of a velocity field in layers adjacent to the phase boundary. The velocity field disturbances influenced by a change of the surface tension gradient either damp or increase with time. Then a problem occurs connected with hydrodynamic instability due to the Marangoni effect. It has been shown [12] that the realization of steady or unsteady hydrodynamic process is influenced by the existence of a critical value of the surface tension gradient. Provided this critical value is exceeded, then slow disturbances of the velocity field caused by local fluctuations of surface tension begin to increase unlimitedly with time. One may characterize the critical value of the surface tension gradient with the so-called Marangoni critical number, the meaning of which will be made clear below by a simple example.

Let us consider a flat ($y=0$) interface "gas-liquid" and suppose that a gas is fixed and a liquid, including a dissolved surface-active substance with bulk concentration c_0, is moving parallel to the phase boundary (towards the x-axis). Furthermore, assume that the convective diffusion is a decelerated stage coming up to the phase boundary of the surface-active substance. In case of a low Reynolds numbers without any external field, Eq. (13) for the balance of tangential forces at the phase boundary becomes more simple:

$$\Gamma M \frac{\partial v^{(s)}}{\partial t} + \mu \frac{\partial v}{\partial y} = \frac{d\sigma}{dc_s} \frac{dc_3}{dx} \quad \text{for} \quad y=0 \tag{20}$$

where c_s is the local value of substance concentration exactly at the interface. This value is to be found from the solution of Eq. (9) describing the convective diffusion along with the boundary conditions (14) for differential mass balance at the phase boundary. It our case this condition can be written as follows:

$$\frac{\partial \Gamma}{\partial t} + \frac{\partial}{\partial x}(v^{(s)}\Gamma) = D \frac{\partial c}{\partial y} \quad \text{for} \quad y=0. \tag{21}$$

Here $c = c(x, y, t, c_0)$ is the solution of Eq. (9) for the liquid phase (the surface diffusion is not considered for simplicity). One may determine the connection of c_s and Γ values in a parametric form (i.e., in terms of c_0), using the following system of equations:

$$c_s = c(x, 0, t, c_0) \tag{22}$$

$$\Gamma = \int_0^\delta (c - c_0)dy \tag{23}$$

where δ is the effective thickness of the boundary diffusive layer where the concentration of the surface-active substance is distinguished from its bulk value. Assuming that $c_s \gg c_0$ and designating as l the characteristic length where the values c_s, v_s, and Γ are changing, let us evaluate the ordering of members in Eqs. (20) and (21) as:

$$\frac{\Gamma M v_s^2}{l} + \mu \frac{\partial v}{\partial y} \sim \frac{c_s}{l} \frac{d\sigma}{dc_s}, \tag{20a}$$

$$\frac{v_s \Gamma}{l} + \frac{v_s \Gamma}{l} \sim \frac{Dc_s}{\delta}. \tag{21a}$$

It is clear that one can neglect the first item on the left hand side of Eq. (20a). Using instead of c_s its ordered value from Eq. (21a) let us introduce this value into the right hand side of Eq. (20a) which leads to:

$$\frac{\partial v}{\partial y} \approx \frac{v_s \Gamma \delta}{D \mu l^2} \frac{d\sigma}{dc_3}. \tag{23a}$$

It follows from Eq. (23) that $\Gamma \approx c_3 \delta$. Furthermore, the values δ and l are physically of the same order (with $l \gg \delta$, changing the concentration field would not change σ, and the alternate version may be realized only in case of very large v_s, that is at variance with the accepted above condition of low Reynolds number). Thus the boundary condition of Eq. (20) in view of the orders of its members can be written as:

$$\frac{\partial v}{\partial y} \approx \frac{v_s}{l} M_a \tag{24}$$

where M_a is the Marangoni number determined as:

$$M_a = \sigma U / \mu D. \tag{25}$$

Here $\Delta \sigma$ is the increment of surface tension over a characteristic length l along the phase boundary.

Solving the problem of velocity field development in time one can write conventionally for v:

$$v = v_0(y) \exp(ik_x x + ik_z z + \beta t) \tag{26}$$

and one of the boundary conditions which the function $v_0(y)$ is to comply with is the condition of Eq. (24). It was shown [11] that there is a critical value of the Marangoni number: if $M_a < M_a^{cr}$ then the disturbances of the velocity field decrease with time, alternately, if $M_a \gg M_a^{cr}$, then the movement becomes unstable (the real part of increment β becomes positive).

A great number of papers [12–23] have been devoted to the problems of linear analysis of the Marangoni instability. These investigations concern the conditions of instability arising from surface tension nonuniformity in systems with various configurations of phase boundaries and various bulk properties of the phases in contact. However, it is beyond the possibilities of linear analysis to solve such fundamental problems such as the formulation of sufficient conditions for regular dissipative structure formation (because the existence of the critical Marangoni number is only a necessary but not a sufficient condition neglecting the systems' initial steady state and its transition to the quasi-steady state), as much as to determine the parameters of quasi-steady states which take place after the system leaves the equilibrium state. These quasi-steady states (usually called the regular dissipative structures) may exist for long enough to be the potential source for the intensification of interfacial heat and mass exchange processes. The problems of the nonlinear theory describing the Marangoni instability are considered in Refs. [24–27].

4. Electro-Hydrodynamic Instability

It is not only the nonuniform surface tension that disturbs the mechanical stability of the mobile interface, but also the existence of free electric charges (ions, for example) at this interface, in view of the external electric field parallel to the boundary. This effect can be illustrated by the example of a two-phase system "liquid-liquid" with a flat charged interface ($y=0$). Let phase 1, occupying the half-space $0<y<\infty$, be the solution of an electrolyte and phase 2 ($-\infty<y<0$) be either an unmixing solution of another electrolyte or a liquid metal. Further, this phase 1 shall have at the interface a specifically adsorbed ion charge with the surface density $ez\Gamma$, where ez is the charge of adsorbed particles and Γ their surface density. If the x-axis is directed along the phase boundary, then Eq. (13) will lead to the following balance condition for tangential forces:

$$\Gamma M \frac{\partial v_s}{\partial t} + \mu_z \frac{\partial v_x^{(2)}}{\partial y} - \mu_1 \frac{\partial v_x^{(1)}}{\partial y} = \frac{1}{4\pi}(\varepsilon, E_x^{(1)}E_y^{(1)} - \varepsilon_2 E_x^{(2)}E_y^{(2)}) \quad \text{for} \quad y=0. \tag{27}$$

From macroscopic electrodynamics laws [3] it follows that at the phase boundary we have:

$$E_x^{(1)} = E_x^{(2)}; \quad \varepsilon_1 E_y^{(1)} - \varepsilon_2 E_y^{(2)} = 4\pi ez\Gamma.$$

In view of these conditions, Eq. (27) is found to be:

$$ez\Gamma E_x^{(s)} = \Gamma M \frac{\partial v_s}{\partial t} + \mu_2 \frac{\partial v_x^{(2)}}{\partial y} - \mu_1 \frac{\partial v_x^{(1)}}{\partial y}. \tag{28}$$

This latter equation shows that the longitudinal electric field, while acting on the charges adsorbed at the interface, will induce an inevitable convective flow both exactly at the boundary and in adjacent areas of each phase. If either the strength E_x of the external field or the surface density Γ of the adsorbed particles is a function of time, one can analyze the stability of the system of the hydrodynamic equations (1)–(2) with an x-component for each phase velocity in the form of Eq. (26). Various kinds of electrohydrodynamic instability result from such an analysis and are described in the literature [28–31]. More details on the problems of electrohydrodynamic instability will be given below using particular model systems as examples.

5. The Linear Analysis of Marangoni Instability

The present section is dedicated to a linear approach to the problem of Marangoni instability at the interface separating two flat, parallel layers of mobile viscous media unlimited in the x- and y-directions and possessing a finite thickness in the z-direction. Let the upper phase ($0 \leq z \leq h_1$) be phase 1, and the lower one ($-h_1 \leq z \leq 0$), phase 2. The lower boundary of phase 2 ($z = -h_2$) will be assumed rigid (with required fulfilment of liquid adhering to the boundary); the upper boundary ($z = h_1$) will be assumed free of tangential tension. The undisturbed interface will be assumed to be plane and horizontal. Such a model of a two-phase system is conventional for experimental designs intended for studying the extraction kinetics [32]. This linear analysis of interface mechanical stability was first carried out in Ref. [22]. Later this analysis was generalized to include systems with distributed surface sources and flows of heat or mass [33] as well as systems with a free boundary for the upper layer [34] (in Refs. [22] and [33] the boundaries $z = h$, and $z = -h_2$ were assumed to be absolutely rigid). In view of the absence of external force fields (including gravity) the system of Eqs. (1)–(2) converts to:

$$\frac{\partial v_i}{\partial t} + v_k \frac{\partial v_i}{\partial x_k} = -\frac{1}{\varrho} \frac{\partial \varrho}{\partial x_i} + v \frac{\partial^2 v_i}{\partial x_k \partial x_k} \quad (i = x, y, z), \tag{29}$$

$$\frac{\partial v_k}{\partial x_k} = 0. \tag{30}$$

The upper integers denoting phase 1 or 2, to which every value belongs, are omitted for simplification.

One can use the dynamic boundary conditions of Eqs. (12)–(13) without considering the influence of the osmotic pressure and the Gibbs adsorption (the case of significant contribution of these values to the interfacial balance of forces is analyzed in Refs. [4, 19]):

$$p_2 - p_1 + \sigma \left(\frac{1}{R_1} + \frac{1}{R_2} \right) = 2\mu_2 \frac{\partial v_z^{(2)}}{\partial z} - 2\mu_1 \frac{\partial v_z^{(1)}}{\partial z}, \tag{31}$$

$$\tau_x \frac{\partial \sigma}{\partial x} + \tau_y \frac{\partial \sigma}{\partial y} = \mu_2 \left\{ \left(\frac{\partial v_z^{(2)}}{\partial y} + \frac{\partial v_y^{(2)}}{\partial z} \right) \tau_y + \left(\frac{\partial v_z^{(2)}}{\partial x} + \frac{\partial v_x^{(2)}}{\partial z} \right) \tau_x \right\}$$

$$- \mu_1 \left\{ \left(\frac{\partial v_z^{(1)}}{\partial y} + \frac{\partial v_y^{(1)}}{\partial z} \right) \tau_y + \left(\frac{\partial v_z^{(1)}}{\partial x} + \frac{\partial v_x^{(1)}}{\partial z} \right) \tau_x \right\}. \tag{32}$$

If the interphase surface tension is not dependent on the concentration of transporting substances and is a function of temperature only, it is necessary to complete the system of Eqs. (29)–(30) with the system of Eqs. (8) and the appropriate boundary conditions:

$$T_1 = T_{10} \quad \text{for} \quad z = h_1; \quad T_2 = T_{20} \quad \text{for} \quad z = -h_2, \tag{33}$$

$$T_1 = T_2; \quad \lambda_1 \frac{\partial T_1}{\partial z} - \lambda_2 \frac{\partial T_2}{\partial z} = -Q(T_s) \quad \text{for} \quad z = 0 \tag{34}$$

where $Q(T_s)$ is a given function of surface temperature T_s.

Linear stability analysis leads to the solution of the system of Eqs. (8), (29), and (30) as a sum of the main and the disturbed states:

$$v_i = \bar{v}_i + \delta v_i; \quad p = \bar{p} + \delta p; \quad T_k = \bar{T}_k + \delta T_k; \quad z = \bar{z} + \delta z \tag{35}$$

where

$$\bar{v}_i = 0, \quad \bar{p} = p_0, \quad \bar{T}_k = \alpha - \beta_k z, \quad \bar{z} = 0,$$

and in view of Eqs. (33)–(34):

$$\alpha - \beta_1 h_1 = T_{10}; \quad \alpha + \beta_2 h_2 = T_{20}; \quad -\lambda_1 \beta_1 + \lambda_2 \beta_2 = \bar{Q}. \tag{36}$$

The values $\bar{Q}$, T_{10}, T_{20}, α, β_1, and β_2 are assumed to be given and constant.

Equations (8), (29), and (30) can be linearized relative to the disturbances δv_i, δp, and δT_k. A standard means of linearization suggested in Ref. [12] results in the following system of equations for functions δv_z, δp, and δT_k:

$$\left(\frac{\partial}{\partial t} - \nu \nabla^2 \right) \nabla^2 \delta v_2 = 0, \tag{37}$$

$$\left(\frac{\partial}{\partial t} - \nu \nabla^2 \right) \delta v_z = -\frac{1}{\varrho} \frac{\partial (\delta_p)}{\partial z}, \tag{38}$$

$$\frac{\partial}{\partial t} (\delta T_k) - \beta_k \delta v_k = \varkappa_k \nabla^2 (\delta T_k). \tag{39}$$

Boundary conditions of Eqs. (11), (31), (32), and (34) at the interface $z = \delta z$ appear as:

$$\Delta \left\{ \delta p - 2\mu \frac{\partial}{\partial z} (\delta v_z) \right\} = -\sigma \left\{ \frac{\partial^2}{\partial x^2} (\delta z) + \frac{\partial^2}{\partial y^2} (\delta z) \right\}, \tag{40}$$

$$\Delta \left\{ \mu \left(\frac{\partial^2}{\partial z^2} - \frac{\partial^2}{\partial x^2} - \frac{\partial^2}{\partial y^2} \right) \delta v_z \right\} = \frac{\partial^2 \sigma}{\partial x^2} + \frac{\partial^2 \sigma}{\partial y^2}, \tag{41}$$

$$\frac{\partial}{\partial t} (\delta z) = v_z^{(1)} = v_z^{(2)}, \tag{42}$$

$$\Delta \{ \delta T \} = (\beta_1 - \beta_2) \delta z. \tag{43}$$

$$\Delta \left\{ \lambda \frac{\partial}{\partial z} (\delta T) \right\} = -\left(\frac{dQ}{dT} \right)_{T = \bar{T}} (\delta T_2 - \beta_2 \delta z). \tag{44}$$

Here Δ is defined by

$$\Delta \{ X \} = X_1 - X_2.$$

At the surfaces $z = h_1$, and $z = -h_2$ the boundary conditions are:

$$\delta v_z^{(1)} = 0; \quad \frac{\partial^2}{\partial z^2} (\delta v_z^{(1)}) = 0; \quad \delta T_1 = 0 \quad \text{for} \quad z = h_1, \tag{45}$$

$$\delta v_z^{(2)} = 0; \quad \frac{\partial}{\partial z} (\delta v_z^{(2)}) = 0; \quad \delta T_2 = 0 \quad \text{for} \quad z = -h_2. \tag{46}$$

The condition for the secondary derivative from $\delta v_z^{(1)}$ in the z-direction follows from the condition of zero tangential force at the free surface $z = h_1$. The condition for the first derivative from $\delta v_z^{(2)}$ in the z-direction follows from the continuity Eq. (30), as much as from the absence of the tangential movement ($\delta v_x = \delta v_y = 0$) at the solid surface $z = -h_2$.

The solution of Eqs. (37)–(39) is:

$$\delta v_z = W(z)\psi(x, y, t); \quad \delta p = \pi(z)\psi; \quad \delta T = \theta(z)\psi; \quad \delta z = \xi\psi, \tag{47}$$

where

$$\psi(x, y, t) = \exp[i(k_x x + k_y y) + \omega t].$$

Substituting Eqs. (47) into Eqs. (37)–(39) and performing a number of simple transformations one obtains:

$$\left[\omega - \nu\left(\frac{d^2}{dz^2} - k^2\right)\right]\left[\frac{d^2}{dz^2} - k^2\right]W = 0, \tag{48}$$

$$k^2\pi + \mu\left[\frac{\omega\varkappa}{\varkappa_1\nu} - \left(\frac{d^2}{dz^2} - k^2\right)\right]\frac{dW}{dz} = 0, \tag{49}$$

$$\left[\omega - \varkappa\left(\frac{d^2}{dz^2} - k^2\right)\right]\theta - \beta W = 0 \tag{50}$$

where $k^2 = k_x^2 + k_y^2$ is a quadratic wavenumber for the movement in the direction parallel to the phase boundary.

One can delete sequentionally the values W, π, δz from Eqs. (48)–(50) and from the boundary conditions of Eqs. (40)–(44) to obtain the system of equations and boundary conditions describing the functions $\theta_1(z)$ and $Q_2(z)$. During these operations it is necessary to use the linear approach for the function $\sigma(\tau)$:

$$\sigma(\tau) = b + cT$$

As shown in Ref. [12], the system of equations for functions $\theta_i(z)$ has the form:

$$\left[\frac{\omega}{\varkappa_i} - \left(\frac{d^2}{dz^2} - k^2\right)\right]\left[\frac{\omega}{\nu_i} - \left(\frac{d^2}{dz^2} - k^2\right)\right]\left(\frac{d^2}{dz^2} - k^2\right)\theta_i = 0 \quad (i = 1, 2). \tag{51}$$

The boundary conditions which the system subject to can be obtained from Eqs. (40)–(46); altogether there are twelve such conditions [much the same as for the system of Eq. (51)]. In the particular case of an undeformed interface (which can be formally realized for $\sigma \to \infty$) these conditions can be expressed as:

$$\Delta\left\{\frac{\varkappa\mu}{\beta}(D^4 - a^4)\theta\right\} = \frac{\varkappa_1\mu_1 a^2}{\beta_1}\theta_1(0)M_a \quad \text{for} \quad z = 0,$$

$$\Delta\left\{\frac{2\varkappa\mu}{\beta}(D^2 - a^2)D\theta - \frac{\varkappa\mu}{a^2\beta}(D^2 - a^2)D\theta\right\} = 0 \quad \text{for} \quad z = 0,$$

$$\theta_1(0) = \theta_2(0);$$

$$\frac{\lambda_1}{\lambda_2}D\theta_1 - \theta_2 = Q^*\theta_2 \quad \text{for} \quad z = 0,$$

$$(D^2 - a^2)\theta_1 = 0 \quad \text{for} \quad z = 0,$$

$(D^2 - a^2)\theta_2 = 0$ for $z = 0$,

$(D^2 - a^2)\theta_1 = 0$ for $z = h_1$,

$(D^2 - a^2)D^2\theta_1 = 0$ for $z = h_1$,

$\theta_1 = 0$ for $z = h_1$,

$(D^2 - a^2)\theta_2 = 0$ for $z = -h_2$,

$(D^2 - a^2)D\theta_2 = 0$ for $z = -h_2$,

$\theta_2 = 0$ for $z = -h_2$.

Here we use the following designations:

$$D = h_1 \frac{d}{dz}; \quad a = kh; \quad M_a = -\frac{d\sigma}{dT} \frac{\beta_1 h_1^2}{\varkappa_1 \mu_1};$$

$$Q^* = -\frac{h}{\lambda_2}\left(\frac{dQ}{dT}\right) \quad \text{for} \quad T = \bar{T}.$$

In the case of $\omega = 0$ with a neutral hydrodynamic stability of the phase boundary, the solution of Eq. (51) is given by:

$$Q_i = \sum_{k=0}^{2} \left\{ A_k^{(i)} \exp\left(\frac{az}{h_1}\right) + B_k^{(i)} \exp\left(-\frac{az}{h_1}\right) \right\} \left(\frac{z}{h_1}\right)^k.$$

There are six arbitrary constants $A_k^{(i)}$ and $B_k^{(i)}$ for each phase ($i = 1, 2$) in this solution. These constants can be found from the above-mentioned twelve boundary conditions. These conditions form a system of twelve linear, uniform, algebraic equations. The nontrivial solution of such a system exists only if its determinant is zero. The latter condition gives the ratio for dispersion including all the characteristic parameters of a system. This ratio allows to interpretate the Marangoni number M_a as a function of all other parameters. In our case of an undeformed interface the expression of the Marangoni number is:

$$M_a = \frac{8a\lambda_1}{\lambda_2 \chi A} \left\{ \chi \left[\frac{\lambda_2}{\lambda_1}(1 - F_1)P + (1 + F_1)(1 - F_2) \right] \left[BU - 4a(1 - F_1)^2(1 + aF_2) \right] \right.$$
$$\left. - \frac{4aG}{A}(1 - F_2)(1 + aF_2)UP \right\} \left\{ \frac{8a\lambda_1 G}{\lambda_2 A}(1 - F_2)(1 + aF_2) - \chi B(1 - F_1) \right\}^{-1} \quad (52)$$

with the following designations:

$$A = -\frac{1}{a}F_1^3 + \left(12a^2 - 4a + 4 - \frac{1}{a} \right)F_1^2 + \left(-4a^2 + 12a - 4 + \frac{5}{a} \right)F_1 - \frac{3}{a};$$

$$B = \left(1 - \frac{4ah_1}{h_2} \right)F_1 - 1;$$

$$F_1 = \exp(-2ah_1/h_2); \quad F_2 = \exp(-2a);$$

$$G = \frac{2h_1 F_1}{h_2}(F_1 - 1) + \frac{4ah_1 F_1}{h_2}\left(1 + \frac{h_1 F_1}{h_2} \right) + \frac{(F_1^2 - 1)(1 - F_1)}{2a};$$

$$U = 2F_2(1 - 2a) - 3 + F_2^2; \quad P = 1 + F_2 + \frac{Q^*}{a}(F_2 - 1); \quad \chi = \frac{\varkappa_1 \mu_1 \beta_2}{\varkappa_2 \mu_2 \beta_1};$$

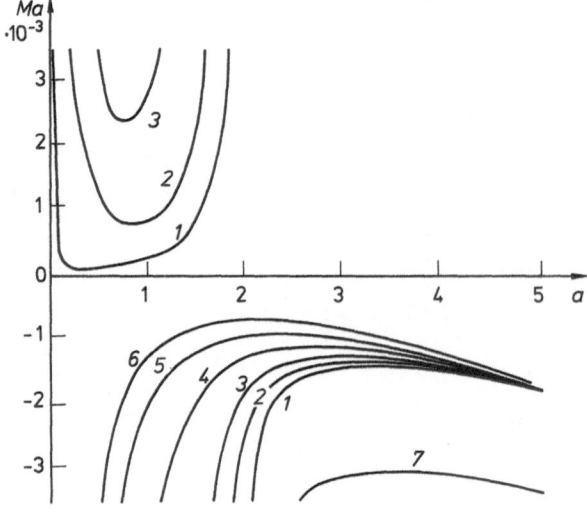

Fig. 1. The neutral-stable states of the liquid-liquid interface with the following parameter values: $\mu^* = \mu_1/\mu_2 = 2.0$; $\varkappa^* = \varkappa_1/\varkappa_2 = 0.5$; $\lambda^* = 0.5$; $\beta^* = 2.0$; $Q^* = 0$; the curves 1–7 correspond to $h^* = 0.1$; 0.5; 0.6; 0.7; 0.8; 1.0; 10

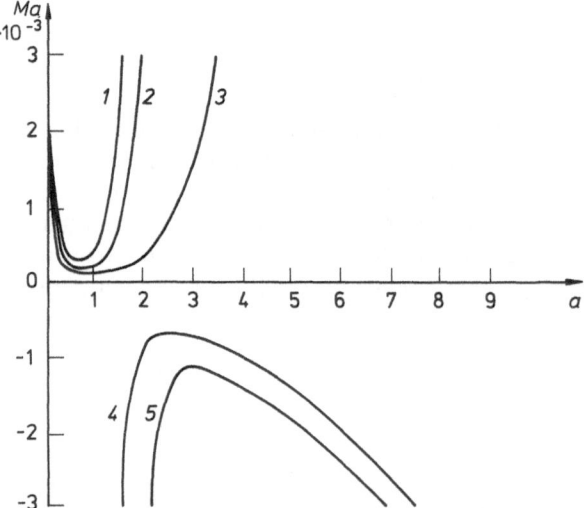

Fig. 2. Influence of heat sources (curve 1) and flows (curves 3, 4) upon the interfacial stability; the parameter values are the following: $\lambda^* = 0.5$; $\mu^* = 2.0$; $\varkappa^* = 0.5$; $h^* = 0.1$; $Q^* = 0$; $\beta_1^* = 1$ (curve 3), 2 (curves 2, 5), and 3 (curve 1)

Both the conditions and the qualitative and the quantitative regularities of the disturbances of mechanical stability at the interface may be illustrated with diagrams for $M_a = M_a(a)$ with the rest of the parameters fixed. Such diagrams are given in Refs. [22, 33] for the case of rigid surfaces $z = h_1$ and $z = -h_2$. Some of these are illustrated in Figs. 1, 2. The areas of positive values of M_a in these diagrams correspond to the heat transfer directed from phase 2 into phase 1, and the areas of negative values correspond to the opposite direction. The curves of Fig. 1, in view of the absence of the distributed surface heat sources ($\beta^* \lambda^* = 1$, $Q^* = 0$, where $\beta^* = \beta_1/\beta_2$, $\lambda^* = \lambda_1/\lambda_2$), coincide with the relevant curves obtained in Ref. [22]. It can be seen that while heat is transported from the phase with a lower $\varkappa$ into the one with a higher $\varkappa$ for any value of ratio $h^* = h_1/h_2$, there exists a critical Marangoni number M_0^{cr}, such that for $|M_a| < |M_a^{cr}|$ all the wavenumbers correspond to steady states, but for $|M_a| > |M_a^{cr}|$ there exists a group of values of a characterizing the unstable wave numbers. Such a feature is typical for the dissipative structures realized in liquid-liquid systems and the instability increase is similar in its nature to the Benard or the Relay-Taylor instabilities [1]. In the case of heat transfer in the opposite direction (for positive M_a) the critical Marangoni number exists only for $h^* < 1$. Figure 2 shows the results of the stability analysis carried out in Ref. [33] for the existence of heat sources with an intensity independent of temperature, ($Q^* = 0$). When the heat transfer is directed from phase 2 into phase 1, then the values of $Q > 0$ relate to a heat source and the ones of $Q < 0$ describe the flow. In the case of the opposite direction of heat transfer the conditions for a source and for a flow are reversed. It can be seen from this figure that the heat flow is the reason for instability, while the presence of a heat source stabilizes the system. The effect of stabilization may be very significant. For example, the curve of neutral stability for $\beta^* = 1$, and for the direction $1 \to 2$ (heat source) lies much lower than the area fulfilled by the graphs in Fig. 2.

6. The Instability Caused by the Electric Forces Acting at the Surface of an Electrolyte Solution

In accordance with the boundary condition of Eq. (13), varying the gradient of surface tension or that of the tangential component of the interfacial electric field strength may induce a macroscopic movement in a system with free charges (ions or electrons). If the local fluctuations are significant, such a movement may increase with time and finally disturb the hydrodynamic stability of the whole interface. There are some experimental data in the literature indicating the existence of such an instability [35–43]. These data cannot be explained in the framework of a Marangoni instability without considering the ionic nature of reagents being transported through the interface or adsorbing at it. For example, the measurement of oscillating movements at the interface of nitrobenzol and water in the presence of highly adsorbing surface-active substances (SAS) [43] has shown that such oscillations can be realized only if the nitrobenzene phase includes the dissolved electrolyte. The cations of the electrolyte go through the interface and substitute for the cations of the water solute (the latter are moving into the nitrobenzene in an equivalent number).

We consider below the simplest model of a system where an instability increase is caused by the electric forces acting at the phase boundary [44]. The system has the form of a two-dimensional, plane, horizontal layer ($0 \leq z \leq h$) occupied by a liquid solution of an electrolyte. The lower boundary ($z = 0$) of the layer is a solid metal wall with potential ϕ_0 (measured from the depth of the solution). The top of the solution ($z = h$) is adjacent to the unrestricted bulk of the second liquid phase. The properties of the second phase can be described with the help of the effective coefficients of mass and charge exchange, with the main phase of the electrolyte solution ($0 \leq z \leq h$), i.e. the value of the mass-transfer coefficient for the ions of the same kind (cations, for instance) and also the density of the electric current at the interface are assumed to be given (and to be constant at all the points on the surface $z = h$). If $c_1(x, y, z)$ and $c_2(x, y, z)$ are the concentration distributions of cations and anions, respectively (the solution is assumed to be binary and symmetrical) and $\phi(x, y, z)$ is the distribution of the electric potential, then in accordance with the above-stated assumption the following conditions can be postulated:

$$\xi_0(c_1 - c_{1\infty}) = \frac{\partial c_1}{\partial z}; \quad i_0 = \frac{\partial \phi}{\partial z} \quad \text{for} \quad z = h \tag{53}$$

where $c_{1\infty}$ is the cation concentration deep in phase 2 ($z \to \infty$), and ξ_0 and i_0 are given constants.

It is assumed finally that the influence of a diffusive double electric layer at the interface of two nonmixing liquids is negligible. This assumption is correct if:

$$\frac{8\pi e^2 c_0}{\varepsilon k T} \ll h^2, \tag{54}$$

where c_0 is the bulk concentration of an electrolyte in the layer $0 \leq z \leq h$. The numerical evaluation shows that if $h = 1$ mm then Eq. (54) is true for practically any concentration of an electrolyte.

Without the influence of a double diffusive layer, the system of equations for functions c_1, c_2, and φ can be expressed as [45]:

$$\frac{\partial c_1}{\partial t} + (v, \nabla c_1) = D_1 \nabla^2 c_1 + \frac{e z_1 D_1}{kT} \operatorname{div}(c_1 \nabla \varphi), \tag{55}$$

$$\frac{\partial c_1}{\partial t} + (v, \nabla c_1) = D_{\text{eff}} \nabla^2 c_1, \tag{56}$$

$$z_1 c_1 + z_2 c_2 = 0 \tag{57}$$

where $D_{\text{eff}} = (z_1 - z_2) D_1 D_2 / (z_1 D_1 - z_2 D_2)$ is an effective coefficient of electrolyte diffusion. According to the above-mentioned assumptions the boundary conditions for Eqs. (55)–(57) are:

$$c_1 = c_0, \quad \phi = \phi_0 \quad \text{for} \quad z = 0,$$

$$\xi_0(c_1 - c_{1\infty}) = \frac{\partial c_1}{\partial z}; \quad i_0 = \frac{\partial \varphi}{\partial z} \quad \text{for} \quad z = h. \tag{58}$$

Equations (55)–(57) are to be completed with the momentum transfer equation:

$$\left(\frac{\partial}{\partial t} - v \nabla^2 \right) \nabla^2 v = 0 \tag{59}$$

and with the boundary conditions

$$v_x = v_y = v_z = 0 \quad \text{for} \quad t = 0,$$

$$\mu \frac{\partial^2 v_z}{\partial z^2} = A \left(\frac{\partial^2 c_1}{\partial x^2} + \frac{\partial^2 c_1}{\partial y^2} \right) + \frac{\varepsilon}{4\pi} \left\{ \frac{\partial \phi}{\partial z} \left(\frac{\partial^2 \varphi}{\partial x^2} + \frac{\partial^2 \varphi}{\partial y^2} \right) \right.$$

$$\left. + \frac{\partial \varphi}{\partial x} \cdot \frac{\partial^2 \varphi}{\partial x \partial z} + \frac{\partial \varphi}{\partial y} \cdot \frac{\partial^2 \varphi}{\partial y \partial z} \right\} \quad \text{for} \quad z = h. \tag{60}$$

The boundary condition for a tangential force at $z=h$ is obtained from the assumption of a linear dependence of surface tension on concentration c_1:

$$\sigma = \sigma_0 + Ac_1. \tag{61}$$

The method of transforming the general boundary condition of Eq. (13) into Eq. (60) is taken from Ref. [12]. Using the boundary conditions of Eq. (60) one can solve Eq. (59) only for the z-component of the velocity field.

According to the procedure of linear analysis described in the previous section, the solution of the problem can be expressed as a number of sums:

$$v_z = \bar{v}_z + \delta v_z, \quad c_1 = \bar{c}_1 + \delta c_1, \quad \phi = \bar{\phi} + \delta\phi.$$

The undisturbed state of a system can be described with the following functions:

$$\bar{v}_z = 0; \quad \bar{c}_1 = \beta z + c_0; \quad \bar{\phi} = i_0\left(h + \frac{c_0}{\beta}\right)\ln\left(\frac{\beta z}{c_0} + 1\right) + \varphi_0 \tag{62}$$

where $\beta = \xi_0(c_0 - c_{1\infty})/(1 - \xi_0 h)$.

Using Eqs. (62), the initial system of equations yield for the undisturbed functions:

$$\left(\frac{\partial}{\partial t} - v\nabla^2\right)\nabla^2(\delta v_z) = 0, \tag{63}$$

$$\frac{\partial}{\partial t}(\delta c_1) + \beta\delta v_z = D_{\text{eff}}\nabla^2(\delta c_1), \tag{64}$$

$$\frac{\partial}{\partial t}(\delta c_1) + \beta\delta v_z = U_1\left\{\beta\frac{\partial}{\partial z}(\delta\phi) + (\beta z + c_0)\nabla^2(\delta\varphi)\right\}$$

$$+ U_1 i_0\frac{\partial}{\partial z}(\delta c_1) - (U_1 i_0\beta/c_0)\delta c_1 + D_1\nabla^2(\delta c_1) \tag{65}$$

where $U_1 = ez, D_1/kT$ is the mobility of cations in the electric field. The boundary conditions for Eqs. (63)–(65) are:

$$\delta v_z = 0; \quad \frac{\partial}{\partial z}(\delta v_z) = 0; \quad \delta c_1 = 0; \quad \delta\phi = 0 \quad \text{for} \quad z=0, \tag{66}$$

$$\delta v_z = 0; \quad \mu\frac{\partial^2}{\partial z^2}(\delta v_z) = A\left(\frac{\partial^2}{\partial x^2} + \frac{\partial^2}{\partial y^2}\right)\delta c_1 + \frac{\varepsilon_{i0}}{4\pi}\left(\frac{\partial^2}{\partial x^2} + \frac{\partial^2}{\partial y^2}\right)\delta\varphi;$$

$$\xi_0\delta c_1 = \frac{\partial}{\partial z}(\delta c_1); \quad \frac{\partial}{\partial z}(\delta\varphi) = 0 \quad \text{for} \quad z=h. \tag{67}$$

The boundary conditions of Eq. (67) suppose that the interface cannot be deformed in the z-direction ($\delta v_z = 0$ for $z=h$). For transforming Eq. (55) into (65), the condition $\beta h \ll c_0$ was used.

The solution of Eqs. (63)–(65) is obtained through:

$$\delta v_z = W(z)\exp[i(k_x x + k_y y) + \omega t],$$

$$\delta c_1 = \theta(z)\exp[i(k_x x + k_y y) + \omega t],$$

$$\delta\varphi = \psi(z)\exp[i(k_x x + k_y y) + \omega t].$$

From Eqs. (63)–(65) one can obtain:

$$\left[\omega - v\left(\frac{d^2}{dz^2} - k^2\right)\right]\left[\frac{d^2}{dz^2} - k^2\right]W(z) = 0, \tag{68}$$

$$\left[\omega - D_{\text{eff}}\left(\frac{d^2}{dz^2} - k^2\right)\right]\theta(z) + \beta W(z) = 0, \tag{69}$$

$$U_1\beta\frac{d\psi}{dz} + U_1(\beta z + c_0)\left(\frac{d^2}{dz^2} - k^2\right)\psi = (D_{\text{eff}} - D)\left(\frac{d^2}{dz^2} - k^2\right)\theta(z)$$

$$- U_1 i_0\left(\frac{d}{dz} - \frac{\beta}{c_0}\right)\theta(z). \tag{70}$$

Here $k^2 = kx^2 + ky^2$.

The boundary conditions for such a system are:

$$W = 0; \quad \frac{dW}{dz} = 0; \quad \theta = 0; \quad \psi = 0 \quad \text{for} \quad z = 0, \tag{71}$$

$$W = 0; \quad \mu\frac{d^2W}{dz^2} = -Ak^2\theta - \frac{\varepsilon i_0 k^2}{4\pi}\psi;$$

$$\xi_0\theta = d\theta/dz, \quad d\psi/dz = 0 \quad \text{for} \quad z = h. \tag{72}$$

The system of Eqs. (68)–(70) are uniform differential equations of the 8th order. Its solution includes unknown coefficients which can be found from the boundary conditions of Eqs. (71)–(72). The nontrivial solution of such a system exists only when the characteristic determinant is zero similar to the case in the previous section. The latter condition is the dispersive ratio connecting the wave vector k, and the frequency ω, while the rest of the parameters are fixed. At the state of neutral stability ($\omega = 0$) the

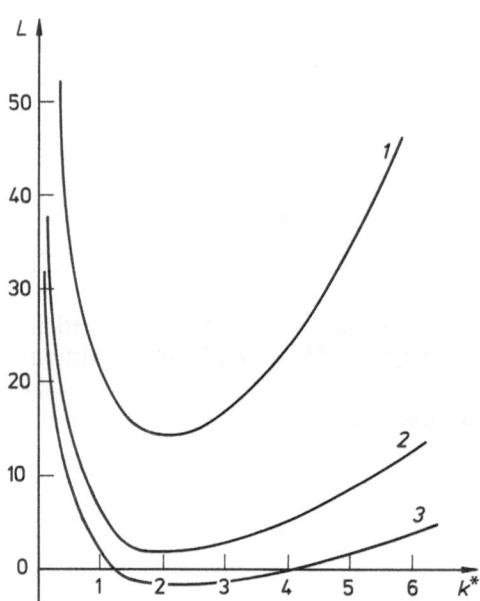

Fig. 3. Influence of the electric field at the interface of two unmixing electrolyte solutions upon the instabilization of phase boundary. The curves 1, 2, 3 correspond to three fixed values of the Marangoni number ($M_a = 5$; 20; 50) for $\xi_0 h = -4$

dispersive ratio can be expressed in form of the dependence:

$$L = L(k^*, \xi_0 k_0 M_a) \tag{73}$$

where

$$L = \frac{\varepsilon z_1 D_1 (D_2 - D_1) i_0}{4\pi (z_1 D_1 - z_2 D_2) U_1 c_0 A}; \quad k^* = kh; \quad M_a = -\frac{\beta h^2 A}{\mu D_{eff}}.$$

Thus in our case of ion mass exchange between two ion mixing solutions of the electrolyte, the hydrodynamic interfacial stability is controlled not only by the Marangoni number but also by the electric parameter L. This parameter may either stabilize or destabilize the interface. Figure 3 illustrates the latter case. It corresponds to a positive z-direction of mass transfer ($\xi_0 < 0$). As follows from the graphs shown in Fig. 3, when the positive values of parameter L are increased with fixed Marangoni number, the range of wavenumbers characterizing the unstable tangential movement at the interface also increases.

7. Nonlinear Methods of Analyzing the Marangoni Instability

As in the previous section, the framework of linear stability analysis allows us to determine the necessary criteria for interfacial destabilization only. Furthermore, it is beyond the scope of the theory to give a quantitative description of new regular states into which a system may transform if the critical value of stability criteria is reached or exceeded. In particular, the linear methods do not permit the determination of the amplitude of the liquid velocity in the Rayleigh-Benard circulational cells, nor do they allow one to establish the steady form of those cells or a number of other physically realizable steady states or to describe the transitions between these states. These problems may be solved, however, with the help of nonlinear methods. Such a method was first applied to the analysis of the Marangoni instability in Ref. [24], where the authors considered the dynamics of forming hexagonal convective cells at a plane interface of two half-infinite media. Later, improved methods were developed for nonlinear analysis [25–27] which helped to extend the theory allowing a detailed hydrodynamic description of the regular circulation cells and to establish the quantitative criteria of some dissipative structures of the phase boundary.

Investigations have been made [26, 27] into the stability of a liquid layer with limited thickness bounded by a cylinder with a circular or rectangular lateral cross section, this layer having an upper interface common with the bulk of an unmovable gas and the side wall of the cylindrical container being heat-proof, provided the bottom wall is maintained at a higher temperature then the upper one. Within the framework of the Boussinesq approximation for a temperature-dependent density it was found that the limit of the capillary number is close to zero $c_a = \mu_0 \varkappa_0 / \sigma_0 d$ (μ_0 is the dynamic viscosity, $\varkappa_0$ is the thermal diffusivity, and σ_0 the surface tension, all at a given standard temperature, and d is the thickness of the undisturbed liquid layer). The unknown distributions of disturbed velocities, pressures, and temperatures were expressed as linear combinations of basic functions of spatial variables with the coefficients depending on time, provided the basic functions are the proper ones of the equations describing the problem in the linear (for the disturbance amplitude) approach. The coefficients depending on time were found from the solution of a system of ordinary nonlinear differential equations derived from the initial system resulting from a modified Galerkin method developed by the authors. The main results of the analysis are the following:

1. The critical Marangoni number is calculated as a function of the Rayleigh, Prandtl, and Biot numbers together with the aspect ratio (i.e., the ratio of the container radius to the mean depth of a liquid).

2. The stability criteria for multiple nonlinear steady states near the critical Marangoni number are determined.

3. The configurations, the characteristic dimensions and the amplitudes of the velocities for axisymmetrical circulational cells, as well as the influence of the Prandtl number upon the stability of the periodic movement in those cells are established.

4. The transitions between two steady convective states of equal probability corresponding to the binary double proper values for the critical Marangoni number are predicted and analyzed.

The results obtained in Refs. [26, 27] give a general procedure for the quantitative analysis of the influence of the self-organizing hydrodynamic dissipative structures upon the kinetics of the interfacial heat and mass transfer. It seems that such an analysis may be carried out in the near future (this requires, however, the development of improved methods to give the solution of the equations with three variables in individual derivatives). The efforts of the investigators are directed at modeling just the structures of Rayleigh-Benard convective cells which do not cover a wide range of regular dissipative structures, and to study the principles of self-organization during the irreversible processes which are interesting in view of theoretical and practical aspects of heat and mass exchange. Some of the model systems examined are discussed below.

8. Models of Systems with Regular Hydrodynamic Dissipative Structures

The physical conditions necessary for regular dissipative structures are: First, the increment of the amplitude temporal evolution of the "most dangerous" disturbance of the hydrodynamic velocity field is to be real and to change its sign from negative to positive, while the criterion characterizing the disturbance (the Marangoni number, for instance) be conserved over the critical point [46]; secondly, the system must simultaneously include a positive and negative feedback between the factors controlling the velocity field evolution [for example, the connection between viscous, electric capillary forces at the mobile interface, which is described by Eq. (13)]. A typical system with positive and negative feedback is considered in Ref. [43] where the authors studied the periodic movement at the interface "benzene-water" with simultaneous adsorption of the surface-active substance (the cetyltrimethylammonium chloride) and ion mass transfer through the interface. The positive feedback in this system is due to the adsorption of SAS, provided the ion mass transfer is responsible for the negative feedback. The basic physical principles of instability for such systems are considered in Ref. [44].

The effect of a regular dissipative structure's appearance near the "liquid-gas" interface due to the Marangoni instability is analyzed in Ref. [18]. The undisturbed system looks like the Quett flow in a plane-parallel layer formed between the immovable "liquid-gas" interface and the flat solid tape positioned inside the liquid at a certain depth under the interface and being moved uniformly and translationally. As a disturbing factor the adsorption of a substance forming an insoluble condensed film at the surface was examined. The analysis was carried out based upon the solution of the nonlinear hydrodynamic equations and upon a model-description of the tangential force arising at the interface as a result of the mutual influence of adsorption and viscous friction. This analysis helped to find the conditions for formation of regular structures appearing like circular cells, perhaps oscillating in time. The characteristic dynamic parameters of these structures were also determined.

The process and the regularities in the formation of dissipative structures induced by the adsorption of the SAS at the "liquid-liquid" interface were analyzed in Ref. [23]. The combined Quett flow was found to be in an initial undisturbed state [18, 23], and propagates between two solid tapes moving in opposite directions being separated by the interface between them (the velocities of the two tapes are chosen such that the undisturbed interface is quiet). The induced Marangoni instability is considered to be a result of the nonuniform distribution of surface tension due to the adsorption controlled by the slowed stage of SAS particles adhering at the interface. The practical prototype of the model considered in Ref. [23] is the electrochemical system "mercury-electrolyte solution with doped SAS" forming a condensed adsorbing phase at the mercury surface. In such a system one may observe so-called polarographic maxima of the third kind which are the sharp increase in the electric current density occurring at certain values of the mercury electrode potential [47–49]. It was found in Ref. [48] that the origin of these maxima is connected with the tangential movement appearing at the surface of a mercury drop: the movement is caused by the formation of areas with increased surface tension at the drop. This can occur when adsorption of the organic substances, with strong mutual attraction, takes place.

The possibility of considering the dissipative structures as circular cells under an oscillating current has been examined in Ref. [23]. In particular, it was shown that the shape of the balanced adsorption isotherm has no influence on the criteria for stability. Comparing this conclusion with the results from Refs. [47–49] concerning the circulational dissipative structures, which are experimentally realized only for systems with surface-active substances (SAS) possessing high attraction constants, one can deduce that: 1) the adsorption process for the systems with polarographic maxima of the third kind is significantly imbalanced, and 2) the quantitative description of the systems with dissipative structures caused by the inbalanced adsorption has to take into consideration the finite velocity of the convective-diffusive SAS transfer onto the interface.

One of the types of regular dissipative structure is a circulational cell. Such cells appear in the space filled with the electrolyte and are located between two plane parallel electrodes, provided the solution is subjected to electrolysis with the upper electrode being the anode. In this case the changing solution density during the electrolysis results in the appearance of natural convection which may induce the self-organizing of the velocity field in a certain regime of the electrochemical mass transfer. In Refs. [50–51] the self-organization regimes were registered by applying the voltage step-

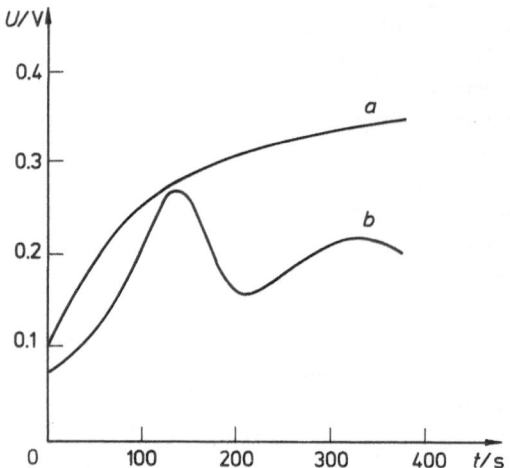

Fig. 4. Typical forms observed for the curves of the voltage dynamic response; a – for the lower anode, b – for the upper anode

pulse to the electrodes and recording the relaxation curve of the current-time response. In the experiments [50–51] the structures of hydrodynamic flows in a cell were not identified visually or by any other means, so it was impossible to give a simple interpretation of the response-curve shape or to establish the quantitative correlation of the electrochemical and hydrodynamical parameters. It was the aim in Ref. [52] to obtain such a correlation. According to this work, the distance between the electrodes, made in the form of two flat copper plates, was variable within a wide range due to a special arrangement (as for the experiments discussed below these distances varied between 0.8–2.5 mm). An aqueous solution of copper sulphate with a concentration of 3.5×10^{-3} M was used in all the experiments. In contrast to the methods of Refs. [51–51], the step-pulse of current was applied to a cell and the "potential-versus-time" curves were recorded. In the case of the lower anode (Fig. 4a), the system was steady in a hydrodynamic sense and the potential was increased monotonically with time up to the stationary value. In the case of the upper anode (Fig. 4b), the recorded "potential-versus-time" curves were nonmonotonic which is indicative of the natural convection appearance and the Benard circulational cell formation. By varying the current pulse value (Fig. 5), the authors could obtain information on the influence of the ion mass transfer velocity upon the liquid circulation intensity in the Benard cells. The experiments carried out with different values of the distance between the electrodes (Fig. 6) show that increasing the distance results in a corresponding increase of the velocity of ion mass transfer, because the potential oscillations with the amplitude proportional to that of the current density signal become more intensive in this case.

However, it is to be noted, that the results of the electrochemical experiments are not sufficient to explain the appearance of Benard cells in the system. In fact, these results are sometimes misinterpreted. For example, in Ref. [51] the monotonic development of the current-versus-time curves, similar to that of the curves "a" and "b" in Fig. 5, was explained as a fact indicative of the absence of any circulational cell. However, optical measurements show that the circulational cell structure of the flux takes place even for a low current density (which possesses certain features connected with the temporal evolution of the velocity field and determining the monotonic

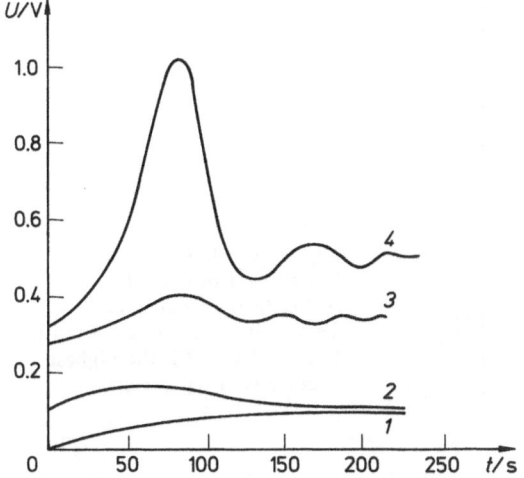

Fig. 5. The form of the response curve versus the current density of a pulse applied; the curves 1–4 correspond to the respective values of i: $6.6 \cdot 10^{-5}$, $11 \cdot 10^{-4}$, $5.5 \cdot 10^{-4}$, and $1.3 \cdot 10^{-3}$ A/cm^2; in every case there were circulational cells near the upper anode (with the distance d between the electrodes equal to 4 mm)

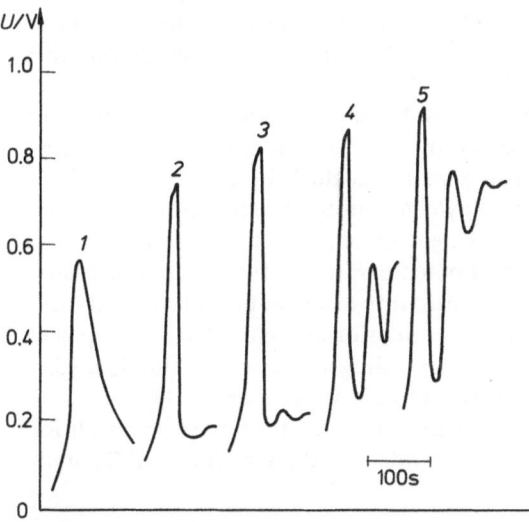

Fig. 6. The response curve's highest amplitude versus the distance between the electrodes with the current pulse value fixed ($i = 1.3 \cdot 10^{-3}$ A/cm^2); the curves 1–5 correspond to $d = 0.8$; 1.0; 1.2; 1.4; and 2.0 mm

response curves). For the above-mentioned reasons, the authors of Ref. [52] developed methods for simultaneously registering the characteristics of the electrochemical mass transfer and the hydrodynamic velocity field. For this purpose the flux visualization method was used by photographing the tracks of the fluorescent particles exposed to an external light source, and by using an argon laser. The fluorescent particles were composed of metacrylate having 10 μm diameter impregnated with rodamine. The local velocities of the liquid flux were determined by measuring the particle track length of the photographs with known exposure time. The highest liquid velocity in the circulational cells varied and depended on the current pulse value in two ways: monotonically if the pulse value is less than its a critical value (Fig. 6); and accompanied with oscillations, if the critical value is exceeded, where the oscillations of the transition process for increasing velocity up to a stationary value are in exact

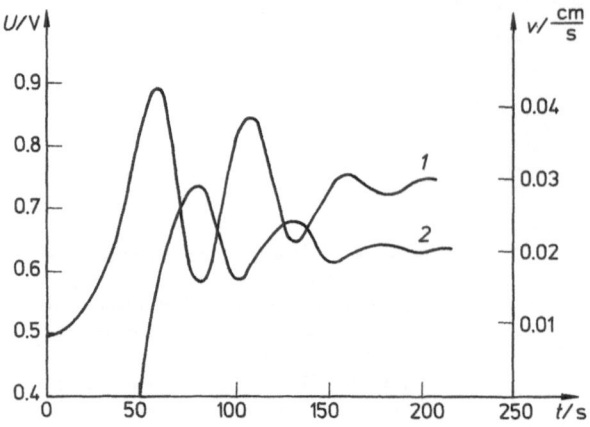

Fig. 7. Correlation of curves U, t and v, t in case of "subcritical current pulse" ($i = 1.3 \cdot 10^{-3}$ A/cm^2, $d = 4$ mm); 1 is a voltage, 2 is the highest circular velocity in cells

correlation with the voltage oscillations (Fig. 7). Such a correlation is of general importance as it is clear that the experimental development of the voltage dependence on time determines not only the presence of the circulational cells (they may arise without any voltage oscillations as follows from Fig. 5) but also the quantitative characteristics of the circulational movement.

The flux between two lateral walls caused by the nonuniformity of the ion concentration profiles in the layers adjacent to the electrodes is of the same nature as the heat convection arising while the bottom wall is heated [2]. In the latter case a disturbance of the steady state occurs if the Rayleigh number reachs a certain (critical) value ($R_a = g\beta d^3 \Delta T / v \varkappa$, where β is the coefficient of bulk heat expansion, d is the distance between the walls, ΔT the increment of temperature, v the dynamic viscosity, and $\varkappa$ the thermal diffusivity); the liquid transforms into a new state with a periodic cell structure in such a way that the circulation in the interior of each cell has an opposite direction compared to that of the adjacent one. According to previous evaluations [53] the critical Rayleigh number in the case of lateral rigid walls is about 1700.

In view of the convection induced by the ion mass transfer, the force responsible for the momentum transfer is proportional to the product of the Grashoff and Schmidt numbers:

$$Gr \cdot Sc = Ra_d = q\beta^*(c_s - c_0)/vD. \tag{74}$$

In this diffusion analogy of the Rayleigh number we have $\beta^* = \varrho^{-1}(\partial \varrho / \partial c)_{\varrho t}$, and the values c_s and c_0 which are the ion concentrations participating in the interfacial mass exchange at the anode surface and in the interior of the solution, respectively.

If the difference $\Delta c = c_s - c_0$ arises at the upper anode due to the electric current, then because of changing the local solution density and with $R_{a_d} > R_{a_d}^{(cr)}$ this difference results in the formation of a hydrodynamic cell structure. The cell convection is the reason why the effective thickness of the diffusive boundary layer decreases [54, 55]. This decrease corresponds to the diminishing part of the U, t-curve (Fig. 4b). As the effective thickness of the diffusive boundary layer δ is connected with the current density i by the relation $i\delta = zFD\Delta c$, then with fixed i the decrease of δ must decrease the force of the ion mass transfer Δc. However, the value Δc changes during a time of about $t_d \approx a^2/D$, where a is the dimension of a circulation cell. The circulational cell dimensions and the circulation velocities have been evaluated by experiment [56]. The characteristic relaxation time for the hydrodynamic cell structure has an order of $t_h \approx a^2/v$, i.e. much less then t_d (because $v \gg D$). According to this the voltage decrease is to be followed by an increase. But this increase would inevitably induce an additional convection and, as a result, the next smallest part of the U, t-curve would appear. This is how the negative feedback between the processes of mass and momentum transfer occurs: the convection caused by the applied current pulse results in such a transformation of the diffusive boundary layer near the electrodes that the motive force of convective mass transfer is diminished. The quantitative characteristic of the mentioned feedback may be established (see Fig. 7) by the simultaneously recording the U, t- and v, t-curves.

The correlation between the behaviour of U, t- and v, t-curves found in Ref. [52] may be the basis for developing quantitative methods controlling the hydrodynamic flux structure near the electrode surfaces, as well as methods for influencing the velocity of the heterogeneous processes by means of controlled changing of the hydrodynamic conditions of interfacial mass exchange.

As mentioned above, the presence of a positive feedback between the interfacial forces and the fluxes of both the momentum and mass initiated by these forces is the requirement for regular dissipative structures to develop. Positive feedback results in an interfacial convective instability. A sufficient condition for transforming the system into the new regular state (either stationary or periodically changable in time) is connected with the existence of the negative feedback. The mechanism of this process counteracts the increase of the velocity field disturbance. For example, in case of self-organization of the interfacial system with a nonuniform distribution of surface tension, the tangential force proportional to the gradient of the surface tension induces the momentum flux which in turn promotes an increase of the extent of surface tension disturbance. That is indicative of the positive feedback existing between the capillary force and the momentum flux induced by it. At the same time there are the viscous friction forces acting near the interface and damping any movement arising. In other words, there exists a negative feedback between the capillary forces and the momentum flux caused by the viscous friction. The regular dissipative structure is in steady state which occurs in a system with wholly balanced positive and negative feedback. An example of such a balanced stationary state is the dissipative structure examined experimentally and theoretically in Ref. [57]. In these experiments [57] the gradient of the surface tension is induced by the air flow over the surface of the liquid filling the cuvette. The liquid is a solution of a surface-active substance in water. Washing away the adsorption film due to the air flow results in SAS concentration near one of the cuvette walls and the surface tension gradient increases. Varying the air flow velocity one may change the value of a capillary force. There exists a negative feedback in the system due to the viscous friction forces which counteracts the otherwise unlimited increase of the velocity field disturbances so that in view of the certain, critical value of the Marangoni number (corresponding to the particular kind of SAS) there are stationary regular fluxes of different configuration at the surface. These fluxes form the periodic configurations of surface structural elements (circles, ellipsoids, rosettes, etc. with the liquid circulating around the axis normal to the surface). By direct measurement, the correlation between the characteristic dimensions l_x and l_y of a surface structure (along the gas flow and normal to it) and the bulk flow velocity v_0 was

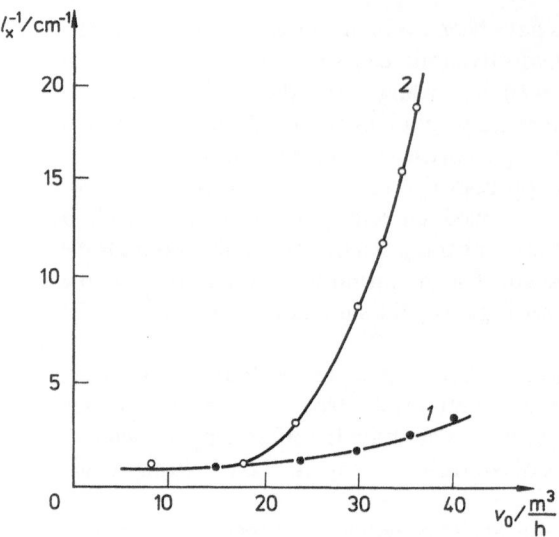

Fig. 8. The characteristic wavenumber for superstructures (curve 1) and substructures (curve 2) forming at the water-solution surface of the caprin acid ($1.5 \cdot 10^{-4}$ M) versus the bulk velocity of the air flow

found [57]. Figure 8 illustrates such a correlation for the l_x-dimension in the direction of a flux. Curve 1 corresponds to so-called "superstructures" with a longitudinal scale of the same order as the length of a cuvette in the x-dimension. Each element of a superstructure has interior "substructures" with elements having similar dimensions positioned at the surface of the superstructure elements in a certain order and with their own velocities of liquid circulation. A decrease of the scale l_x following the increase of the air velocity is a much more drastic for the substructures than for the superstructures (Fig. 8, Curve 8). When the flux velocity increases, the divergence of the observed values is also extended and this is indicative of the tendency to disturb the regular structure: the circulational cells become less stable. The theoretical analysis in Ref. [57] allows one to reproduce the form of the measured dependence illustrated in Fig. 8. According to this analysis, the critical Marangoni number $M_a^{cr} = (\partial\sigma/\partial x)_{cr} h^2/\varrho v^2$ $\left[\text{ here } \left(\dfrac{\partial\sigma}{\partial x}\right)_{cr} \text{ is}\right.$ the value of the surface tension gradient for the first appearance of the regular dissipative structures; h is the height of the gas layer above the liquid, ϱ and v are the density and the dynamic viscosity of a gas $\Big]$ is independent both of the liquid properties, and of the equilibrium adsorptional characteristics of the SAS and has a value of about 6000.

The dissipative structures arising as a result of the disturbances of the interfacial hydrodynamic stability may be either steady or unsteady. An example of a significantly nonstationary dissipative structure is the system studied experimentally in Ref. [58]; when a mercury drop is introduced into a field with an unstable gradient of the SAS concentration, oscillatory motion is observed at the interface. The concentration gradient was due to the immediate contact of two solutions with different concentrations of SAS (the latter being sucrose). The mercury drop was slowly squeezed out of the narrow capillary into the contact area of the two solutions, the upper one containing the lower SAS concentration, the lower density and the stronger surface

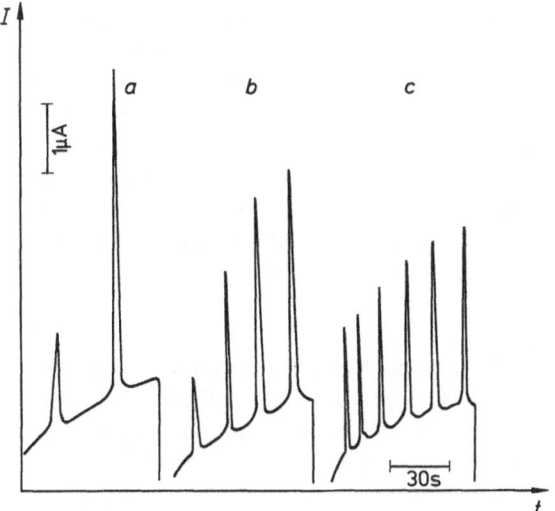

Fig. 9. The current oscillation versus the position of a mercury drop in the field of the SAS-concentration gradient (the initial gradient was induced by the contact of two water solutions: 0.1 H.KCl and 0.1 H.KCl + 25 weight % saccharose). The intervals a, b, c correspond to the drop center displacement for 0.6 mm in vertical direction towards the increase of the SAS concentration

tension at the mercury interface. Besides SAS there was an electrolyte in the solution with its cations (Tl^+-ions, as a rule) being the depolarizer. The current of the electrical reduction of the cations at the mercury dropping electrode was a measure of the interfacial mass exchange velocity. At the same time an optical system with fluorescent particles was used in a laser radiation field for registering the tangential motion at the mercury surface.

One of the typical results illustrating the existence of the unstable dissipative structure is shown in Fig. 9. Here the current I of the electroreduction is given as a function of time for three different positions of a drop center, removed at 0.6 mm from each other in a vertical direction along with the SAS-concentration increases. As it is shown, the removal of a drop in the direction of higher concentration gradients is followed by current oscillations; the increase of the concentration gradient results in an increased oscillation frequency and decreased amplitude. The optical measurements carried out at the same time have shown that the electrochemical regularities found are a reflection of the hydrodynamic picture of the oscillating tangential flows with the typical scale and velocity being a function of the gradient of the interfacial surface tension.

9. Regular Circulational Fluxes Caused by Hydrodynamic Instability and their Role in Interfacial Mass Exchange

Experimental data in the literature are indicative of the fact that the hydrodynamic instability caused by mass transfer, by homogeneous or heterogeneous chemical reactions, by heat exchange and the adsorption of the surface-active-substance may increase, under some circumstances, the interfacial mass exchange (sometimes by an order of magnitude). The possibility of such an intensification was first demonstrated in Refs. [59, 60]. Later experiments [61–69] and theoretical approaches [12–23, 55,

70–73] helped to establish and to analyze in a theoretical framework, the quantitative regularities of the mass transfer with respect to the hydrodynamic instability.

A very successful effort to settle the quantitative correlation between the velocity of the interfacial mass exchange and the Marangoni number in case of significantly nonuniform distribution of the interfacial surface tension was undertaken in Ref. [68]. The authors examined the kinetic rectification of binary mixtures with widely differing coefficients of surface tension of their components. The investigation was carried out for seven different mixtures in the film and shaft rectificational columns resulting in the following correlations:

$$\beta_x/\beta_x^0 = 1 + B(R)M_a \tag{75}$$

where β_x and β_x^0 are the values of the mass-output coefficient in the liquid phase with and without the influence of the surface tension gradient, respectively; the value $R = m\beta_y/\beta_x^0$ is the ratio of mass-exchange resistances in a vapour and a liquid phase (β_y is a mass-output coefficient in a vapour phase, m is a tangent to the line of the equilibrium at the diagram $y^* = f(x)$). The Marangoni number is determined as $M_a = \Delta\sigma l/\mu D$, where $\Delta\sigma$ is the difference between the true surface tension at the interface and that corresponding to the liquid in the flux core; in such a case the characteristic length l would be the thickness of the liquid film $l = (v^2/g)^{1/3}$; D is a diffusion coefficient in the liquid phase; μ the dynamic viscosity of the liquid. Equation (75) suggests the following procedure for calculating the influence of the surface tension gradient on the rectification kinetics [68]:

1. The coefficients of mass transfer in contacting phase are determined experimentally (or if possible with known correlations or adequate theoretical expressions).

2. The quantity R is calculated as the ratio of phase resistance to mass transfer.

3. The quantity B is determined by using the experimental data of Ref. [68].

4. The Marangoni number is calculated.

5. The resulting coefficient of the mass transfer is determined by using the rule of phase resistance additivity.

An equation similar to Eq. (75) was proposed [70] based upon the dynamic analysis of the mass transfer in a thin film possessing the given difference $\Delta\sigma$ of the surface tension at the gas-liquid interface. Thus a semi-empirical procedure proposed in [68] was theoretically developed.

There is as yet no completed quantitative theory of the mass exchange at unstable hydrodynamic interfaces. However, the steps directed to constructing such a theory show that unsteady behaviour of convective-diffusive transportation in the systems with regular dissipative structures may be described by exact quantitative relations in the near future and that methods for calculating the practical mass-transfer intensification could soon be developed. An example of such a predictive method is given by the analytic and numerical results of Refs. [71–73]. Their authors have calculated the intensification effect of the mass exchange in a two-phase system consisting of two flat parallel layers of nonmixing liquids for a regular circulational structure of hydrodynamic fluxes in each phase. The circulational structures have been described analytically based on the solution of the system of the Navier-Stokes linearized equations with the given tangential force at the phase boundary. Figure 10 illustrates the dependence of the intensification power of mass exchange η versus the Péclet

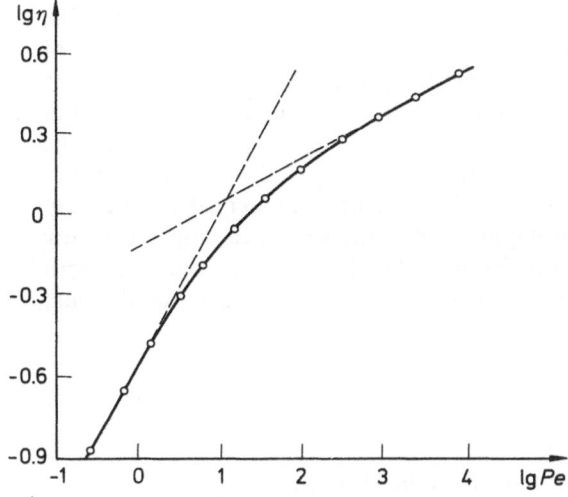

Fig. 10. The intensification power of the interfacial mass exchange η as a function of the Péclet number Pe based on the typical circulational velocity and the typical scale of a circulational cell

number Pe, obtained in Refs. [71–73]. The value η was determined as

$$\eta = (J - J_0)/J_0 \tag{76}$$

where J and J_0 are the diffusive fluxes through the interface with and without the Marangoni convection, respectively. The Péclet number was written as:

$$Pe = \frac{v_0 l}{D} \tag{77}$$

where v_0 is the highest velocity of liquid circulation in a cell in the interior of the phase limiting the velocity of the interfacial mass exchange; l is a characteristic length of a circulational cell; D is the coefficient of diffusion. From Fig. 10 it follows, in particular for low Péclet numbers, that the relation $\eta = \text{const} \cdot Pe^2$ is true in spite of the asymptotic behaviour (where $Pe \to \infty$) of a curve describing the intensification effect corresponding to a low $\eta \sim Pe^{1/2}$. Since v_0 (and consequently Pe) is proportional to the difference of the surface tension $\Delta\sigma$, the results given in Fig. 10 show that it is, in principal, possible to obtain a significant intensification of the mass exchange by controlled variation of the surface tension.

10. Conclusion

It is often pointed out in the literature [74] that one of the most important directions in the field of theoretical analysis of chemical and electrochemical technology is connected with the investigation of the extremal behaviour of the interfacial heat and mass exchange (extremal with respect to the thermodynamic balance); this behaviour allows the formation of self-organizing dissipative hydrodynamic structures. Other important ways of using the dissipative structures include the possible realization and controlled variation of the velocity of lateral ion transportation along the interface of two nonmixing electrolyte solutions, as well as the processes of light ion transport through the interface in systems with biologically active components [31, 75]. The realization of interfacial ion transfer controlled by the external electric fields is one of the currently important directions of modern bioelectrochemistry [76]. The examples considered above show that in some cases it is possible to influence the formation of dissipative structures by various external parameters (such as a polarization in the external electric field, thermal or concentrational influence upon the interfacial surface tension, the surface heat sources or flows, etc.). The prospect of practical utilization of the structures has stimulated the development of a theory concerning the highly unbalanced processes of momentum transfer, described by nonlinear equations with solutions that permit the existence of stationary or oscillating quasistable circulational flows.

References

1. Glensdorf, P., Prygozhin, I.: Termodin. Teor., M., Mir, 357 (1973)
2. Ebeling, V.: Obrazovanie Structur., M., Mir, 256 (1979)
3. Landau, L.D., Lifshitz, E.M.: Electrodinamica Sploshnykh Sred., M., Phismatgiz., 93 (1959)
4. Frumkin, A.N.: Zhurn. Chist. i Prikl. Zn., Otdel. Phis.-Mat. i Tekhn. Nauk. *1*, 1 (1921)
5. Levich, V.G., Krylov, V.S.: in "Annual Review of Fluid Mechanics," Palo Alto, California: Ann. Revs. Inc. *1*, 293 (1969)
6. Melcher, J.R., Taylor, G.I.: in "Annual Review of Fluid Mechanics," Palo Alto, California: Ann. Revs. Inc. *1*, 111 (1969)
7. Squire, H.B.: J. Mech. Appl. Math. *4*, 321 (1951)
8. Knowles, C.P., Gebhart, B.: Fluid Mech. *34*, 657 (1968)
9. Dzalurija, Y.: Estestv. Convectia, M., Mir. 296 (1983)
10. Benney, D.J., Lin, C.C.: Phys. Fuilds *3*, 656 (1960)
11. Audunson, T., Gebhart, B.: Int. J. Heat Mass Transfer *19*, 737 (1976)
12. Pearson, J.R.A.: J. Fluid Mech. *4*, 489 (1958)
13. Sterling, C.V., Seriven, L.E.: A.I.Ch.E.Journal. *6*, 514 (1959)
14. Scriven, L.E., Sterling, C.V.: J. Fluid Mech. *19*, 321 (1964)
15. Mield, D.A.: ibid. *19*, 341 (1964)
16. Whitaker, S.: Industr. Engng. Chem., Fundam. *3*, 132 (1964)
17. Smith, K.A.: J. Fluid Mech. *24*, 401 (1966)
18. Linde, H., Schwartz, P.: Teor. Osnovi Chim., Teknol., *5*, 401 (1971)
19. Brian, P.L.T., Smith, K.A.: A.I.Ch.E.Journal *18*, 231 (1972)
20. Zeren, R., Reynolds, W.C.: J. Fluid Mech. *53*, 305 (1972)
21. Sorensen, T.S., Hennenberg, M.: in "Lecture Notes in Physics. Dynamics and Instability of Fluid Interfaces" (Sorensen, T.S. Ed.), Springer Verlag, 276 (1979)
22. Reichenbach, J., Linde, H.: J. Colloid Sci. *84*, 433 (1981)
23. Schwartz, P., Vilke, G., Krylov, V.S.: Teor. Osnovi Chim. Teknol. *16*, 777 (1982)
24. Scanlon, J.W., Segel, L.A.: J. Fluid Mech. *30*, 149 (1967)
25. Kraska, J.R., Sani, R.L.: Int. J. Heat Mass Transfer *22*, 535 (1979)
26. Rosenblat, S., Davis, S.H., Homsy, G.M.: J. Fluid Mech. *120*, 91 (1982)
27. Rosenblat, S., Homsy, G.M., Davis, S.H.: ibid. *120*, 123 (1982)
28. Malcus, W.V.R., Veronis, G.: Phys. Fluids *4*, 13 (1961)
29. Choi, H.Y.: Trans. ASME, J. Heat Transfer *90*, 98 (1968)
30. Melcher, J.R., Firebaugh, M.S.: Phys. Fluids *10*, 1178 (1967)
31. Galler, D., Sanfeld, A., Bisch, P.: Physicochem. Hydrodynamics, *3*, 1 (1982)
32. Yagodin, G.A., Kagan, S.Z., Tarasov, B.B. et al.: Osnovi Extract. Zhidkost. M., Chimia 399 (1981)
33. Frentzel, G., Linde, H.: Teor. Osnovi Chim. Tekn. *20*, 28 (1986)
34. Mitjushev, P.V., Krylov, V.S.: Electrochimiya *21*, 22, 1604 (1986)
35. Dupeyrat, M., Michel, J.: J. Exp. Suppl. *18*, 269 (1971)

36. Dupeyrat, M., Nakache, E.C.R.: Acad. Sci., Ser. C. 277, 599 (1973)
37. Flett, D.S., Okuhara, D.N., Spink, D.R.: J. Inorg. Nucl. Chem. 35, 2471 (1973)
38. Flett, D.S., Cox, M., Heels, D.: in "Proc. Int. Solvent Extraction Conf." (Soc. Chem. Ind., London) 3, session 24, p. 2560 (1974)
39. Whewell, R.J., Hugues, M.A., Hanson, C.J.: Inorg. Nucl. Chem. 37, 2303 (1975)
40. Price, R., Tumelty, J.: Ind. Eng. Chem., Symp. Ser. 45, 18 (1975)
41. Dobson, S., Van der Zecuw, A.: J.Chem. Ind. 5, 176 (1976)
42. Nitsh, W., Navazio, L.: in "Proc. Interfacial Solvent Extraction Conf. 80, 220 (1980)
43. Nakache, E., Dupeyrat, M., Vignes-Adler, M.: Faraday Discuss., Chem. Soc. 77, paper 13, 1 (1984)
44. Mitjushev, P.V., Krylov, V.S.: Electrochimiya, 22, 552 (1986)
45. Newman, G.: Electrochim. Sistemy, M., Mir., 456 (1977)
46. Vidal, A., Acrivos, A.: Phys. Fluids 9, 615 (1966)
47. Frumkin, A.N., Stenina, E.V., Fedorovich, N.V.: Electrochimiya 6, 1572 (1970)
48. Frumkin, A.N., Fedorovich, N.V., Stenina, E.V., Damaskin, B.B., Krylov, V.S.: J. Electroanalyt. Chem. 50, 103 (1974)
49. Frumkin, A.N., Fedorovich, N.V., Stenina, E.V.: in "Itogi Nauki i Tekniki", ser. Electrochimiya, M., VINITI, 13, 5 (1978)
50. Baranowski, B., Kawczynski, A.L.: Roczn. Chemii 44, 2447 (1970)
51. Baranowski, B., Kawczynski, A.L.: Electrochim. Acta 17, 695 (1972)
52. Vessler, G.R., Schwartz, P., Krylov, V.S., Linde, H.: Electrochimiya 22, 623 (1986)
53. Pellew, A., Southwell, R.V.: Proc. Roy. Soc. A 176, 312 (1940)
54. Isaeva, L.A., Poljakov, P.V., Michalev, J.G., Rogozin, J.N.: Electrochimiya 18, 1697 (1982)
55. Slinko, M.G., Dilman, V.V., Rabinovich, L.M.: Teor. Osnovi Chim. Tekn. 17, 10 (1983)
56. Sawistowski, G.: in "Posledn. Dostidg. v Oblasti." (Hanson, K., Ed.), M., Chimiya 204 (1974)
57. Schwartz, P., Biletzky, J., Krylov, V.S., Linde, H.: Electrochimiya (in press)
58. Vessler, G.R., Hirhe, Hr., Krylov, V.S., Schwartz, P., Linde, H., Ering, H.: ibid. (in press)
59. Linde, H.: Fette, Seifen, Anstrichmittel 60 (826) 1053 (1985)
60. Linde, H., Mber, D.T.: Akad. Wiss. Berl. 1, 699 (1959)
61. Orell, A., Westwater, J.M.: Chem. Eng. Sci. 16, 127 (1961)
62. Orell, A., Westwater, J.M.: A.I.Ch.E.Journal 8, 350 (1962)
63. Sawistowski, H., Goltz, G.E.: Trans. Inst. Chem., Engrs. 41, 174 (1963)
64. Sawistowski, H., James, B.R.: Chem.-Ing.-Techn. 35, 175 (1963)
65. Bakker, C.A.P., Van Huytenen, P.M., Beek, W.J.: Chem. Eng. Sci. 21, 1039 (1966)
66. Sawistowski, H., Austin, L.J.: Chem. Ing.-Techn. 39, 224 (1967)
67. Dilman, V.V.: Teor. Osnovi Chim. Tekn. 9, 844 (1975)
68. Grymzin, J.N., Kwashnin, S.J., Lothov, V.A., Maljusov, V.A.: ibid. 16, 251 (1982)
69. Poljakov, P.V., Isaeva, L.A., Michalev, Y.G.: Electrochimiya 16, 1132 (1980)
70. Krylov, V.S., Maljusov, V.A., Nitshke, U., Lothov, V.A.: Teor. Osnovi. Chim. Tekn. 19, 10 (1985)
71. Nitshke, U., Schwartz, P., Krylov, V.S., Linde, H.: ibid. 19 (1985)
72. Nitshke, U., Schwartz, P., Krylov, V.S., Linde, H.: ibid. 19, 134 (1985)
73. Nitshke, U., Schwartz, P., Krylov, V.S., Linde, H.: ibid. 19, 245 (1985)
74. Krylov, V.S.: ibid. 17, 15 (1983)
75. Van Lamsweerde-Gallez, D., Meessen, A.: J. Membrane Biol. 23, 103 (1975)
76. Koryta, J.: Electrochim. Acta 29, 445 (1984)

Galvani and Volta Potentials at the Interface Separating Immiscible Electrolyte Solutions

Zbigniew Koczorowski

Abstract

Recent progress and main problems of the study of electrochemical equilibrium properties are reviewed for interfaces between two immiscible liquid electrolyte solutions. The discussed properties are mainly described in terms of the Galvani, Volta, zero charge, and surface (dipolar) potentials at the liquid–liquid interfaces and free liquid surfaces. Different galvanic and voltaic cells with liquid–liquid, mainly water–nitrobenzene interfaces, are described. These interfaces may be polarizable or reversible with respect to one or several ions simultaneously.

1. Introduction

An interface between two immiscible liquid electrolyte solutions is common in nature. First electrical and electrochemical investigations dealt with tissues and cells of live organisms, i.e. with systems containing such interfaces. As a result, the present state of knowledge of metal-solution electrochemistry is on a high level. On the contrary, the development of interfacial electrochemistry dealing with boundaries between liquid phases has been slow and practically limited only to liquid galvanic cells. It was only within the last ten years that significant progress has been made, in particular, in the area of ion and electron transfer through interfaces separating immiscible solutions. During this period of time, however, developments in the field of electrochemical equilibria and their representative potentials were much less pronounced. The reason is that the latter relies on a precise description and arrangement of earlier results, an accurate definition of reproducible experimental conditions and reference systems, as well as broadening of the interpretation of the investigated phenomena. This interpretation requires an adaptation of the theory and experimental techniques used in electrochemistry of metal–electrolyte solution systems. Hence, the field pays off its debt for the assistance rendered at its origin.

The state of art and achievements in the area of electrochemical equilibria at the interfaces between immiscible electrolyte solutions up to the late sixties are best described in the reviews by Rideal and Davies [1, 2], Kortüm [3], and Sollner [4]. More recent achievements have also been reviewed [5–15]. Here, the articles by Koryta should be mentioned, as they resulted in the development and popularization of the present research in the field of immiscible electrolyte solutions.

The Interface Structure and Electrochemical Processes
at the Boundary Between Two Immiscible Liquids
Editor: V. E. Kazarinov
© Springer-Verlag Berlin, Heidelberg 1987

Knowledge of ion distribution equilibria at liquid interfaces is of primary importance to various fields of science and technology, not only to the investigation of reactions of ion and electron transfer but also to the theory and practice of various processes examined in research and industrial laboratories. To begin with, one should mention bioelectrochemistry, liquid–liquid extraction, ion-selective electrodes, emulsification and some reactions of interest to organic chemistry, e.g. interfacial processes catalyzed in the presence of tetraalkylammonium or tetraalkylphosphonium salts. It should be noted here that a water–immiscible organic phase forms an interfacial liquid membrane; such liquid membranes can be applied for the purpose of selective separation of ions [16]. A potential difference between two aqueous solutions, separated from each other by a liquid or other membrane, is defined as the membrane potential.

It was Frumkin who pointed out that the investigations of interfacial phenomena of immiscible electrolyte solutions are also very important from the theoretical point of view [17]. They provide convenient approaches to the determination of various physicochemical parameters, such as transfer and solvation energy of ions, partition and diffusion coefficients, as well as interfacial potentials [7]. It should be remembered here that at equilibrium, either in the presence or absence of an electrolyte, solvents forming the discussed system are saturated in each other. Therefore, these two phases, in a sense, constitute two mixed solvents.

A model polar nonaqueous solvent that has been most commonly applied hitherto is nitrobenzene. The choice of nitrobenzene mainly results from the following reasons:

– very low solubility of water in nitrobenzene (0.34% by weight),
– low electrical resistance,
– good solubility of various chemical compounds,
– high permittivity, 35.96, practically providing almost complete dissociation of many salts within a wide range of concentrations,
– large specific density, $1.2 \text{ g} \cdot \text{cm}^{-3}$, and low vapour pressure.

The aim of this chapter is to provide a short review covering the present state of research and knowledge, as well as the problems concerning electrochemistry of liquid interfaces at equilibrium. These systems are best described by mutually related chemical parameters, such as ion transfer energy, $\Delta_w^s G_i$, and electrical parameters, such as Galvani and Volta potentials, $\Delta_w^s \varphi$ and $\Delta_w^s \psi$, where s and w refer to the system consisting of organic and aqueous phases mutually saturated. It is well known that both these potentials can be correlated with the difference of surface potentials of the s

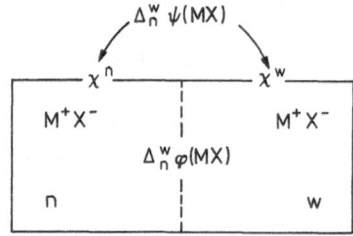

Fig. 1. Sketch illustrating Galvani, Volta, and surface potentials in water (w)-nitrobenzene (n) system

and w phases, $\Delta_w^s \chi$, according to the following equation:

$$\Delta_w^s \varphi = \Delta_w^s \psi + \Delta_w^s \chi \tag{1}$$

schematically illustrated in Fig. 1.

It should be noted that in the present review the results of the investigations performed with the use of voltaic cells of the type water–nonpolar solvent, e.g. water–decane [1, 2, 18] or water–octane [19–23] will be intentionally omitted. In the case of such interesting and important systems the organic phase does not constitute a solution of a dissociated electrolyte; hence here, as justified in detail by Davis and Rideal [1, 2], only changes of surface potentials could be examined in practice.

2. Liquid Galvanic Cells – a Historical Survey

An electrochemical galvanic cell consisting of at least one interface separating two immiscible electrolyte solutions is called for short a *liquid cell*. Such cells are sometimes, although less precisely, referred to as phase, oil or membrane cells [4, 24–27].

Research dealing with liquid cells originates directly from biological, or more precisely – bioelectrochemical, interest. First investigations and theoretical descriptions of liquid cells come from Nernst and Riesenfeld [28–32]. The idea of the liquid cells in which an organic phase models biological cell membranes is due to Cremer [33].

Results of investigations of liquid cells constituted a subject of many discussions or even violent disputes. Historically, the most controversial was the question of the conditions of EMF measurements, as well as the Beutner [26, 34] – Baur [35] dispute concerning the principle and origin of the electrical potentials under investigation.

It was promptly realized [1, 36] that liquid cells, being in the state of complete equilibrium, i.e. containing reversible electrodes in both phases, cannot constitute any subject of research interest. According to the Second Law of Thermodynamics such systems, irrespective of their composition, would have zero or constant value of EMF [8, 24, 25, 37]. The first case concerns reversible electrodes in respect to the same ion; the second, however, takes place when one electrode is reversible in respect to the cation and the second one in respect to the anion. It is well known that application of salt bridges, providing practically constant or negligible diffusion potentials, constitutes a non-thermodynamic procedure essential in many electrochemical experiments. Also in the case of investigation of interfaces between immiscible electrolyte solutions only a liquid (galvanic) cell with transport allows the characterization of the interface.

Practically all liquid cells with reversible interfacial equilibria examined to date can be considered as liquid galvanic cells of the Nernst, Haber or intermediate type [4].

Galvanic cells of the Nernst type are also termed cells with "dissolution membranes" or "solvent type membranes" [4, 28–32]. Such systems are defined by the distribution equilibria in which all ions, present in aqueous and in organic solvents, participate.

The second type of liquid cells is characterized by the presence of a liquid organic phase containing a liquid exchanger. Cremer [38] and particularly Haber [38, 39], who together with Klemensiewicz discovered the glass electrode, should be credited with the above idea. Thus, they have started research in the area of solid and liquid ion-selective electrodes.

Investigations of cells of the Nernst and Haber types have progressed for many years practically independently [4]. Despite doubtless differences, especially in their various ways of application at the present time there is a possibility of a uniform description of the theory and properties of such cells.

2.1 Investigations of the Nernst Type Cells

Investigations of liquid cells containing an organic phase permeable to all ions present were initiated by Nernst and Riesenfeld [28–32]. In such systems it is necessary to distinguish concentration cells and chemical galvanic cells.

In the case of concentration cells that, in general, can be represented in the form of the scheme:

$$
\text{SCE} \;\big|\big|\;
\begin{array}{c} \text{aqueous} \\ \text{solution} \\ M^+A^-, c_1 \end{array}
\;\big|\;
\begin{array}{c} \text{organic} \\ \text{solution} \\ M^+A^- \\ Bc_1\ Bc_2 \\ (c_1 \neq c_2) \end{array}
\;\big|\;
\begin{array}{c} \text{aqueous} \\ \text{solution} \\ M^+A^-, c_2 \end{array}
\;\big|\big|\; \text{SCE}
\qquad\qquad (\text{I})
$$

where the completely dissociated MA electrolyte, with the distribution coefficient B (concentrations $c_1 \neq c_2$), generates an EMF practically equal to the diffusion potential in the organic phase. As predicted by Nernst, independently of concentration, the distribution potentials of both liquid interfaces should be identical [29, 32]. In the above scheme SCE represents a saturated calomel electrode. The distribution potential is a Galvani potential resulting from the different distribution coefficients of the ions (see Sect. 3.2). The $|$ symbol represents the liquid interface, and the symbol $||$ designates the salt bridge eliminating the diffusion potential.

Chemical cells of the Nernst type can be represented by the following scheme:

$$
\text{SCE} \;\big|\big|\;
\begin{array}{c} \text{aqueous} \\ \text{solution} \\ M^+A^-, c_1 \end{array}
\;\big|\;
\begin{array}{c} \text{organic} \\ \text{solution} \\ M^+A^- \quad M^+L^- \\ B_1c_1 \quad B_2c_2 \end{array}
\;\big|\;
\begin{array}{c} \text{aqueous} \\ \text{solution} \\ M^+L^-, c_2 \end{array}
\;\big|\big|\; \text{SCE}
\qquad\qquad (\text{II})
$$

In such cells, aqueous solutions contain electrolytes with a common cation M^+. The EMF of this cell is equal to the difference of distribution potentials of both electrolytes and to the diffusion potential in the organic phase. According to Nernst, under appropriate conditions the EMF depends only on the difference of distribution potentials. It should be noted that cells functionally resembling type II can also contain many various ions in both phases.

Investigations of liquid cells, differing in their ion content and containing various organic solvents, brought up a dispute concerning the origin of EMF of type II cells. Beutner, after Nernst, thought that the difference of distribution coefficients of ions constituted the main source of EMF [26, 34]. On the contrary, Baur presumed that it originated predominantly from the adsorption of ions [35]. This idea was developed by Ehreneward and Sillen [40]. According to Cremer, the crucial contribution came from diffusion potentials. Finally, the problem was solved by the Rideal's school [1, 2, 41–43]. According to them, in the state of equilibrium and at the interface of immiscible electrolyte solutions, when both phases contain ions the distribution potential is the decisive factor. On the other hand, the surface potential related to the orientation and adsorption of dipole constituents of the system starts to dominate when a solution in nonpolar solvent constitutes one of the phases.

Numerous investigations of liquid cells of type I and II, including the works by Alleman [44], Kamieński and Karczewski [45–47], showed additivity of the distribution potentials in the presence of various ions. Such solvents as cresol, quinoline, ethyl acetate, n-butanol, n-pentanol, furfural were used for the organic phase; nitrobenzene appeared in later works.

Works by Karpfen and Randles [48], Boenhoffer, Kahlweit, and Strehlow [49–54, 3], Dupeyrat [55–57], Gavach and coworkers [58–61], Joos and Verburgh [62–63], as well as by Koczorowski et al. [64–67] allowed quantitative verification of the Nernst theory and practical application of liquid cells for various physicochemical purposes. In this research, in particular in Refs. [48, 52, 64], strictly reproducible experimental reference systems for the determination of distribution potentials were defined (see Sect. 4).

With the same view Minc and Koczorowski formulated the concept of voltaic cells with a liquid interface and measured a number of differences of distribution potentials in the water–nitrobenzene system [68–70] (see also Sects. 4 and 5). It should be noted that numerous investigations of this type were also performed by Kamieński et al. [71–73], and in particular by Frumkin, Boguslavsky, and Gugeshashvili with coworkers [74–78].

2.2 Investigations of the Haber Type Cells

Liquid cells containing an ion exchanger in the organic phase, e.g. a salt with a highly hydrophobic anion or cation R^+, could be considered as concentration cells [4, 54]:

$$
\text{SCE} \left|\left| \begin{array}{c} \text{aqueous} \\ \text{solution} \\ M^+X^- \\ c_1 \end{array} \right| \begin{array}{c} \text{organic} \\ \text{solution} \\ R^+X^- \\ c_1 \; c_2 \end{array} \right| \begin{array}{c} \text{aqueous} \\ \text{solution} \\ M^+X^- \\ c_2 \end{array} \left|\left| \text{SCE} \right.\right. \tag{III}
$$

or as chemical cells:

$$
\text{SCE} \left|\left| \begin{array}{c} \text{aqueous} \\ \text{solution} \\ M^+X^- \\ c_1 \end{array} \right| \begin{array}{c} \text{organic} \\ \text{solution} \\ R^+X^- \end{array} \right| \begin{array}{c} \text{aqueous} \\ \text{solution} \\ M^+L^- \\ c_2 \end{array} \left|\left| \text{SCE} \right.\right. \tag{IV}
$$

There are large cations in these cells, e.g. tetraalkylammonium cations in the organic phase; and the interfacial ion exchange involves only so-called critical ions, here X^- and L^-. M^+ ions are practically not transferred through the organic phase. Both liquid interfaces are reversible with respect to the appropriate anion, X^- or L^-.

In general, at least in the ideal case, EMF of cells of type III and IV can be described by the Nernst equation, but in the first case it depends on the logarithm of the ratio of activity coefficients of the same electrolyte, and in the second case on that of two different electrolytes [4].

EMF is, in practice, also influenced by the diffusion potential in the organic phase, and in the case of cells of type IV – by the difference of standard transfer energies of both ions (see Sect. 3.3).

Cells of type III represent the simplest case of an ion-selective liquid cell; its EMF is often called a membrane [79], or monoionic [80], potential. The first term is too narrow due to the fact that the membrane potential corresponds to the behaviour of a number of cells, both to those of type I and IV, and to the cells with solid membranes and with Donnan equilibria.

Cells of type IV represent the simplest case of cells with a biionic potential [4, 81]. Hence, in the case of a larger number of ions transferred through the organic phase a multi- or polyionic potential should be considered possible.

Liquid ion-selective electrodes operate on the basis of cells of type III; their selectivity can be examined with the use of type IV and polyionic cells.

It should be noted that, although it is justified, the terms mono-, bi-, and multiionic practically are not used in the case of liquid Nernst cells of type I or II. Such terms are, however, used in the case of solid, e.g. porous, membranes [79].

Research on Haber-type cells is very popular. Here, works by Sollner [4, 81, 82], Beutner, Boenhoffer and coworkers [49–54], Dupeyrat [55–57], Gavach et al. [58–61] and by Italian investigators [80, 83–85] should be mentioned. Works by Dupeyrat and Gavach also include investigations of the liquid Nernst-type cells (Sect. 2.1), thus resulting in an unequivocal description of both types of cells (Sect. 3).

The above works can be considered as complementary to the main subject of the characterization of ion-selective electrodes [86–91]. In this context, the recent achievements are presented in a review by Koryta [8].

Systematic investigations of cells of type IV, containing an ion-exchange system in the form of trioctylopropyl ammonium salts in o-dichlorobenzene were performed by Shean and Sollner [4, 82]. In the case of such anions as IO_3^-, Cl^-, Br^-, NO_3^-, and SCN^-, a very good additivity of biionic potentials has been found. These potentials form a series resembling the Hofmeister series, and they do not depend on the activity of the salt in the aqueous solution or on the concentration of exchanger in the organic phase.

The Italian scientists [80, 84–85] have made an important contribution to the investigations of type-III and -IV cells. Danesi et al. examined the correlation of membrane potentials with ion-exchange and association constants as well as with ion mobilities [80]. Tetraheptyl ammonium salts in benzene, chlorobenzene, o-dichlorobenzene, and nitrobenzene were used. Fabiani et al. broadened the interpretation of the results of this research of the works cited herein [87–89]. Scibona et al. [84] started original and promising temperature investigations of monoionic potentials of systems reversible with regard to Cl^-. This approach was employed by Fabiani et al. [85] in the analysis of the selectivity of ion-exchange phases.

Yoshida [92] published a semiempirical dependence describing the biionic potential with account of the influence of diffusion layers formed at the interface between the aqueous solution and the organic ion-exchange phase. This dependence was verified for the cells containing salts of crystal violet in nitrobenzene. The importance of ion association in the organic phase and its influence on selectivity of the ion-exchange phase was investigated by Iyo et al. [93].

2.3 Other Types of Liquid Cells

Cells of the Nernst and Haber type constitute two extreme cases of liquid cells with a reversible interface of immiscible electrolyte solutions. Intermediate cases are, however, common; e.g. in analytical applications, ion-selective systems with non-ideal (lower than 100%) selectivity are used.

Cells with liquid polarizable interfaces that are vital to kinetic studies, as well as voltaic cells, will be discussed in Sects. 4 and 5.

In modern investigations of the electrochemical properties of immiscible electrolyte solutions mixed cells are used, i.e. cells containing one interface, e.g. that under investigation – Nernstian or polarizable, and a second reference interface of the Haber type (Sect. 3.3).

Liquid cells are largely analogous to the cells containing molten salts or metals. Possibly it was the research in the area of those liquid cells that later inspired further investigations of cells with liquid metallic membranes, or "molten bridges" [4]. The main works to be noted here were performed by Lemke [94] and Holmberg [95] and they dealt both with concentration systems and chemical ones.

3. Galvani Potential at the Interface of Immiscible Electrolyte Solutions

Works by Nernst [28, 32] constitute a fundamental contribution to the electrochemical analysis of the phase equilibrium between two immiscible electrolyte solutions. According to these works, in the above system electrical potentials originate from the difference of distribution coefficients of ions of the electrolyte present in both phases. Verwey and Niessen [96] described this problem in the following manner: "Excluding the case when one of the liquids is nonpolar and solubility or dissociation is equal to zero, each of the electrolytes added to the system, even in very small amounts, acts as the potential-forming electrolyte. The electrolyte, despite that it is unevenly distributed between the two phases, forms at the interface, as a result of the unequal distribution coefficients of cations and anions, an electrical double layer".

Nernst's approach as well as Verwey and Niessen's interpretation, was supported in the work by Randles and Karpfen [48], Boenhoffer et al. [49–54], Overbeek [37], Kortüm [3], and in particular by Davies and Rideal [1, 2]. They consolidated the prevailing description of Galvani potential in the case of the 1:1 electrolyte distribution equations [see Eqs. (14–17)] below. Karpfen and Randles [48], as well as Joos and Verburgh [62] described the distribution systems containing 3 and 4 monovalent ions, [Eq. (36)].

3.1 General Approach of Le Hung

Le Hung presented a general theoretical approach for calculating the Galvani potential $\Delta_s^w \varphi$ at the interface of two immiscible electrolyte solutions, e.g. aqueous (w) and organic solvent (s) [97]. Le Hung's approach allows the calculation of $\Delta_s^w \varphi$ when the initial concentration, activity coefficients and standard energies of transfer of ions are known in both solutions. This can be considered as a quantitative realization of the idea implied in the above excerpt from the work by Verwey and Niessen [96].

According to Le Hung, one deals, in general, with a system composed of two liquid immiscible phases containing $I_i^{z_i}$ ions; this can be described as follows:

$$
\begin{array}{c|c}
w & s \\
I_i^{z_i} & I_i^{z_i} \\
V_w \rightleftarrows V_s
\end{array}
\tag{V}
$$

where z_i represents the charge of the $I_i^{z_i}$ ion, and V_w and V_s are the volumes of phases w and s, respectively.

In the equilibrium state, the electrochemical potentials of each $\bar{\mu}_i$ ion are identical in both phases:

$$
\bar{\mu}_i^w = \bar{\mu}_i^s \tag{2}
$$

Using the definition of the electrochemical potential this can be rewritten in the form:

$$
\mu_i^{0,w} + RT \ln a_i^w + z_i F \varphi^w = \mu_i^{0,s} + RT \ln a_i^s + z_i F \varphi^s \tag{3}
$$

where $\mu_i^{0,s}$, $\mu_i^{0,w}$ stand for standard chemical potentials of ions, a_i^w and a_i^s represent their activities, φ^w and φ^s are inner potentials of both phases, and R, F, and T have usual meanings and stand for the gas constant, Faraday constant and temperature (K).

From Eq. (3) one can derive the dependence of the Galvani potential at the interface

$$
\Delta_s^w \varphi = \varphi^w - \varphi^s = \Delta_s^w \varphi_i^0 + \frac{RT}{z_i F} \ln \frac{a_i^s}{a_i^w}. \tag{4}
$$

From Eq. (4) it follows that the Galvani potential can be calculated from the values of the standard potential and the activity of any ion participating in the equilibrium distribution. The standard Galvani potential is defined as follows:

$$
\Delta_s^w \varphi_i^0 = \Delta_w^s G_i^0 / z_i F. \tag{5}
$$

The standard transfer energy of an ion from the aqueous phase (w) to the nonaqueous phase (s), $\Delta G_{tr,i}^{0; w \rightarrow s}$, denoted in abbreviated form by the symbol $\Delta_w^s G_i^0$ is the difference of standard chemical potentials of the ions, i.e. of the standard Gibbs energies of solvation in both phases

$$
-\Delta_s^w G_i^0 = \Delta_w^s G_i^0 = \mu_i^{0,s} - \mu_i^{0,w}. \tag{6}
$$

Equation (4) can be rewritten in the form:

$$
\Delta_s^w \varphi = \varphi_s^w \varphi_i^0 + \frac{RT}{z_i F} \ln \frac{\gamma_i^s c_i^s}{\gamma_i^w c_i^w} \tag{7}
$$

where c_i^w and c_i^s represent the concentrations of $I_i^{z_i}$ ions; γ_i^w and γ_i^s stand for the activity coefficients of the ions in the phases w and s, respectively. It is apparent from the mass conservation law that:

$$V_w c_i^w + V_s c_i^s = m_i \tag{8}$$

where m_i represents the number of moles of $I_i^{z_i}$ ions in both phases. For these phases the condition of electroneutrality is still valid:

$$\sum_1^j z_i c_i^w = 0; \quad \sum_1^j z_i c_i^s = 0; \quad \sum_1^j z_i m_i = 0. \tag{9}$$

Using Eqs. (7–9), Le Hung derived the general dependence describing the difference of Galvani potentials between the phases w and s:

$$\sum_1^j z_i m_i \bigg/ \left(V_w + V_s \frac{\gamma_i^w}{\gamma_i^s} \exp\left[\frac{z_i F}{RT} (\Delta_s^w \varphi - \Delta_s^w \varphi_i^0) \right] \right) = 0 \tag{10}$$

or in the form:

$$\sum_1^h z_i c_i^{0,w} \bigg/ \left(1 + \frac{V_s}{V_w} \frac{\gamma_i^w}{\gamma_i^s} \exp\left[\frac{z_i F}{RT} (\Delta_s^w \varphi - \Delta_s^w \varphi_i^0) \right] \right)$$

$$+ \sum_h^j z_i c_i^{0,s} \bigg/ \left(1 + \frac{V_w \gamma_i^w}{V_s \gamma_i^s} \exp\left[\frac{z_i F}{RT} \Delta_s^w \varphi - \Delta_s^w \varphi_i^0 \right] \right) = 0. \tag{11}$$

Ions having initial concentrations $c_i^{0,w}$ and $c_i^{0,s}$, marked with indices from 1 to h and from h to j, occur in the phases w and s, respectively. When the values of z_i, V_w, V_s, γ_i^w, γ_i^s, $c_i^{0,w}$, $c_i^{0,s}$, $\Delta_s^w \varphi_i^0$, and T are known, one can find $\Delta_s^w \varphi$.

The above equations allow the calculation of Galvani potentials at the interfaces of immiscible electrolyte solutions in the presence of any number of ions with any valence, also including the cases of association or complexing in one of the phases. Marklik [98] described the cases of association and formation of complexes with participation of one of the ions but in both phases. In a later work [99] Le Hung extended his approach and also considered any mutual interaction of ions and molecules present in both phases.

Detailed solutions of the Le Hung's equation, for practically important and simple systems, are discussed below.

3.2 Distribution Potentials for Binary Electrolytes

In the case of an electrolyte of the type $M_m X_x$, in the state of distribution equilibrium, i.e. for the system:

$$
\begin{array}{c|c}
\text{w} & \text{s} \\
M_m X_x & \rightleftharpoons M_m X_x \\
V_w & V_s
\end{array}
\tag{VI}
$$

Equation (10) can be written as follows:

$$Z_{M^+} m_{M^+} \Bigg/ \left(V_w + V_s \frac{\gamma_{M^+}^w}{\gamma_{M^+}^s} \exp\left[\frac{z_{M^+} F}{RT} (\Delta_s^w \varphi - \Delta_s^w \varphi_{(M^+)}^0) \right] \right)$$

$$+ Z_{X^-} m_{X^-} \Bigg/ \left(V_w + V_s \frac{\gamma_{X^-}^w}{\gamma_{X^-}^s} \exp\left[\frac{z_{X^-} F}{RT} (\Delta_s^w \varphi - \Delta_s^w \varphi_{(X^-)}^0) \right] \right) = 0. \tag{12}$$

From the dependence of Eq. (12) and Eq. (9) one can obtain:

$$\Delta_s^w \varphi^0 = \frac{z_{M^+} \Delta_s^w \varphi_{M^+}^0 - z_{X^-} \Delta_s^w \varphi_{X^-}^0}{-z_{X^-} + z_{M^+}}$$

$$+ \frac{RT}{(-z_{X^-} + z_{M^+})F} \ln \frac{\gamma_{M^+}^s \gamma_{X^-}^w}{\gamma_{M^+}^w \gamma_{X^-}^s}. \tag{13}$$

When the distribution equilibrium refers to the $1:1$ valent electrolyte, e.g. MX, where $z_{X^-} = -1$, i.e. for the system

$$\begin{array}{cc} w & \vert & s \\ M^+ X^- & \rightleftarrows & M^+ X^- \end{array} \tag{VII}$$

Eq. (13) appears in the form:

$$\Delta_s^w \varphi_{(MX)} = \frac{\Delta_s^w \varphi_{(M^+)}^0 + \Delta_s^w \varphi_{(X^-)}^0}{2} + \frac{RT}{2F} \ln \frac{\gamma_{M^+}^s \gamma_{X^-}^w}{\gamma_{M^+}^w \gamma_{X^-}^s}. \tag{14}$$

The above dependence, which has been known for a long time, can be directly derived from Eq. (3) and the electroneutrality condition of Eq. (9) which for that case are in the form:

$$\Delta_s^w \varphi = \Delta_s^w \varphi^0(M^+) + \frac{RT}{F} \ln (a_{M^+}^s / a_{M^+}^w), \tag{15}$$

$$\Delta_s^w \varphi = \Delta_s^w \varphi^0(X^-) - \frac{RT}{F} \ln(a_{X^-}^s / a_{X^-}^w), \tag{16}$$

and

$$c_{M^+}^s c_{X^-}^w / c_{M^+}^w c_{X^-}^s = 1. \tag{17}$$

The formal Galvani potential, described by Eq. (14), does not depend on the concentration of ions of the electrolyte MX. Since the term containing the activity coefficients of ions in both solutions is, as experimentally shown, equal to zero [48, 69, 75, 76, 65] (Sect. 3.4) it may be neglected. This results predominantly from the cross-symmetry of this term and is even more evident when the ion activity coefficients are replaced by their mean values. A decrease of the difference in the activity coefficients in both phases is, in addition, favored by partial hydration of the ions in the organic phase. Thus, a liquid interface is practically characterized by the standard Galvani potential, usually known as the distribution potential [1, 2, 48]. It can be expressed in three identical forms; for example, for the system s/w they are as follows:

$$\Delta_w^s \varphi^0(MX) = \tfrac{1}{2}[\Delta_w^s \varphi^0(M^+) + \Delta_w^s \varphi^0(X^-)], \tag{18}$$

$$\Delta_w^s \varphi^0(MX) = \frac{1}{2F}[\Delta_s^w G_{M^+}^0 - \Delta_s^w G_{X^-}^0], \tag{19}$$

$$\Delta_w^s \varphi^0(MX) = \frac{RT}{2F} \ln \frac{B_w^{s,0}(M^+)}{B_w^{s,0}(X^-)}. \tag{20}$$

It is obvious from Eq. (13) that Eqs. (14, 18, 19, and 20) apply to all symmetrical electrolytes, i.e. to electrolytes dissociating into the same number of cations and anions.

According to Eq. (18), which directly ensues from Eq. (14), the distribution potential is the arithmetic mean of the Galvani potentials of cation and anion. These potentials are the ionic constituents of the distribution potential, and in fact, according to Eq. (5) they can be considered as electrical representations of the ionic transfer energies ΔG_i or limiting distribution coefficients of the ions, B_i [64, 65]. Here, the reader is referred to the following equations:

$$F\Delta_s^w \varphi^0(M^+) = -\Delta_s^w G_{M^+}^0 = -RT \ln B_w^{s,0}(M^+), \tag{21}$$

$$F\Delta_s^w \varphi^0(X^-) = \Delta_s^w G_{X^-}^0 = RT \ln B_w^{s,0}(X^-). \tag{22}$$

It is noteworthy [64] that for the case where:

$$B_w^{s,0}(M^+) = [B_w^{s,0}(X^-)]^{-1} \tag{23}$$

it appears that

$$\Delta_w^s \varphi^0(MX) = \Delta_w^s \varphi^0(M^+) = \Delta_w^s \varphi^0(X^-). \tag{24}$$

Equation (23) is true only when the standard chemical potentials (chemical solvation energies) of cations and anions are identical in both phases. Indeed, this occurs when two solutions in the same solvent are separated by a membrane. Hence, the Donnan equilibrium expressed in the form of Eq. (24) can be considered as a particular case of the Nernst distribution equilibrium; these phenomena are, however, not identical [64].

The distribution coefficients or distribution constants of the ions, $B_w^{s,0}(M^+)$ and $B_w^{s,0}(X^-)$, are related to the extraction constant $K_{MX}[100]$ and to the distribution coefficient of the B_{MX}^0 electrolyte in the manner described below.

The standard distribution constant describing the equilibrium in system VII,

$$M^+(w) + X^-(w) \rightleftarrows M^+(s) + X^-(s) \tag{VIIa}$$

can be expressed, in the equilibrium state, as follows:

$$K_{MX}^0 = \exp[-\Delta G_{MX}^0/RT] = \frac{a_{M^+}^s a_{X^-}^s}{a_{M^+}^w a_{X^-}^w}. \tag{25}$$

Expressing the distribution constant in terms of the mean electrolyte activities one obtains:

$$K_{MX} = \left(\frac{a_\pm^s}{a_\pm^w}\right)^2. \tag{26}$$

From the above dependence, and from the definition of the limiting or activity distribution coefficient of electrolyte, i.e. from the equation:

$$B_{w(MX)}^{s,0} = \frac{a_{\pm}^s}{a_{\pm}^w} \tag{27}$$

a relationship between the above-mentioned functions can be derived as follows:

$$B_{w(MX)}^{s,0} = K_{MX}^{1/2} = [B_{w(M^+)}^{s,0} \cdot B_{w(X^-)}^{s,0}]^{1/2}. \tag{28}$$

The transfer energies and distribution coefficients refer to two mutually saturated, i.e., in a sense, mixed solvents. It should be noted that this is a case where, under conditions of distribution equilibria, the quantities in question can be experimentally measured; this would not be possible with mutually miscible solvents [101]. There have also been numerous attempts, for example in Refs. [102, 103] or recently in Ref. [104], at theoretical calculations of transfer energies.

3.3 Interfaces Reversible with Respect to Single Ions

The interface between w and s containing electrolytes of the MX_1 and MX_2 type, respectively, and represented in the form:

$$\begin{array}{c|c} w & s \\ M^+X_1^- & M^+X_2^- \end{array} \tag{VIII}$$

constitutes an interface reversible in regard to the common cation M^+ under the following conditions [6–12, 97]:

$$\Delta_w^s G_{X_1^-}^0 \gg 0, \tag{29}$$

$$\Delta_w^s G_{X_2^-}^0 \ll 0, \tag{30}$$

$$\Delta_w^s G_{X_2^-}^0 < \Delta_w^s G_{M^+}^0 < \Delta_w^s G_{X_1^-}^0. \tag{31}$$

It is obvious from the above conditions that the transfer of strongly hydrophilic X_1^- anions from phase w to s and of strongly hydrophobic X_2^- anions from phase s to w is much more difficult compared to the transfer of the common hydrophilic-hydrophobic M^+ cations. In the equilibrium state, the Galvani potential is defined in terms of the Nernst equation [see Eqs. (4, 7, 15, and 16)]:

$$\Delta_w^s \varphi = \Delta_w^s \varphi(M^+) = \Delta_w^s \varphi_{(M^+)}^0 + \frac{RT}{F} \ln \frac{a_{M^+}^w}{a_{M^+}^s}. \tag{32}$$

When the concentration of M^+ ions the same in both phases of system VIII, the interface is characterized by the formal Galvani potential of the M^+ ions, i.e. by the equation:

$$\Delta_w^s \varphi_{(M^+)}^f = \Delta_w^s \varphi_{(M^+)}^0 + \frac{RT}{F} \ln \frac{\gamma_{M^+}^w}{\gamma_{M^+}^s}. \tag{33}$$

Equations (32) and (33) have been known and used for a long time, both as special relations describing Haber's ion-selective systems (Sect. 2.2) and modern investigations of the interfaces of immiscible electrolyte solutions.

The system of VIII, usually in the form shown below:

$$
\begin{array}{c|c}
\text{w} & \text{n} \\
\text{TBA}^+\text{Cl}^- & \text{TBA}^+\text{TPhB}^- \\
0.01\ \text{M} & 0.01\ \text{M}
\end{array}
\qquad\qquad\qquad\text{(VIIIa)}
$$

is a typical reference interface commonly used in the above investigations. It follows from Le Hung's analysis [97] that in the presence of tetrabutylammonium chloride in water and of tetraphenylborate (TBATPhB) in nitrobenzene, the interface appears to be reversible with respect to TBA$^+$ ions.

Interfaces of type VIII are used as liquid ion-selective electrodes. It is apparent from Le Hung's analysis that they constitute a special case of Nernstein distribution systems reversible in regard to two or more ions. Here, Le Hung's equation allows quantitative evaluation of the influence of the presence of other ions on the selectivity of these systems [7, 105, 106].

Kheifets and coworkers [107–112] showed the practical importance of the application of the potentials and Eq. (32) for the analysis and characterization of ion-exchange extraction. They also proposed an approach to calculate the $\Delta_w^s \varphi_i$ distribution potentials from the interface tensions, namely using dependences of Gibbs' isotherm type and the Lippmann equation.

Standard ionic potentials $\Delta_w^s \varphi_i^0$ can be calculated from the ionic distribution coefficients or transfer energies, [Eqs. (21) and (22)]. In order to perform such calculations, an appropriate non-thermodynamic assumption that allows division of the $B_{(MX)}^0$ or ΔG electrolyte function into ionic constituents, has to be made. At the present time, the assumption about the equality of the transfer energies of tetraphenyl-arsonium cations (TPhAs$^+$) and tetraphenylborate anions (TPhB$^-$) [113–114] is considered most appropriate; and it can be presented in the following form:

$$\tfrac{1}{2}\Delta G^0(\text{TPhAsTPhB}) = \Delta G^0(\text{TPhAs}^+) = \Delta G^0(\text{TPhB}^-). \qquad (34)$$

Using the above assumption, Rais [115] divided into components the radiometrically measured distribution coefficients and energies of transfer of various salts from water saturated with nitrobenzene to nitrobenzene saturated with water.

The ionic potentials $\Delta_w^s \varphi_i^0$ can be experimentally determined either with the use of galvanic cells containing interfaces of type VIII or electroanalytically [65]. During the reaction of transfer of a particular ion, e.g. Cs$^+$ through a liquid interface, one can find, using the electroanalytical methods, by polarography the characteristic half-wave ($E_{1/2}$) potential by chronopotentiometry, the potential $E_{\tau/4}$, or by voltammetry the peak potential E_p, and from these the appropriate $\Delta_w^s \varphi_i^0$ potentials [6–14]. The above measurements require application of suitable supporting electrolytes, as well as a low concentration of the ion being investigated.

3.4 Experimental Investigations of Galvani Potentials of Nonpolarizable Interfaces

For many years there existed a widely accepted view in the literature that Galvani potentials of interfaces between two immiscible electrolyte solutions could not be measured; only the potential differences resulting from the changes in the ionic composition were measurable [1, 2, 5, 6–12]. Hence, the differences of potentials, of the distribution potentials in particular, that correspond to the substitution of electrolyte M_1X_1 by M_2X_2 were investigated [48]:

$$\Delta_{M_2X_2}^{M_1X_1}\Delta_w^s\varphi^0 = \Delta_w^s\varphi^0(M_1X_1) - \Delta_w^s\varphi^0(M_2X_2)$$

$$= \frac{RT}{F}\ln\frac{B_w^{s,0}(M_1X_2)}{B_w^{s,0}(M_2X_1)}. \tag{35}$$

It could be concluded from the above dependence that when two electrolytes contain a common ion, then the value of $\Delta_{M_2X_2}^{M_1X_1}\Delta_w^s\varphi^0$ does not depend on that ion. Equation (35) was experimentally verified in several works, among which the investigations by Karpfen and Randles [48], Kahlweit and Strehlow [52], and later by Minc and Koczorowski [68–70] and Boguslavsky et al. [74–78] should be mentioned.

Karpfen and Randles investigated liquid galvanic cells of the type:

$$\text{SCE} \left|\begin{array}{c|c|c|c|c} w & \text{DIPK} & \text{DIPK} & w \\ MX & MX & \text{TEAPi} & \text{TEAPi} \end{array}\right| \text{SCE} \tag{IX}$$

where MX represents an investigated electrolyte in the state of distribution equilibrium, and DIPK stands for di-isopropyl ketone. It should be noted that the experimentally found equality of the transport numbers of tetraethylammonium (TEA^+) and picrate (Pi^-) ions in DIPK allowed this solution to be used for the preparation of the liquid junction that eliminated the contribution of the diffusion potential from the nonaqueous phase side.

A direct relationship between the differences of distribution potentials and the values of $B^0(MX)$ was also verified for a few electrolyte mixtures in this work with the use of the dependence [48]:

$$\Delta_w^s\varphi^0(MA + MX) - \Delta_w^s\varphi^0(MA) = \frac{RT}{2F}\ln\frac{1}{(1-n) + n\dfrac{B_w^{s,0}(X^-)}{B_w^{s,0}(A^-)}} \tag{36}$$

where n is a fraction of the total anion concentration in the aqueous phase at equilibrium, provided by X^-.

Cell IX that still contained the water–DIPK distribution system with TEAPi as a salt bridge, was also employed by Karpfen and Randles to determine the distribution potentials of the water–nitrobenzene systems [48]. The cells of this type were also used by Gugeshashvili et al. [74, 75], Dupeyrat [55–57], Gavach et al. [58–61] and by Joos et al. [62–63]; in the later works of the last-mentioned authors the bridging system water–nitrobenzene with TEAPi was employed. Kahlweit and Strehlow [52] studied the water–quinoline system with the use of a 0.07 M LiCl solution in quinoline as a salt bridge.

Equations [35] was verified in Refs. [69–78] with the use of voltaic cells (Sect. 5). Boguslavsky et al. also employed galvanic cells for direct measurements of the distribution potentials of two electrolytes using a salt bridge containing TEAPi in nitrobenzene [74, 76]; these were as follows:

$$\text{SCE} \left|\begin{array}{c|c|c|c|c|c} w & n & n & n & w \\ M_1X & M_1X & \text{TEAPi} & M_2X & M_2X \end{array}\right| \text{SCE}. \tag{X}$$

Alternatively, the above authors used a cell with direct contact of two nitrobenzene solutions containing a large common ion, e.g. M^+ [55, 61, 74, 76]:

$$SCE \left|\left| \begin{array}{c} w \\ MX_1 \end{array} \right| \begin{array}{c} n \\ MX_1 \end{array} \right|\left| \begin{array}{c} n \\ MX_2 \end{array} \right| \begin{array}{c} w \\ MX_2 \end{array} \left|\left| SCE \right.\right.\right. \qquad (XI)$$

It is apparent from the above studies that the existing interfaces are typical both for Nernst and Haber cells.

Koczorowski pointed out that despite the views existing till then it was possible to measure the values of $\Delta_w^s \varphi(MX)$ and $\Delta_w^s \varphi_b$, and not only those of $\Delta_{M_2X_2}^{M_1X_1}\Delta_w^s\varphi$ [64–65]. The proposed method originated from numerous previous experiments. It relied on the application of symmetrical cells with transport, namely of type [64]:

$$SCE \left|\left| \begin{array}{c} w \\ MX \end{array} \right| \begin{array}{c} s \\ MX \end{array} \right|\left| \begin{array}{c} n \\ TEAPi \end{array} \right| \begin{array}{c} w \\ TEAPi \end{array} \left|\left| SCE \right.\right.\right. \qquad (XII)$$

Three diffusion potentials and two distribution potentials appear in this cell besides two identical potentials of the reversible electrode, for example of aqueous calomel electrodes. If phase s contains nitrobenzene, all the diffusion potentials in system XII appear at the contact of solutions in the same solvent. The method is based on the following two assumptions [64]:

1. That the TEAPi solution in some non-aqueous solvents, in particular in nitrobenzene, eliminates the diffusion potential at the contact of two solutions in nitrobenzene or between nitrobenzene and other solvent, like KCl or KNO_3 in the case of aqueous solutions;

2. That the distribution potential of the water–immiscible solvent (in particular, nitrobenzene) system, in the presence of TEAPi distributed between two phases, is close to zero. It can be concluded from the above assumptions that the EMF of cell XII is equal to the value of the distribution potential (defined by Eqs. (14, 18–20) of the salt MX. It is noteworthy that salt bridges containing TEAPi were often used, and as mentioned above, both in direct studies of the distribution systems or in measurements of the ion transfer energy, e.g. of Ag^+, between two miscible or immiscible solvents [116–119]. In some works a bridge with TEAPi in di-isopropyl ketone was most frequently used; but unfortunately, different reference interfaces or electrodes, e.g. calomel and picrate electrodes, were employed [48, 62].

The method that assumes a negligible influence of the diffusion potential has often been criticized [120–122], mainly due to its weak theoretical foundation. However, it has to be emphasized that this is an experimentally simple approach, very well justified in the cases of interest to us. The above assumption (1) is supported by the results of equivalent conductivity measurements of TEA^+ and Pi^-. According to Krumgaltz [123], there is only a difference in these results of ca. 3.4% in non-aqueous nitrobenzene; probably in nitrobenzene saturated with water this difference is even lower. Further, the original assumption (2) was supported by the results of measurements by Rais [115], as well as by Koczorowski and Geblewicz [64, 66]. EMF measurements of type-XII cells were performed in glass vessels where the contact of miscible solutions was made via a glass sinter, as shown in Fig. 2 [65], or via a fine capillary to avoid mixing [67].

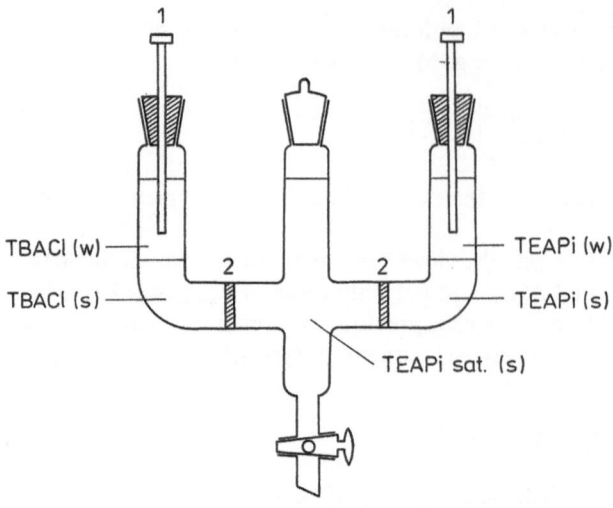

Fig. 2. Diagram of the measuring vessel with an example of an electrolyte [65]; *1* – represents saturated calomel electrodes, and *2* – glass sinters

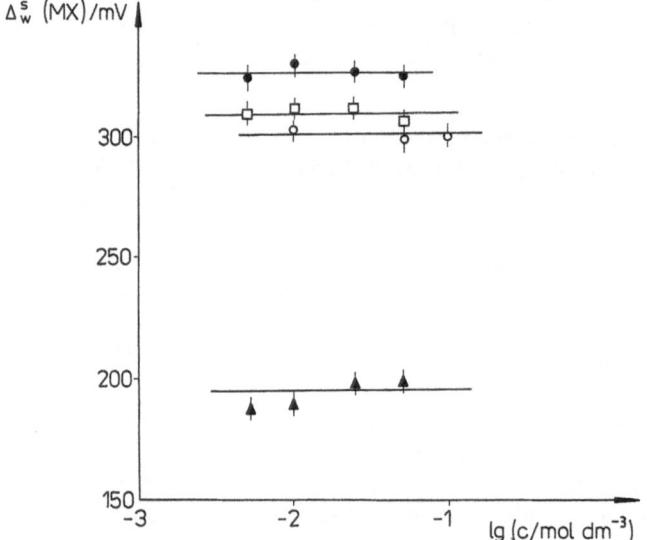

Fig. 3. The dependence of the distribution potential on the logarithm of its initial concentration in the aqueous phase [65]: for (●) TBACl and (▲) TEACl electrolyte in the water–nitrobenzene system; as well as for (□) TBACl and (○) TBABr in the water–1,2-dichloroethane system

Results of measurements with inorganic electrolytes [64], particularly with electrolytes containing one hydrophobic and one hydrophilic ion [65] have shown, in principle, that both nonthermodynamic assumptions, namely about the reference electrolyte (TPhAsTPhB) method (Sect. 3.4), and about the method of the cell with transport containing a TEAPi bridge, are equivalent.

The fact that the distribution potential is independent of the electrolyte concentration, postulated by Eqs. (18–20), can be considered as a criterion of the correctness of the measurements of $\Delta_w^s \varphi^0(MX)$ potentials (Fig. 3). Possible limitations of the constancy of the potential result from experimental conditions. The low concentration limit is mainly due to the interference of alien ions originating from impurities present in the investigated salt, or from solvent dissociation. On the other hand, the high concentration limit is due to influence of the differences of the ionic activity coefficients [Eq. (14)], which are more conspicuous at higher concentrations, and also to nonfulfillment of the assumptions of the method; for example when the ion concentration in the non-aqueous phase is not sufficiently lower than the TEAPi concentration of 0.2 to 0.4 mol $\cdot$ dm^{-3} in the bridge solution. Here values of 0.5 and of 0.01 M are the optimal initial concentrations for electrolytes containing hydrophilic ions and for salts with at least one hydrophobic ion, respectively.

Application of assumptions (1) and (2) and of cells of type XII allows construction of a common potential scale for various interfaces of immiscible electrolyte solutions. To achieve this goal in practice it is necessary to determine experimentally the standard distribution potential of TEAPi, or the formal potential of TBA$^+$ or any other reference standard in the water–given solvent system (w/s), versus the zero TEAPi water–nitrobenzene system. For this purpose, the cells schematically shown below can be used [64–67, 124–125];

$$\text{SCE} \bigg|\bigg|\begin{array}{c} w \\ \text{TEAPi} \end{array}\bigg|\begin{array}{c} s \\ \text{TEAPi} \end{array}\bigg|\bigg|\begin{array}{c} n \\ \text{TEAPi} \end{array}\bigg|\begin{array}{c} w \\ \text{TEAPi} \end{array}\bigg|\bigg|\text{SCE} \qquad\qquad \text{(XIII)}$$

and

$$\text{SCE}\bigg|\bigg|\begin{array}{c} w \\ \text{TBACl} \\ c^w \end{array}\bigg|\begin{array}{c} s \\ \text{TBATPhB} \\ c^s \end{array}\bigg|\bigg|\begin{array}{c} n \\ \text{TEAPi} \end{array}\bigg|\begin{array}{c} w \\ \text{TEAPi} \end{array}\bigg|\bigg|\text{SCE}. \qquad\qquad \text{(XIV)}$$

Using the above cells, the potentials of the reference systems have been determined, e.g. for water-nitrobenzene and water-1,2 dichloroethane interfaces [65, 66, 125]:

$$\begin{array}{cc} w & n \\ \text{TBACl} & \text{TBATPhB}; \\ 0.01\text{ M} & 0.01\text{ M} \end{array} \quad \Delta_n^w\varphi_{(\text{TBA}^+)}^f = -258\pm5\text{ mV}, \qquad\qquad \text{(XIVa)}$$

$$\begin{array}{cc} w & n \\ \text{TBACl} & \text{TBATPhB}, \\ 0.05\text{ M} & 0.05\text{ M} \end{array} \quad \Delta_n^w\varphi_{(\text{TBA}^+)}^f = -265\pm5\text{ mV}, \qquad\qquad \text{(XIVb)}$$

$$\begin{array}{cc} w & d \\ \text{TBACl} & \text{TBATPhB}, \\ 0.01\text{ M} & 0.01\text{ M} \end{array} \quad \Delta_d^w\varphi_{(\text{TBA}^+)}^f = -245\pm5\text{ mV}, \qquad\qquad \text{(XIVc)}$$

$$\begin{array}{cc} w & d \\ \text{TEAPi} & \text{TEAPi'} \end{array} \quad \Delta_d^w\varphi_{(\text{TEAPi})}^0 = +15\pm5\text{ mV}. \qquad\qquad \text{(XIIIa)}$$

3.5 Ion Transfer Energies and Galvani Potentials

Data on the distribution coefficients and transfer energies of electrolytes for water–nitrobenzene systems mainly originate from the works by Davies [126] and Rais [115]. Although the results of the first of the cited works have been used to test Eq. (35), the publication itself [126] has been somewhat forgotten. In practice, however, data by

Table 1. Standard Gibbs energies of ion transfer and corresponding standard Galvani potentials in the water–nitrobenzene systems

Ion	$\Delta_s^w G_i^0/$ kJ·mol^{-1} [115, 14]	$\Delta_s^w \varphi_i^0/V$ [115, 14]	$\Delta_s^w G_i^0/$ kJ·mol^{-1}	$\Delta_s^w \varphi_i^0/V$
Li$^+$	−38.4	0.398		
Na$^+$	−34.5	0.358	−29.9 [65]	0.310 [65]
H$^+$	−32.5	0.337		
NH$_4^+$	−27.4	0.284		
K$^+$	−24.3	0.252		
Rb$^+$	−19.9	0.206		
Cs$^+$	−15.5	0.161		
TMA$^+$	− 3.4	0.035		
TEA$^+$	5.8	−0.060	3.4 [66]	−0.035 [66]
TPA$^+$	15.5	−0.161		
TBA$^+$	24.2	−0.251	24.6 [65]	−0.255 [65]
TPhAs$^+$	35.9	−0.372		
Cl$^-$	−30.5	−0.316	−37.1 [65]	−0.385 [65]
Br$^-$	−28.5 [60]	−0.295 [60]	−34.3 [65]	−0.355 [65]
J$^-$	−18.8	−0.195		
ClO$_4^-$	− 8.7	−0.090		
Pi$^-$	4.6	0.048		
I$_3^-$	23.4	0.242		
NO$_3^-$	−24.4	−0.253		
TPhB$^-$	35.9	0.372		
Crystal Violet	39.5 [133]	−0.410 [133]		

Table 2. Standard Gibbs energies of ion transfer and corresponding standard Galvani potentials in the water–1,2-dichloroethane system

Ion	$\Delta_s^w G_i^0/$ kJ·mol^{-1} [141, 14]	$\Delta_s^w \varphi_i^0/V$ [141, 14]	$\Delta_s^w G_i^0/$ kJ·mol^{-1}	$\Delta_s^w \varphi_i^0/V$
Cs$^+$			−17.9 [65]	+0.185 [65]
TMA$^+$	−17.6	0.182	19.1 [65]	−0.198 [65]
TEA$^+$	− 4.2	0.044	− 4.1 [66]	+0.042 [66]
TPA$^+$	8.8	−0.091		
TBA$^+$	21.8	−0.225	19.9 [65]	−0.206 [65]
TPhAs$^+$	35.1	−0.364		
Cl$^-$	−46.4	−0.481		
Br$^-$	−38.5	−0.399		
J$^-$	−26.4	−0.273		
ClO$_4^-$	−17.2	−0.178		
TPhB$^-$	35.1	0.364		
Pi$^-$			− 2.9 [66]	−0.030 [66]

Rais are widely used and applicable to many ions. It should be noted that Rais, on the basis of Eq. (34), divided the quantities into G_i^0 ionic components. These data are cited almost completely in numerous review articles and publications [6, 9, 12, 97, 128]; More recently they have been supplemented with the results of electroanalytical determinations obtained by other authors [6–15, 60, 65, 66, 129–137].

Table 3. Standard Gibbs energies of ion transfer and corresponding standard Galvani potentials in the water–dichloromethane system [139, 14]

Ion	$\Delta_s^w G_i^0/kJ \cdot mol^{-1}$	$\Delta_s^w \varphi_i^0/V$
TMA$^+$	−18.8	0.195
TEA$^+$	− 4.2	0.044
TPA$^+$	8.8	−0.091
TBA$^+$	22.2	−0.230
Cl$^-$	−46.4	−0.481
Br$^-$	−39.3	−0.408
J$^-$	−26.4	−0.273
ClO$_4^-$	−21.3	−0.221
Pi$^-$	− 6.7	−0.069

Table 4. Standard Gibbs energies of ion transfer and corresponding standard Galvani potentials in the water–acetophenone system [140]

Ion	$\Delta_s^w G_i^0/kJ \cdot mol^{-1}$	$\Delta_s^w \varphi_i^0/V$
TBA$^+$	15.63	−0.162
TPhAs$^+$	16.21	−0.168
J$^-$	−12.35	−0.128
SCN$^-$	−11.00	−0.114
JO$_4^-$	− 1.45	−0.015
ClO$_4^-$	− 1.83	−0.019
ClO$_3^-$	−19.30	−0.200
NO$_3^-$	−20.36	−0.211
TPhB$^-$	16.21	0.168

Table 5. Standard Gibbs energies of ion transfer and corresponding standard Galvani potentials in the water–isobutyl-methyl ketone system [67]

Ion	$\Delta_s^w G_i^0/kJ \cdot mol^{-1}$	$\Delta_s^w \varphi_i^0/V$
Li$^+$	−20.3	0.21
TEA$^+$	− 8.7	0.09
TBA$^+$	19.3	−0.20
Cl$^-$	−50.2	−0.52
Pi$^-$	−11.6	−0.12

Table 1 lists the values of $\Delta_n^w G_i^0$ and $\Delta_n^w \varphi^i$ for more common ions, studied in the water–nitrobenzene system. Columns 1 and 2 contain values from Koryta's [14] review based on Rais' data [115, 138], but columns 3 and 4 contain values that also seem to merit special attention. It should be noted that many data are rather controversial [65]. For example, Samec and coworkers consider the value of the very important standard potential of TBA$^+$ ions to be equal to −0.275. This value is, however, rather unlikely.

The values of $\Delta_n^w G_i^0$ and $\Delta_n^w \varphi_i^0$, obtained with the use of electrochemical methods for the water–1,2-dichloroethane [65–66, 134, 138], water–dichloromethane [139], water–acetophenone [140] and water–methylisobutyl ketone [67] systems are shown in Tables 2 to 5. In the case of the systems with 1,2-dichloroethane the data obtained from

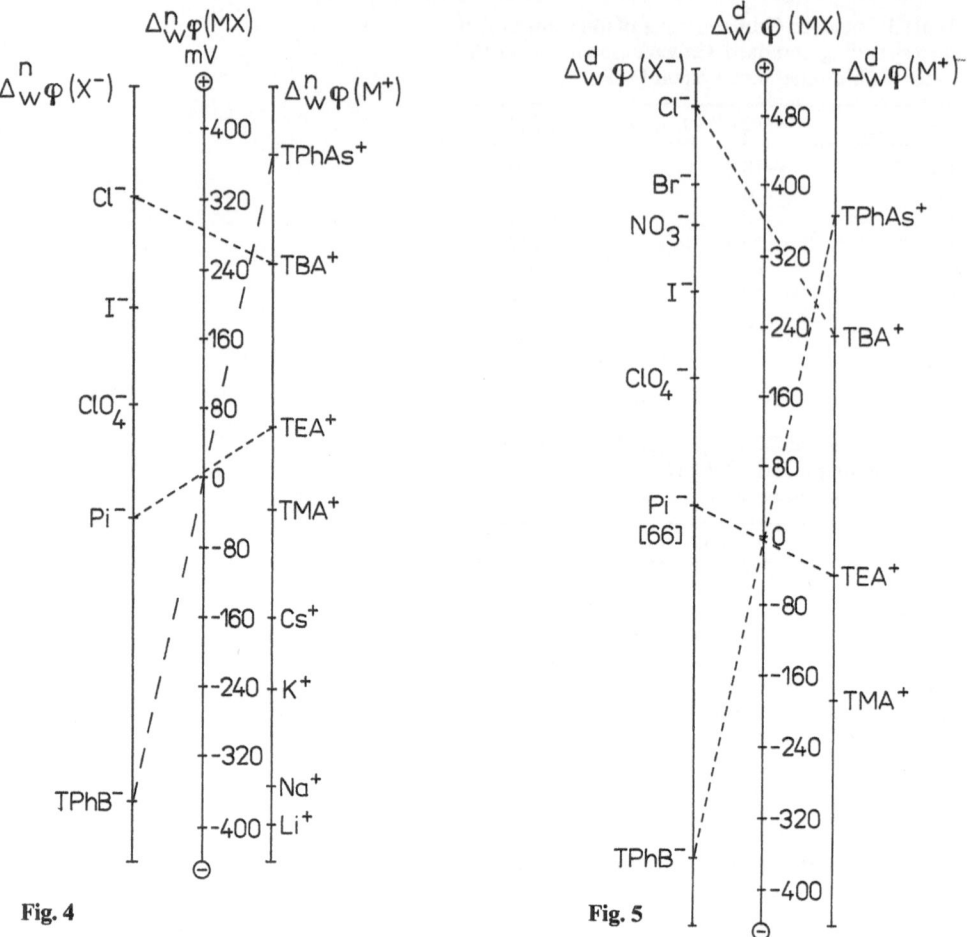

Fig. 4. Nomogram of the relationship [18] between the distribution potential of a salt and its ionic components for the water–nitrobenzene system [64]; data from Table 1 [115]

Fig. 5. Nomogram of relationship [18] between the distribution potential of a salt and its ionic components for the water–1,2-dichloroethane system; data from Table 2 [141]

the studies of distribution coefficients are also included (see Table 2, columns 1 and 2) [141, 14]. The data on ion transfer energies for pure 1,2-dichloroethane and water, i.e. for solvents not mutually saturated, can be found in Ref. [142]. Some results included in the tables are presented as nomograms (Figs. 4 and 5). Here, Eq. (18), the additivity of ionic potentials, and the physical meaning of the $\Delta_w^s\varphi^0(MX)$ distribution potentials, are graphically illustrated. The potentials that characterize the interfaces of systems containing TEAPi, TPhAsTPhB, and TBACl, are marked on the nomograms.

4 Polarizable Interface of Immiscible Electrolyte Solutions

The interface between an aqueous solution containing a strongly hydrophilic electrolyte (e.g. LiCl) and a nitrobenzene solution containing a strongly hydrophobic salt [e.g. tetrabutylammonium tetraphenylborate (TBATPhB)], schematically shown below:

$$
\begin{array}{c|c}
\text{w} & \text{n} \\
\hline
\text{LiCl} & \text{TBATPhB}
\end{array}
\tag{XV}
$$

constitutes a very important system in the studies of immiscible electrolyte solutions. Since the distribution coefficients of the above electrolytes can be defined by the inequalities:

$$
B_w^{n,\,0}(\text{LiCl}) \ll 1 \quad \text{and} \quad B_w^{n,\,0}(\text{TBATPhB}) \gg 1
\tag{37}
$$

the interface XV at least in certain potential limits, behaves as an ideally polarizable one, like mercury in an aqueous KF solution. A detailed analysis of this behaviour, as well as its analogy to the mercury–KF solutions system, can be found in the works by Koryta [7, 9]. Electrolyte ions, existing in the system XV, are practically present only in one of the phases, which allows them to function as supporting electrolytes in both solvents. Hence, the above system is necessary to study electrical double layers and the kinetics of ion and electron reactions at interface between immiscible electrolyte solutions.

The degree of polarizability of system XV can be found from the data calculated by Le Hung [97] with the use of Eq. (11). In the state of equilibrium of the solutions of 0.05 M LiCl in water and 0.05 M TBATPhB in nitrobenzene, the concentrations of Li^+ and Cl^- in the organic phase are equal to $1 \cdot 10^{-7}$ M and $4 \cdot 10^{-9}$ M, respectively, and the concentrations of TBA^+ and $TPhB^-$ in the aqueous phase are about $3 \cdot 10^{-7}$ M each. These concentrations are too low to establish permanent reversible equilibria. They are, however, significantly higher compared to those of the components present in the mercury–aqueous KF solution system [143]. According to Frumkin [144] and Parsons [143], an ideally polarizable interface constitutes, in a sense, a limiting case of a charge-transfer system. The limits, however, are markedly different for the two interfaces being compared here.

At the present time the electrochemistry of immiscible electrolyte solutions is developing mainly on the basis of the studies of the water–nitrobenzene and water–1,2-dichloroethane interfaces. The polarizability ranges of these interfaces in the presence of typical electrolytes (system XV) are about 0.30 V. Extension of these ranges towards negative currents has been achieved by substitution of TBA^+ ions by tetraphenylarsonium ions [145], and by crystal violet cations [133, 134]. But the substitution of $TPhB^-$ ions by dicarbollyl cobaltate (III) anions leads to an extension of the above range towards positive currents [127].

Investigations of the properties of polarizable interfaces of two immiscible electrolyte solutions, as well as of their double-layer structures and zero-charge potentials were undertaken in numerous works [61, 145, 148]. Samec et al. [149] on the basis of the pioneer works by Verwey and Niessen [96], as well as by Gross et al. [61], formulated a model of the double layer. In this model a layer of oriented solvent molecules (the inner layer) separates two diffuse layers of the Gouy-Chapman type.

This approach is, of course, to some extent controversial. Girault and Schiffrin question the applicability of the inner-layer model to liquid–liquid interfaces due to mixed-solvation and interfacial mixing phenomena [148, 151, 152]. Under these conditions it is very difficult to define the charge density. In addition, the influence of ion-pair formation, being a feature of specific adsorption, could not be neglected [152, 153]. Experimental investigations did not result in an unequivocal conclusion, though according to Samec et al. [149], all considered approaches are compatible.

It seems that certain additional information on this subject could be obtained from analysis and correlation of Volta and zero-charge potentials [155]. It is noteworthy that such an approach resulted in significant achievements in the study of metal–solution interfaces [156–159].

The Galvani potential of a liquid interface can be considered, similarly to the case of the metal–solution interface [37, 157], as a sum of dipole and ionic components [64]:

$$\Delta_n^w \varphi = \Delta_n^w g(\text{dipole}) + \Delta_n^w g(\text{ion}). \tag{38}$$

At the zero-charge potential the ionic component is equal to zero, thus the zero-charge potential itself is equal to the dipole component. It can be presumed on the basis of the water and nitrobenzene surface potentials (Sect. 5) that the value of $\Delta_n^w g(\text{dipole})$ is negligible, and likely to be close to zero [155] (Sect. 5). This is also supported by the previously determined values of zero-charge potentials [61, 137, 145, 149, 150, 160–162]. Dispersion of these data (from 0 to about 60 mV – Table 6) is not only

Table 6. Galvani zero charge potentials, $\Delta_n^w \varphi_{pzc}$, at water–nitrobenzene interfaces [155]

	Salt composition of system (mol · dm³)		$\Delta_n^w \varphi_{pzc}$ mV	Method of measurement	Reference
	in water	in nitrobenzene			
1	NaBr $3 \cdot 10^{-2}$ + TAABr var.	TAATPhB 10^{-2}	1 ± 5	interfacial tension	[61]
2	LiCl var.	TBATPhB var.	42 (15)	capacity	[145] ([150])
3	LiCl var.	TBATPhB var.	20 (−7)	interfacial tension	[161] ([150])
4	LiCl 10^{-1}	TBATPhB 10^{-1}	58	capacity	[162]
5	MgCl₂ $5 \cdot 10^{-3}$	TBATPhB 10^{-2}	27	capacity	[149]
6	LiCl 10^{-2}	TPhAsTPhB 10^{-2}	46	capacity	[149]
7	LiCl 10^{-2}	TBATPhB 10^{-2}	32	capacity	[149]
8	NaBr $3 \cdot 10^{-2}$ + TAABr var.	TAATPhB 10^{-2}	0 ± 5	interfacial tension	[160]
9	NaBr 10^{-2}	TBATPhB 10^{-2}	0	capacity	[150]

var. – variable concentrations
TAABr – tetraalkylammonium bromide
TAATPhB – tetraalkylammonium tetraphenylborate
TBATPhB – tetrabutylammonium tetraphenylborate
TPhAsTPhB – tetraphenylarsonium tetraphenylborate

evidence of experimental errors but also attests to existence of various charges at the interfaces of immiscible electrolyte solutions [155]. Probably, it will be necessary to distinguish between free and total charges, just as it was done in the case of metal–solution interfaces [158].

5 Volta Potentials at the Interfaces of Immiscible Electrolyte Solutions

A gas gap is a necessary phase in voltaic cells. This phase separate two condensed phases, e.g. metals, liquid electrolyte solutions or solid electrolytes. Investigations of these cells allow the determination of Volta potentials, i.e. differences of the outer potentials of the phases, separated by the gas gap. Volta potentials are directly related to such quantities as real potentials of charged species in investigated phases and surface potential changes.

The Volta potential, $\Delta_w^s\psi(MX)$, can also be considered in the case of immiscible electrolyte solutions, e.g. for the nonpolarizable water–nitrobenzene interface in the state of distribution equilibrium of the MX electrolyte (Fig. 1). The above potential can be operationally defined as the compensating voltage of the voltaic cell [163]:

$$M \left| \begin{array}{c} w \\ MX \end{array} \right| N_2 \left| \begin{array}{c|c} n & w \\ MX & MX \end{array} \right| M, E_{16} \tag{XVI}$$

where M stands for metal, and N_2 represents a chemically inert gas, e.g. nitrogen or pure air. Such cells can be investigated with the use of the jet method. However, it is easier to apply the ionization or, especially, the condenser methods. In the latter case, the value of $\Delta_w^s\psi^0(MX)$ is obtained from the difference of the compensation voltages of the following two cells [163]:

$$M \left| \begin{array}{c} \text{Vibrating} \\ \text{plate} \end{array} \right| N_2 \left| \begin{array}{c|c} n & w \\ MX & MX \end{array} \right| SCE \left| M, E_{17}, \right. \tag{XVII}$$

$$M \left| \begin{array}{c} \text{Vibrating} \\ \text{plate} \end{array} \right| N_2 \left| \begin{array}{c} w \\ MX \end{array} \right| SCE \left| M, E_{18}, \right. \tag{XVIII}$$

The voltaic cells of type XVII and XVIII are shown schematically in Fig. 6.

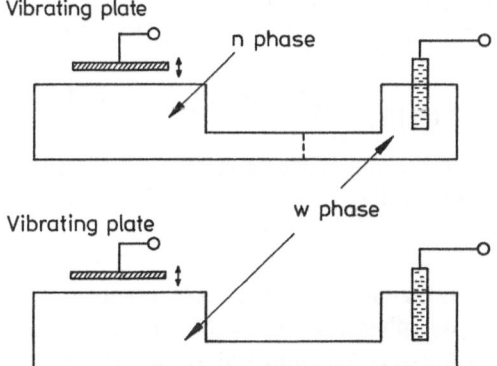

Vibrating plate

n phase

w phase

Vibrating plate

Fig. 6. Schematic illustration of an experimental realisation of the Volta potential measurement by a dynamic condenser method; a – cell (XVII), b – cell (XVIII)

For the same reasons in the case of the Galvani potential [Eqs. (18–20)], the obtained quantity $\Delta_n^w \psi^0(MX)$ also represents the standard Volta potential. The dependence:

$$\Delta_n^w \psi^0(MX) = E_{16} = E_{17} - E_{18} \tag{39}$$

is true if during the E_{17} and E_{18} measurements the following conditions are satisfied [163]:

1. the constancy of the surface potential of the vibrating plate, e.g. the gold plate, protected with a layer of permaflon [163];
2. the constancy of the surface potential of the water–saturated nitrobenzene phase χ^n, i.e. its independence of the presence of electrolytes dissolved therein;
3. the constancy of the surface potential of the nitrobenzene–saturated water phase, χ^w, i.e. its independence of the presence of electrolyte therein;
4. the constancy of the reference electrode (SCE) potential, and the constancy of the possible diffusion potential.

Conditions (1) and (4) are typical for all investigations of the Volta potentials. There are no serious problems with realizing condition (2). It is also important to satisfy this condition in the studies of the differences of distribution potentials with the use of Eq. (35), i.e. by the method of voltaic cells [68–78]. In the above investigations, type-XVII cells are employed. As it can be easily proved, substitution of the electrolyte M_1X_1 by M_2X_2, when the conditions (1), (2), and (4) are satisfied, leads to a change of the compensation voltage, ΔE_{17}, defined by the equation:

$$\Delta E_{17} = \Delta_{M_2X_2}^{M_1X_1} \Delta_n^w \psi^0 = \Delta_{M_2X_2}^{M_1X_1} \Delta_n^w \varphi^0 . \tag{40}$$

In the case of many salts including all those containing hydrophobic ions, it is not possible, however, to satisfy directly condition (3), For such systems it is necessary to determine experimentally the correction that corresponds to the change of the surface potential, caused by the presence of a given electrolyte in cell XVIII; or to use the value of E_{18} for a non-adsorbing salt, e.g. NaCl [163].

The performed measurements led to determination of the Volta potentials for several inorganic (KCl, NaCl, NaBr) and organic (TEACl, TBACl, TPhAsCl, TEAPi, NaTPhB) salts [163, 164]. Differences of these values are equal, as it follows from Eq. (40), to the respective differences of Galvani potentials (Sect. 3.4).

Utilization of the properties of the TEAPi system used in cell XII, Sect. 3.4, offers interesting possibilities. Investigations of the following voltaic cells [163, 164, 119]:

$$\text{M} \left| \begin{array}{c} \text{vibrating} \\ \text{plate} \end{array} \right| \text{N}_2 \left| \begin{array}{c} n \\ \text{MX} \end{array} \right|\left| \begin{array}{c} n \\ \text{TEAPi} \end{array} \right| \begin{array}{c} w \\ \text{TEAPi} \end{array} \left|\right| \text{SCE} \left| \text{M}, E_{19} \right. \tag{XIX}$$

and of type-XVIII cells allow determination of the difference of surface potentials between water saturated with nitrobenzene and nitrobenzene saturated with water:

$$\Delta_n^w \chi = E_{18} - E_{19} \tag{41}$$

of course, on condition that assumptions (1) and (2), mentioned in Sect. 3.4, are satisfied. When in cells XVIII and XIX, the investigated phases with the MX supporting electrolyte do not contain mutually saturated nitrobenzene and water, then Eq. (41)

allows the determination of $\Delta_{n0}^{wo}\chi$ for pure solvents. It should be noted that a similar approach to measure the surface potential differences has been used by Damaskin and Kaganowicz [165]; in all the investigated solvents, however, salt bridges with aqueous KCl have been employed.

It has been established that the values of $\Delta_{n0}^{wo}\chi$ and $\Delta_n^w\chi$ are equal to 0.24 ± 0.01 V [119] and 0.10 ± 0.02 V [163], respectively. The above data suggest that the presence of water in nitrobenzene and, in particular, of nitrobenzene in water strongly alters their surface structure. It can be supposed that once mutually saturated water and nitrobenzene achieve a direct contact, their surface structures become even more similar. Such a case would correspond to a lower difference of surface potentials, $\Delta_s^w g(\text{dipole})$ (Sect. 4). Hence the experimentally justified inequality can be written as follows [155]:

$$\Delta_{n0}^{wo}\chi > \Delta_n^w\chi > \Delta_n^w g(\text{dipole}) \tag{42}$$

on condition that the dipole potential corresponding to the zero charge potential (Sect. 4), should be negligible, close to zero. The equation written below, namely the following differences:

$$\Delta_n^w g(\text{dipole}) - \Delta_n^w\chi = \delta\chi^w - \delta\chi^n = \Delta_n^w\delta\chi \tag{43}$$

contain formal information about the structure of liquid interfaces. The value of $\Delta_n^w\delta\chi$, just as in the case of the metal–solution interface, expresses the change of the surface potentials χ^w and χ^n by bringing both liquids into contact [165]. The experimentally proved statement that in the presence of tetraalkylammonium picrate the distribution potential of the water–nitrobenzene interface is equal to zero (Sect. 3.5) [64, 65, 115] indicates that the $\Delta_n^w\delta\chi$ potential of many systems is also close to zero.

The value of the standard Volta potential, obtained for TEAPi, $\Delta_n^w\varphi^0(\text{TEAPi})$ $= -106\pm15$ mV [164], is almost exactly equal, but with opposite sign, to the value $\Delta_n^w\chi$ $= +105\pm20$ mV [163]. Substituting these data into the Eq. (1)-type relationship:

$$\Delta_n^w\varphi^0(\text{TEAPi}) = \Delta_n^w\psi^0(\text{TEAPi}) + \Delta_n^w\chi \tag{44}$$

yields $\Delta_n^w\varphi^0(\text{TEAPi}) = 0\pm20$ mV. This result is in agreement with the previously estimated value [64, 65].

Subtraction of the already mentioned value of $\Delta_n^w\chi$ from the standard values of ionic Galvani potentials (Table 1) allows the calculation with the use of an Eq. (44)-type dependence of the standard Volta potentials of the ion in the water–nitrobenzene systems.

In real systems, i.e. in the presence of a supporting electrolyte (Sect. 4, system XV), the structure of the interface is also affected by TBA^+ and TPhB^- ions. According to Gross et al. [61], these ions are adsorbed at the water–nitrobenzene interface. However, taking into account very strong adsorption of TBA^+ [163] and TPhB^- on the free water surface and on that saturated with nitrobenzene, the above conclusion is to be considered as approximate.

It is interesting to note the markedly different surface behaviour of TBACl and NaTPhB in water saturated with nitrobenzene. Investigation of the surface potentials and surface tensions of such solutions suggests the anti-parallel orientation of nitrobenzene and Na^+TPhB^- dipoles unlike the parallel orientation of nitrobenzene and TBA^+Cl^- dipoles [164]. This problem, as well as many others discussed above, requires further studies. In this context the seldom considered works by Ohlenbush [166] and Krylov et al. [167] would be helpful.

6 Final Observations

The review presented here certainly does not cover completely the broad subject of the equilibrium phenomena at the interfaces between immiscible electrolyte solutions. Also many of the discussed questions are still open, and to be better or more precisely explained, require further studies. To achieve this goal, it is necessary to develop new measurement techniques, e.g. the jet method for zero-charge potential [168] and spectral approaches to investigation of the composition of liquid interfaces. A very important issue is the ability to better define the conditions of the measurement of liquid interfaces. The results of various authors are controversial, perhaps due to the difference in the purity and methods of preparation of solutions and their constituents, as well as due to time-dependent chemical processes and emulsification. It is also well known that the presence of electrolyte salts alters the solubility of organic compounds, e.g. of nitrobenzene in water. Inorganic salts, e.g. NaBr, lower (salting-out effect), and organic salts, e.g. TBABr, increase (salting-in effect) the solubility [169]. Finally, one should pay attention to the recently initiated electrochemical studies of the interfaces between water and a mixed organic solvent [124, 170].

Note

Three out of six figures in this chapter originate from my published papers, namely:
   – Figs. 2 and 3 are taken from Ref. [65] by Koczorowski Z. and Geblewicz G., J. Electroanal. Chem. *152*, 55 (1983).
   – Fig. 4 is taken from Ref. [64] by Koczorowski Z., J. Electroanal. Chem. *127*, 11 (1981).
   As far as Figs. 1, 5, and 6 are concerned, they have been prepared especially for this chapter.

Warsaw, 17th February 1986   (Zbigniew Koczorowski)

References

1. Davies, J.T., Rideal, E.K.: Can. J. Chem. *33*, 947 (1956)
2. Davies, J.T., Rideal, E.K.: Interfacial Phenomena, Academic Press, N.Y. – London, p. 56, 1961
3. Körtum, G.: Treatise on Electrochemistry, Elsevier, Amsterdam – London – N.Y., 1965
4. Sollner, K.: in: Sherwood, J.N. et al. (Eds.) Diffusion Process, Proceedings of the Thomas Graham Memorial Symposium, Vol. 2, Gordon and Breach, London, p. 656, 1971
5. Boguslavsky, L.I., Yagushinsky, L.S.: in: Electrosynthesis and Bioelectrochemistry (in Russian), Nauka, Moskva, p. 305, 1975
6. Koryta, J.: Electrochim. Acta *24*, 293 (1979)
7. Koczorowski, Z.: Wiad. Chem. *33*, 279 (1979)
8. Koryta, J.: Hung. Sci. Instr. *49*, 25 (1980)
9. Koryta, J., Vanysek, O.: in: Gerisher, Tobias, (Eds.), Advances in Electrochemistry and Electrochemical Engineering, Wiley, N.Y., Vol. 12, p. 113, 1981
10. Koryta, J.: Anal. Chim. Acta. *111*, 1 (1979)
11. Koryta, J.: Ion-Selective Electrode Rev. *5*, 131 (1983)

12. Koryta, J.: Ions, Electrodes, Membranes (in Russian), Mir, Moskva, p. 201, 1983
13. Koczorowski, Z.: in: Galus, Z. (Ed.) Theoretical Fundamentals of Applied Electrochemistry (in Polish), Ossolineum, Wrocław, p. 77, 1983
14. Koryta, J.: Electrochim. Acta 29, 445 (1984)
15. Vanysek, P., Buck, R.P.: J. Electroanal. Chem. 163, 1 (1984)
16. Gökalp, M., Hodgson, K.T., Cussler, E.L.: AIChEJ 29, 144 (1983)
17. Frumkin, A.N.: Zh. Fiz. Khim. 35, 2163 (1961)
18. Minc, S., Koczorowski, Z.: Roczniki Chem. 39, 469 (1965)
19. Frumkin, A.N., Gugeshasvili, M.I., Boguslavsky, L.I.: Dokl. ANSSSR 198, 1452 (1971)
20. Gugeshasvili, M.I., Boguslavsky, L.I., Frumkin, A.N.: ibid. 206, 985 (1972)
21. Volkov, A.G., Bibikova, M.A., Mironov, A.F., Boguslavsky, L.I.: Bioelectrochem. Bioenerg. 10, 477 (1983)
22. Volkov, A.G., Gugeshasvili, M.I., Mironov, A.F., Boguslavsky, L.I.: ibid. 10, 485 (1983)
23. Gugeshasvili, M.I., Volkov, A.G., Yaguzhinsky, L.S., Mironov, A.F., Boguslavsky, L.I.: ibid. 10, 493 (1983)
24. Brodsky, A.I.: Physical Chemistry (in Russian), Goskhimizdat, Moskva, Vol. 4, 1948
25. Izgaryshev, N.A., Gorbachov, S.V.: Course of Theoretical Electrochemistry (in Russian), Goskhimizdat, Moskva, 1951
26. Beutner, R.: Physical Chemistry of Living Tissues and Life Processes, London, 1933
27. Butler, J.A.: Electrocapillarity, The Chemistry and Physics of Electrodes and Other Charged Surfaces, London, 1940
28. Nernst, W.: Z. Phys. Chem. 6, 16 (1890); 8, 110 (1891); 9, 137 (1892)
29. Nernst, W., Riesenfeld, E.H.: Ann. Physics 8, 600 (1902)
30. Riesenfeld, E.H.: ibid. 8, 616 (1902)
31. Riesenfeld, E.H., Reinhold, B.: Z. phys. Chem. 68, 459 (1905)
32. Nernst, W.: Theoretische Chemie, Verlag Enke, Stuttgart 1926
33. Cremer, M.: Z. Biol. 47, 562 (1906)
34. Beutner, R.: Z. Elektrochem. 19, 319, 467 (1913)
35. Baur, E., et al.: ibid. 32, 547 (1926)
36. Adam, N.K.: The Physics and Chemistry of Surface, Oxford, 1941
37. Overbeek, J.T.G., in Kruyt, H.R. (Ed.): Colloid Sciences, Vol. 1, Elsevier Publ. Comp., Amsterdam 1952
38. Haber, F.: Ann. Phys. 26, 927 (1908)
39. Haber, F., Klemensiewicz, Z.: Z. Phys. Chem. 67, 385 (1909)
40. Ehrensward, G., Sillen, L.G.: Nature 141, 788 (1938)
41. Craxford, S.R., Gatty, O., Rotschild, I.: ibid. 141, 1098 (1938)
42. Dean, R.B.: ibid. 144, 32 (1939)
43. Dean, R.B., Gatty, O., Rideal, E.K.: Trans. Faraday Soc. 35, 161 (1940)
44. Alleman, E.: Z. Elektrochem. 34, 373 (1928)
45. Kamieński, B., Karczewski, K.: Roczniki Chem. 14, 375, 384, 394 (1934)
46. Kamieński, B.: ibid. 17, 497 (1937); 18, 600 (1938)
47. Karczewski, K.: ibid. 16, 69, 126, 254, 348, 560 (1936); 17, 9, 61 (1937)
48. Karpfen, F.M., Randles, J.E.B.: Trans. Faraday Soc. 49, 823 (1953)
49. Bonhoeffer, K.F., Kahlweit, M., Strehlow, H.: Z. Elektrochem. Ber. Bunsenges. 57, 614 (1953)
50. Bonhoeffer, K.F., Kahlweit, M., Strehlow, H.: Z. Phys. Chem., (N.F.) 1, 21 (1954)
51. Bonhoeffer, K.F., Kahlweit, M., Strehlow, H., Hocking, C.S.: ibid. 4, 212 (1955)
52. Kahlweit, M., Strehlow, H.: Z. Elektrochem. 58, 658 (1954)
53. Bonhoeffer, K.F.: Angew. Chem. 67, 1 (1956)
54. Kahlweit, M.: Pfügers Archiv. ges. Physiol. 271, 139 (1960)
55. Dupeyrat, M.: Compt. Rend. 249, 102 (1959); 252, 269 (1961)
56. Dupeyrat, M.: J. Chim. Phys. 61, 306 (1964)
57. Dupeyrat, M.: Rev. Gen. Electr. 76, 1273 (1967)
58. Gavach, C., Seta, P.: Anal. Chim. Acta 50, 470 (1970)
59. Gavach, C., Davion, N.: Electrochim. Acta 18, 649 (1973)
60. Gavach, C., Savajols, A.: ibid. 19, 573 (1974)
61. Gross, M., Gromb, S., Gavach, C.: J. Electroanal. Chem. 89, 29 (1978)
62. Joos, P., Verburgh, Y.: Bull. Soc. Chim. Belg. 87, 737 (1978)

63. Joos, P., Jansegers, L., Verburgh, Y.: J. Colloid Interface Sci *63*, 27 (1978)
64. Koczorowski, Z.: J. Electroanal. Chem. *127*, 11 (1981)
65. Koczorowski, Z., Geblewicz, G.: ibid. *152*, 55 (1983)
66. Geblewicz, G., Koczorowski, Z.: ibid. *158*, 37 (1983)
67. Koczorowski, Z., Geblewicz, G., Paleska, I.: ibid. *172*, 327 (1984)
68. Minc, S., Koczorowski, Z.: Roczniki Chem. *34*, 349 (1960)
69. Minc, S., Koczorowski, Z.: Electrochim. Acta *8*, 575 (1963)
70. Koczorowski, Z., Minc, S.: ibid. *8*, 645 (1963)
71. Kamieński, B., Kulawik, I., Kulawik, J., Mikulski, J., Pawełek, J.: Bull. Acad. Polon. Sci., Ser. Sci. Chim. *15*, 249, 253 (1967)
72. Kamieński, B., Kulawik, I., Kulawik, J., Pawełek, J., Mikulski, J.: ibid. *16*, 57 (1968)
73. Kulawik, I., Baumgartner, T.: Zeszyty Nauk. Uniw. Jagiellońskiego, prace chem. *489*, 153 (1978)
74. Gugeshasvili, M.I., Lozhkin, B.T., Boguslavsky, L.I.: Elektrokhimiya *10*, 1272 (1974)
75. Gugeshasvili, M.I., Manvelyan, M.A., Boguslavsky, L.I.: ibid. *10*, 819 (1974)
76. Kharkats, Yu.I., Volkov, G., Boguslavsky, L.I.: Dokl. ANSSSR *220*, 1441 (1975)
77. Manvelyan, M.A., Neugodova, G.I., Boguslavsky, L.I.: Elektrokhimiya *12*, 309 (1976)
78. Boguslavsky, L.I., Frumkin, A.N., Manvelyan, M.A.: Dokl. ANSSSR *233*, 144 (1977)
79. Nasim Beg, M., Siddgi, F.A., Singh, S.P., Gupta, V., Prakash, P.: Bull. Chem. Soc. Jpn. *52*, 2696 (1979)
80. Danesi, P.R., Salvemini, F., Scibona, G., Scuppa, B.: *75*, 554 (1971)
81. Sollner, K.: J. Phys. Chem. *53*, 1211, 1226 (1949)
82. Shean, G., Sollner, K.: J. Membr. Biol. *9*, 297 (1972)
83. Fabiani, C., Danesi, P.R., Scibona, G., Scuppa, B.: J. Phys. Chem. *78*, 2370 (1974)
84. Scibona, G., Magini, M., Scuppa, B., Castagnola, A., Fabiani, C.: Anal. Chem. *49*, 212 (1977)
85. Fabiani, C., Scuppa, B., Scibona, G., Pizzichini, M.: Gazetta Chim. Ital. *109*, 409 (1979)
86. Nikolsky, B.P.: Acta Physicochem. (USSR) *7*, 597 (1937)
87. Eisenman, G.: Biophys. J. *2*, 259 (1962)
88. Sandblom, J.P., Eisenman, G., Walker, J.L. (Jr.): J. Phys. Chem. *71*, 3862 (1967)
89. Sandblom, J.P., Eisenman, G.: Biophys. J. *7*, 217 (1967)
90. Srinivasan, K., Rechnitz, G.A.: Anal. Chem. *41*, 1230 (1969)
91. Lesourd, J.B.: J. Electroanal. Chem. *86*, 81 (1978)
92. Yoshida, N.: Bull. Chem. Soc. Jpn. *52*, 3139 (1979)
93. Jyo, A., Mihara, H., Ishibash, N.: Denki Kagaku *44*, 268 (1976)
94. Lemke, C.H.: J. Electrochem. Soc. *101*, 203 (1954)
95. Holmberg, B.: Acta Chem. Scand. *27*, 875, 3550, 3557 (1973); A *28*, 284 (1974); A *30*, 641, 680, 797 (1976)
96. Verwey, E.J.W., Niessen, K.F.: Phil. Mag. *28*, 435 (1939)
97. Le Hung, Q.: J. Electroanal. Chem. *115*, 159 (1980)
98. Marklik, E.: Electrochim. Acta *28*, 573 (1983)
99. Le Hung, Q.: J. Electroanal. Chem. *149*, 1 (1983)
100. Irving, H.M.N.H.: Pure Appl. Chem. *21*, 111 (1970)
101. Popovych, O.: Crit. Rev. Anal. Chem. *1*, 73 (1970)
102. Nermeijer-Denessen, H.J.M., de Ligny, C.L., Remijnse, A.G.: *77*, 153 (1977)
103. Abraham, M.H., Liszi, J.: J. Inorg. Chem. *43*, 143 (1981)
104. Kornyshev, A.A., Volkov, A.G.: J. Electroanal. Chem. *180*, 363 (1984)
105. Kihara, S., Yoshida, Z.: Talanta *31*, 789 (1984)
106. Hundhammer, B., Seidlitz, H.J., Becker, S., Dhawan, S.K.: J. Electroanal. Chem. *180*, 355 (1984)
107. Kheifets, V.L., Gindin, L.M., Shneerson, A.A., Volkov, L.V.: Izv. SO ANSSSR, Ser. Khim. Nauk *14*, 9 (1969)
108. Kheifets, V.L., Gindin, L.M.: ibid. *7*, 155 (1970)
109. Kheifets, V.L., Shneerson, A.A.: Zh. Prikl. Khim. *50*, 528 (1977)
110. Shneerson, A.A., Kheifets, V.L.: ibid. *54*, 1250 (1981)
111. Shneerson, A.A., Kheifets, V.L.: Kolloidnyi Zh. *42*, 935 (1980)
112. Gindin, L.M.: Processes of Extraction and Their Application (in Russian), Nauka, Moskva 1984

113. Grunwald, E., Baughman, G., Konstan, G.: J. Am. Chem. Soc. 82, 5801 (1960)
114. Alexander, R., Parker, A.J.: ibid. 89, 5549 (1967)
115. Rais, J.: Collect. Czech. Chem. Commun. 36, 3253 (1971)
116. Parker, A.J., Alexander, R.: J. Am. Chem. Soc. 90, 3313 (1968)
117. Alexander, R., Parker, A.J., Sharp, J.H., Waghorne, W.E.: ibid. 94, 1148 (1972)
118. Cox, B.G., Parker, A.J., Waghorne, W.E.: ibid. 96, 1010 (1974)
119. Koczorowski, Z., Zagórska, I.: J. Electroanal. Chem. 193, 113 (1985)
120. Popovych, O.: Crit. Rev. Anal. Chem. 1, 73 (1970)
121. Bauer, D., Breant, M.: in: (Bard, A.J., Ed.) Electroanalytical Chemistry, Dekker, N.Y. 281, 1975
122. Popovych, O.: in: (Kolthof, I.M., Elving, P.J., Eds.) Treatise on Analytical Chemistry, Wiley-Interscience, N.Y., I, 711, 1978
123. Krumgalz, B.S.: Elektrokhimiya 8, 1320 (1972)
124. Koczorowski, Z., Paleska, I., Geblewicz, G.: J. Electroanal. Chem. 164, 201 (1984)
125. Geblewicz, G., Figaszewski, Z., Koczorowski, Z.: ibid. 177, 1 (1984)
126. Davies, J.T.: J. Phys. Chem. 54, 185 (1950)
127. Koryta, J., Vanysek, P., Brezina, M.: J. Electroanal. Chem. 75, 211 (1977)
128. Marcus, Y.: Pure Appl. Chem. 55, 977 (1983)
129. Samec, Z., Marecek, V., Weber, J., Homolka, D.: J. Electroanal. Chem., 99, 385 (1979)
130. Samec, Z., Marecek, V., Weber, J.: ibid. 96, 245 (1979)
131. Samec, Z., Marecek, V., Weber, J.: ibid. 100, 841 (1979)
132. Homolka, D., Marecek, V.: ibid. 112, 91 (1980)
133. Vanysek, P.: ibid. 121, 149 (1981)
134. Hundhammer, B., Solomon, T., Alemu, H.: ibid. 159, 19 (1983)
135. Osakai, T., Kakutani, T., Nishivaki, Y., Senda, M.: Bunseki Kagaku, 32, E81 (1983)
136. Yoshida, Z., Freiser, H.: J. Electroanal. Chem. 162, 307 (1984)
137. Koczorowski, Z., Geblewicz, G.: ibid. 139, 177 (1982)
138. Rais, J., Selucki, P., Kyrs, M.: J. Inorg. Nucl. Chem. 38, 1376 (1976)
139. Samec, Z., Homolka, D., Marecek, V., Kavan, L.: J. Electroanal. Chem. 145, 213 (1983)
140. Solomon, T., Alemu, H., Hundhammer, B.: ibid. 169, 303 (1984)
141. Czapkiewicz, J., Czapkiewicz-Tutaj, B.: J. Chem. Soc. Faraday Trans., I, 76, 1663 (1980)
142. Abraham, M.H., Danil de Namor, A.F.: ibid. I, 72, 955 (1976)
143. Parsons, R.: in: (Bockris, J.O'.M., Convay, B.E., Yeager, E., Eds.) Comprehensive Treatise of Electrochemistry, Plenum Press, N.Y., Vol. 1, 1, 1980
144. Frumkin, A.N.: Phil. Mag. 40, 363 (1920)
145. Samec, Z., Marecek, V., Homolka, D.: J. Electroanal. Chem. 126, 121 (1981)
146. Kakiuchi, T., Senda, M.: Bull. Chem. Soc. Jpn. 56, 1322 (1983)
147. Reid, J.D., Vanysek, P., Buck, R.P.: J. Electroanal. Chem. 161, 1 (1984)
148. Girault, H.H.J.: ibid. 170, 127 (1984)
149. Homolka, D., Haykova, P., Samec, Z.: ibid. 159, 233 (1983)
150. Samec, Z., Marecek, V., Homolka, D.: Faraday Discuss. Chem. Soc. 77, 10 (1984)
151. Girault, H.H.J., Schiffrin, D.J.: J. Electroanal. Chem. 150, 43 (1983)
152. Girault, H.H.J., Schiffrin, D.J.: in: (Ellen, M.J., Usherwood, P.N.R., Eds.) Charge and Field Effects in Biosystems, Abacus Press, 171, 1984
153. Haykova, P., Homolka, D., Marecek, V., Samec, Z.: J. Electroanal. Chem. 151, 277 (1983)
154. Silva, F., Moura, C.: ibid. 177, 317 (1984)
155. Koczorowski, Z.: ibid. 190, 257 (1985)
156. Frumkin, A.N., Iofa, Z.A., Gerovich, M.A.: Zh. Fiz. Khim. 30, 1455 (1956)
157. Parsons, R.: in: (Bockris, J.O'M., Convay, B.M., Eds.) Modern Aspects of Electrochemistry, Butterworths, London, Vol. I, Chapt. 3, 1954
158. Frumkin, A.N., Petrii, O.A., Damaskin, B.B.: in: (Bockris, J.O'M., Convay, B.E., Yeager, E., Eds.) Comprehensive Treatise of Electrochemistry, Plenum Press, N.Y., Vol. 1, 221, 1980
159. Trasatti, S.: in: (Bockris, J.O'M., Convay, B.E., Yeager, E., Eds.) Comprehensive Treatise of Electrochemistry, Plenum Press, N.Y., Vol. 1, 45, 1980
160. Reid, J.D., Melroy, O.R., Buck, R.P.: J. Electroanal. Chem. 147, 71 (1983)
161. Kakiuchi, T., Senda, M.: Bull. Soc. Jpn. 56, 1322 (1983)
162. Kakiuchi, T., Senda, M.: ibid. 56, 1755 (1983)

163. Koczorowski, Z., Zagórska, I.: J. Electroanal. Chem. *159*, 183 (1983)
164. Zagórska, I., Koczorowski, Z.: ibid. *204*, 273 (1986)
165. Damaskin, B.B., Kaganovich, R.I.: Elektrokhimiya *13*, 293 (1977)
166. Ohlenbusch, Z.D.: Z. Electrochem. *60*, 607 (1950)
167. Krylov, V.S., Myamlin, V.A., Boguslavsky, L.I., Manvelyan, M.A.: Elektrokhimiya *13*, 834 (1977)
168. Girault, H.H.J., Schiffrin, D.J.: J. Electroanal. Chem. *161*, 415 (1984)
169. Iwamoto, E., Hiyama, Y., Yamamoto, Y.: J. Solution Chem. *6*, 371 (1977)
170. Solomon, T., Alemu, H., Hundhammer, B.: J. Electroanal. Chem. *169*, 311 (1984)

Acknowledgement. The author is very grateful to Pawel J. Kulesza of the Department of Chemistry, University of Warsaw for his assistance in preparing the English version of this text.

Electrocapillarity and the Electric Double Layer Structure at Oil/Water Interfaces

Mitsugi Senda, Takashi Kakiuchi, Toshiyuki Osakai, and Tadaaki Kakutani

Summary

Oil/water interfaces are classified into the ideal-polarized interface and the non-polarized interface. The interface between a nitrobenzene solution of tetrabutylammonium tetraphenylborate and an aqueous solution of lithium chloride behaves as an ideal-polarized interface in a certain potential range. Electrocapillary curves of the interface were measured. The results are analyzed using the electrocapillary equation of the ideal-polarized interface and the Gouy-Chapman theory of diffuse double layers. The electric double layer structure consisting of the inner layer and the two diffuse double layers on each side of the interface is discussed. Electrocapillary curves of the nonpolarized oil/water interface are discussed for two cases of a nonpolarized nitrobenzene/water interface.

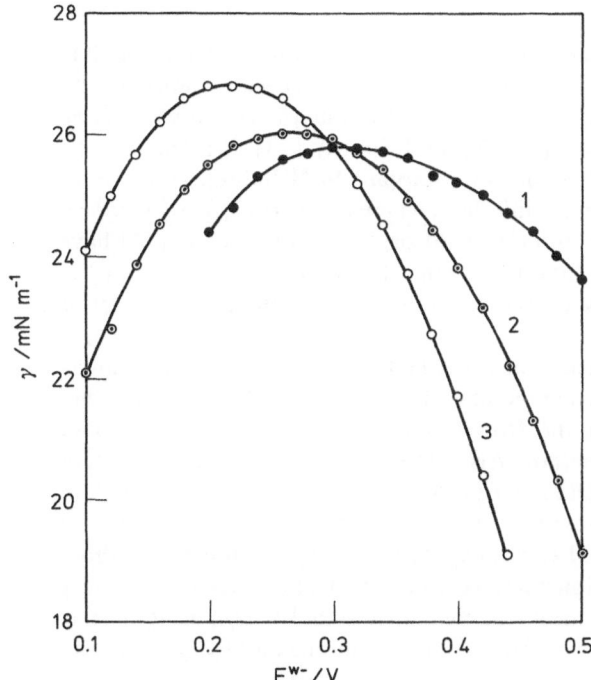

Fig. 1. Electrocapillary curves of the interface between nitrobenzene solution of $0.1 \, mol \, dm^{-3}$ TBATPB and aqueous solution of (1): 0.01, (2): 0.10, and (3): $1.00 \, mol \, dm^{-3}$ LiCl at 25°C

The Interface Structure and Electrochemical Processes
at the Boundary Between Two Immiscible Liquids
Editor: V. E. Kazarinov
© Springer-Verlag Berlin, Heidelberg 1987

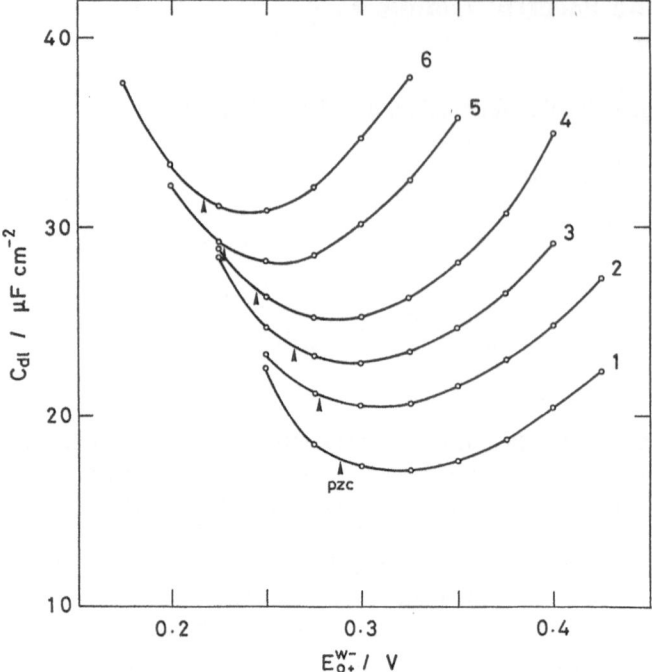

Fig. 2. Differential capacity curves of the interface between nitrobenzene solution of 0.1 mol dm^{-3} TBATPB and aqueous solution of (1): 0.02, (2): 0.05, (3): 0.10, (4): 0.20, (5): 0.50, and (6): 1.00 mol dm^{-3} LiCl at 25°C. Arrows indicate the potential of zero charge

Electrocapillary phenomena at the interface between two immiscible electrolyte solutions, which we will call the oil/water (O/W) interface for short, were studied first by Guastalla [1], then by Blank and Feig [2, 3], Watanabe et al. [4, 5], Dupeyrat et al. [6, 7], Joos et al. [8, 9], Gavach et al. [10–12], and Spurny [13]. Watanabe et al. applied the electrocapillary equations such as the Lippmann-Helmholtz equation to elucidate the double layer structure of the interface, whereas others [2, 3, 6, 7] made a distinction between electrocapillarity and electroadsorption. Koryta et al. [14] have discussed the electric polarizability of the oil/water interface on the basis of the transfer Gibbs energies of ions from one solvent (the aqueous phase) to the other (the oil or organic phase).

In recent years the interfacial-tension measurements of nitrobenzene/water interfaces under exact control of the electric potential difference across the interface were made by Gavach et al. [15, 16] using the drop weight method and by Kakiuchi and Senda [17, 18] using the drop time method. Also, Reid et al. [19] used the maximum bubble pressure method, while Girault et al. [20–24] used the drop shape method, in which a drop image processing technique was developed to measure the interfacial tension. These authors have obtained electrocapillary curves very much like those obtained at mercury/electrolyte solution interfaces (see Fig. 1). The differential capacity of nitrobenzene/water interfaces was measured by Samec et al. [25–27] using the a.c. polarographic technique and the Galvanostatic pulse technique and by Osakai et al.

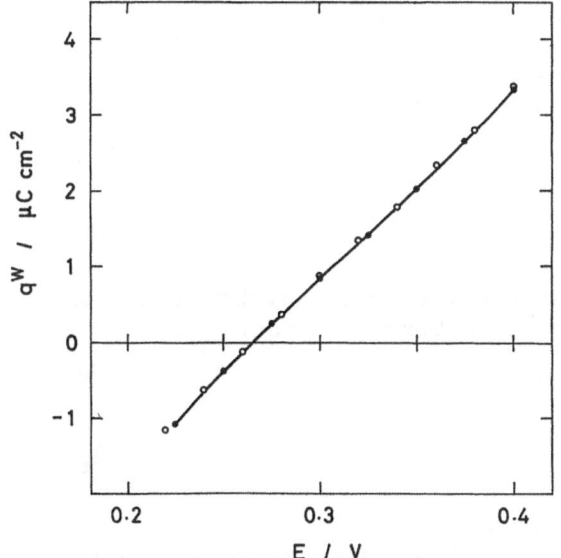

Fig. 3. Comparison of the surface charge densities in the aqueous phase obtained by differentiation of the electrocapillary curve (o) and by integration of the differential capacity curve (●) for the interface between 0.1 mol dm^{-3} TBATPB nitrobenzene solution and 0.1 mol dm^{-3} LiCl aqueous solution

[28, 29] using the a.c. polarographic technique (see Fig. 2). The values of the surface charge density at nitrobenzene/water interfaces obtained by differentiation of their electrocapillary curves were compared with the surface charge density data obtained by integration of the differential capacity curve [17, 29] (see Fig. 3).

Since it is essential to distinguish between ideal-polarized and nonpolarized interfaces [30] to elucidate the double layer structure of the oil/water interface on the basis of electrocapillary curve and capacity curve measurements, we first discuss the polarizability of the oil/water interface [17, 31].

Ideal-Polarized and Nonpolarized Oil/Water Interfaces

The Galvani potential difference across the oil/water interface $\Delta\phi_O^W$ (the potential of the water (W) phase referred to that of the oil (O) phase) when both the O and W phases contain a common ion M which is transferable across the interface is given by the Nernst-Donnan equation [31, 39–42],

$$\Delta\phi_O^W = \Delta\phi_M^\phi + \frac{RT}{z_M F} \ln \frac{a_M^O}{a_M^W}, \tag{1}$$

where a_M^O and a_M^W are the activities of M ion in the phases O and W, respectively, z_M the number of charges of the M ion, and F, R, and T have their usual meanings. $\Delta\phi_M^\phi$ is the standard potential of the transfer of the M ion and related to the standard Gibbs energy of the transfer of the M ion from O to W, $\Delta G_{tr,M}^{\phi,O\to W}$, by

$$\Delta\phi_M^\phi = -\Delta G_{tr,M}^{\phi,O\to W}/z_M F. \tag{2}$$

The $\Delta\phi_O^W$ given by Eq. (1) may be called the distribution potential of the M ion at the oil/water interface.

Let us first discuss the interface between an oil phase containing a highly lipophilic electrolyte B_1A_1, e.g., tetrabutylammonium tetraphenylborate (TBATPB) in nitrobenzene, and an aqueous phase containing a highly hydrophilic electrolyte B_2A_2, e.g., LiCl in water (hereafter $B_j (j = 1, 2, ...)$ represents a cation and $A_j (j = 1, 2, ...)$ an anion and, unless otherwise stated, it is supposed that $z_M = +1$ for $M = B_j$ and -1 for $M = A_j$), as shown in Cell Ia:

$$\alpha_1 \begin{vmatrix} B_1A_1 & B_2A_2 \\ (O) & (W) \end{vmatrix} \alpha_2. \tag{Ia}$$

In this cell α_1 and α_2 represent the reference electrodes reversible to either cation (B_1 or B_2) or anion (A_1 or A_2). The situation is described by the equations:

$$\Delta\phi^\phi_{A_1} \gg 0 \quad \text{and} \quad \Delta\phi^\phi_{B_2} \gg 0, \tag{3a}$$

$$(\Delta G^{\phi, O \rightarrow W}_{tr, A_1} \gg 0 \quad \text{and} \quad \Delta G^{\phi, O \rightarrow W}_{tr, B_2} \ll 0),$$

$$\Delta\phi^\phi_{B_1} \ll 0 \quad \text{and} \quad \Delta\phi^\phi_{A_2} \ll 0, \tag{3b}$$

$$(\Delta G^{\phi, O \rightarrow W}_{tr, B_1} \gg 0 \quad \text{and} \quad \Delta G^{\phi, O \rightarrow W}_{tr, A_2} \ll 0).$$

Here $\gg 0$ and $\ll 0$ mean "positively very large" and "negatively very large", respectively. Thus, as long as $\Delta\phi^W_O$ does not exceed a certain range (positive and negative) of potential difference, the transfer of the ions B_1, A_1, B_2, and A_2 across the interface is negligibly small, that is, B_1A_1 is confined to phase O and B_2A_2 to W. There exists a range of $\Delta\phi^W_O$ where determination of the potential difference by the activities of the ions present in the system according to Eq. (1) is fictious, since $a^W_{B_1}$, $a^W_{A_1}$, $a^O_{B_2}$, and $a^O_{A_2}$ are extremely small. Then $\Delta\phi^W_O$ is determined much more by applying an electric field to the interface via the two reference electrodes α, and α_2 from an external source than by equations like (1). The situation is completely analogous to that of an ideal-polarized metal/electrolyte solution interface. We call this case the ideal-polarized oil/water interface.

Let us discuss another case shown in Cell IIa;

$$\alpha_1 \begin{vmatrix} B_3A_1 & B_3A_2 \\ (O) & (W) \end{vmatrix} \alpha_2. \tag{IIa}$$

Both the O and W phases have in common the ion B_3 which is transferable across the oil/water interface, but the transfer of ion A_1 from O to W and that of ion A_2 in the opposite direction are negligible in the potential range considered. The situation is described by the equations:

$$\Delta\phi^\phi_{A_1} \gg 0, \tag{4a}$$

$$\Delta\phi^\phi_{A_2} \ll 0, \tag{4b}$$

$$\Delta\phi^\phi_{A_2} \ll \Delta\phi^\phi_{B_3} \ll \Delta\phi^\phi_{A_1}. \tag{4c}$$

Under these conditions the electric potential difference between O and W is determined practically only by the activities of B_3 ion in both phases according to Eq. (1), if the concentrations of the common ion B_3 in O and W have comparable values;

$$\Delta\phi^W_O = \Delta\phi^\phi_{B_3} + \frac{RT}{z_{B_3}F} \ln \frac{a^O_{B_3}}{a^W_{B_3}}. \tag{5}$$

We call this case the nonpolarized oil/water interface. Similarly the oil/water interfaces as shown in Cells IIIa and IVa are also nonpolarized interfaces;

$$\alpha_1 \begin{array}{|c|c|} B_3A_1 & B_3A_2, B_2A_2 \\ (O) & (W) \end{array} \alpha_2 ,$$ (IIIa)

$$\alpha_1 \begin{array}{|c|c|} B_1A_1, B_3A_1 & B_3A_2, B_2A_2 \\ (O) & (W) \end{array} \alpha_2 .$$ (IVa)

In these cases the common ion B_3 is transferable across the interface, whereas the other ions B_1, A_1, B_2, and A_2 are practically not transferable in the potential range considered. Then the potential difference across the interface is determined by Eq. (5).

In the last decade increasing attention has been paid to the electeochemical studies on the kinetics and equilibrium of ion transfer at the oil/water interfaces that are polarized in the sense that the interface is of the ideal-polarized nature for all ions (here B_1, A_1, B_2, and A_2) except transferring ion or ions (here B_3) (for reviews see [31–38].) In such cases the nontransferable ions serve as the supporting electrolytes for the transport process of the transferable ions across the interface. It has been shown that the oil/water interface works as an ion-selective electrode surface for both vol-tammetric (and amperometric) and potentiometric measurements of ions, both based on the same electrochemical principle of ion transfer across the interface [43, 44]. There it is essential that the oil/water interface is of the ideal-polarized nature for all ions (such as counter ion and supporting electrolyte ion; so-called potential window) except the monitored ion or ions (voltammetry).

Electrocapillary Curves of Ideal-Polarized Oil/Water Interfaces

We first discuss the electrocapillary phenomenon at the interface between a nitro-benzene (NB) solution of tetrabutylammonium tetraphenylborate (TBATPB) and an aqueous (W) solution of LiCl, which corresponds to the above system Ia. The interface was studied using the following cell at 25°C [17, 18];

$$\text{Ag} \left| \text{AgCl} \right| \begin{array}{c|c|c} a \,\text{mol dm}^{-3} & b \,\text{mol dm}^{-3} & c \,\text{mol dm}^{-3} \\ \text{TBACl (W)} & \text{TBATPB (NB)} & \text{LiCl (W)} \end{array} \left| \text{AgCl} \right| \text{Ag} .$$ (Ib)

The interface to be investigated is marked "x". In this cell the interface between a mol·dm^{-3} TBACl (tetrabutylammonium chloride) in water (W) and b mol·dm^{-3} TBATPB in nitrobenzene (NB) is a nonpolarized interface and the potential difference across the interface is determined by the distribution potential of TBA$^+$ ion (Eq. (5)). Thus the left-hand half-cell Ag/AgCl/TABCl(W)/TBATPB(NB) constitutes a reference electrode reversible to TBA$^+$ ion in NB, whereas the right-hand half-cell Ag/AgCl/LiCl(W) constitutes a reference electrode reversible to Cl$^-$ ion in W. The cell potential E is the potential of the right-hand terminal of the cell, E_{right}, referred to that of the left-hand, E_{left}, that is, $E = E_{\text{right}} - E_{\text{left}}$, and is related to the potential drop across the nitrobenzene/water interface to be investigated, $\Delta\phi_O^W$, by:

$$E = \Delta E_{\text{ref}} + \Delta\phi_O^W .$$ (6)

Table 1a. Coordinates of electrocapillary maximum for the interface between 0.10 mol dm^{-3} TBATPB(NB) and 0.01 to 1.00 mol dm^{-3} LiCl(W) obtained with Cell Ib with $a=0.1$ at 25°C

$\dfrac{c_{LiCl}^{W}}{mol\,dm^{-3}}$	$a_{LiCl}^{\pm,W}$	$\dfrac{\gamma_{ecm}}{mN\,m^{-1}}$	$\dfrac{E_{ecm}}{V}$
0.01	0.009	25.8	0.309
0.02	0.017	25.7	0.289
0.05	0.041	25.8	0.278
0.10	0.079	26.0	0.265
0.20	0.151	26.2	0.245
0.50	0.369	26.5	0.227
1.00	0.776	26.8	0.217

Table 1b. Coordinates of electrocapillary maximum for the interface between 0.01 to 0.17 mol dm^{-3} TBATPB(NB) and 0.10 mol dm^{-3} LiCl(W) obtained with Cell Ib with $a=0.005$ at 25°C

$\dfrac{c_{TBATPB}^{NB}}{mol\,dm^{-3}}$	$a_{TBATPB}^{\pm,NB}$	$\dfrac{\gamma_{ecm}}{mN\,m^{-1}}$	$\dfrac{E_{ecm}}{V}$
0.01	0.008	25.9	0.167
0.02	0.014	25.9	0.153
0.03	0.021	26.0	0.142
0.05	0.033	26.0	0.131
0.07	0.044	26.0	0.129
0.10	0.061	26.1	0.122
0.17	0.099	26.3	0.108

Here ΔE_{ref} is a constant which is determined by the difference of the potential between the two reference electrodes, $E_{ref,right}$ and $E_{ref,left}$;

$$\Delta E_{ref} = E_{ref,right} - E_{ref,left}. \tag{7}$$

Electrocapillary curves were measured [18] for seven different concentrations of LiCl(W), $c=0.01$ to 1.0, at $a=0.1$ and $b=0.1$ and for seven different concentrations of TBATPB(NB), $b=0.01$ to 0.17, at $a=0.005$ and $c=0.1$ in Cell Ib. The coordinates of the electrocapillary maximum are given in Table 1a, b. Some representative curves are shown in Fig. 1. Also, differential capacity vs. potential curves were measured [28, 29, 45] for six different concentrations of LiCl(W), $c=0.02$ to 1.0, at $a=0.1$ and $b=0.1$ and for three different concentrations of TBATPB(NB), $b=0.05$ to 0.17, at $a=0.1$ and $c=0.1$ in Cell Ib. Some results are shown in Fig. 2 [45], in which the positions of the electrocapillary maximum potential are indicated by arrows on the curves.

The interface between TBATPB(NB) and LiCl(W) behaves as an ideal-polarized interface in a certain range of the potential difference across the interface [17]. Thus the electrocapillary equation of this system at constant temperature and pressure is given by [18, 46]:

$$-(a\gamma)_{T,p} = q^{W}dE_{O+}^{W-} + \Gamma_{Li/W}d\mu_{LiCl}^{W} + \Gamma_{TPB/NB}d\mu_{TBATPB}^{NB}, \tag{8}$$

where γ is the interfacial tension; q^W is the surface charge density in the aqueous phase and is defined by:

$$q^W = F(\Gamma_{Li} - \Gamma_{Cl}) = F(\Gamma_{TPB} - \Gamma_{TBA}) = -q^O, \tag{9}$$

where Γ_j is the surface amount of j-component per unit interface area (the surface concentration of j-component, $j = Li^+$, Cl^-, TBA^+, and TPB^- ions, and W and NB) and q^O is the surface charge density in the nitrobenzene phase; the super- and subscripts W$-$ and O$+$ on the cell potential E indicate that the reference electrodes in the W and NB phases are reversible to an anion (Cl^-) and a cation (TBA^+), respectively; $\Gamma_{Li/W}$ and $\Gamma_{TPB/NB}$ are the relative surface excesses (the relative surface concentrations) of Li^+ and TPB^- ions, respectively, which are defined by the following Eq. (10a, b) and μ_i^α is the chemical potential of i-component ($i = LiCl$ and $TBATPB$) in the phase α ($\alpha = W$ and NB).

$$\Gamma_{Li/W} = \Gamma_{Li} - \frac{\chi_{LiCl}^W}{\chi_W^W} \Gamma_W, \tag{10a}$$

$$\Gamma_{TPB/NB} = \Gamma_{TPB} - \frac{\chi_{TBATPB}^{NB}}{\chi_{NB}^{NB}} \Gamma_{NB}. \tag{10b}$$

Here χ_i^α is the mole fraction of i-component in α phase. In Eq. (10a, b) and the following the correction terms for the mutual solubilization of nitrobenzene with water are neglected for simplification [18, 30].

The surface charge density can be obtained by differentiation of the electrocapillary curve:

$$q^W = -\left(\frac{\partial\gamma}{\partial E_{O+}^{W-}}\right)_{T, p, \mu_{LiCl}^W, \mu_{TBATPB}^{NB}} \tag{11}$$

while the differential capacity C_{dl} is given by:

$$C_{dl} = \left(\frac{\partial q^W}{\partial E_{O+}^{W-}}\right)_{T, p, \mu_{LiCl}^W, \mu_{TBATPB}^{NB}} \tag{12}$$

The values of surface charge density obtained by numerical differentiation of the electrocapillary curve agreed well with those obtained by numerical integration of the differential capacity curve [17, 29] (Fig. 3). These results indicate that the interface between a nitrobenzene solution of TBATPB and an aqueous solution of LiCl actually behaves as an ideal-polarized interface in a certain potential range and also that the differential capacity measurements should give essentially the same information on the electrocapillarity and the double layer structure of nitrobenzene/water interfaces as the electrocapillary curve measurements, provided that their electrocapillary maximum potential which is now equal to the potential of zero charge (pzc) and interfacial tension at the pzc (γ_{pzc}) are known.

According to Verwey [47], the Galvani potential difference across the oil/water interface may be divided into a surface potential jump (the chi potential) caused by orientation of dipoles etc. and a double layer potential related to ionic displacement. Verwey and Niessen [47] investigated the distribution of this double layer potential over the two phases, assuming that the double layer charge is situated in two diffuse double layers on both sides of the interface and calculating the electric potential

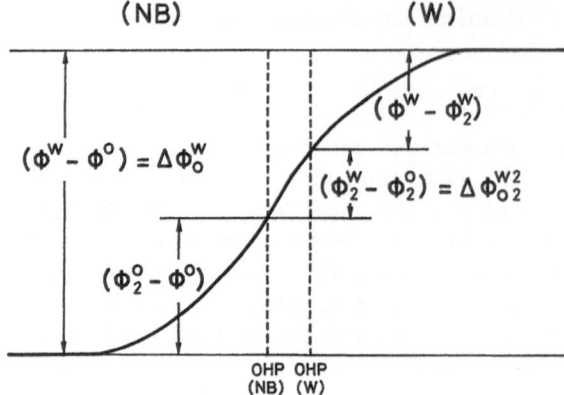

Fig. 4. The electric potential distribution at the oil/water interface

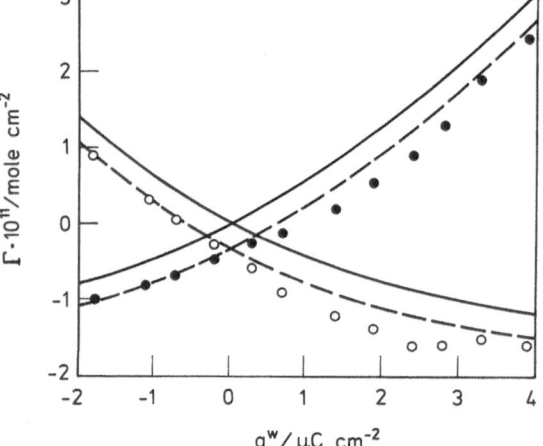

Fig. 5. Relative surface excesses of Li^+ (●) and Cl^- (○) ions as functions of the surface charge density in aqueous phase at the interface between 0.1 mol dm^{-3} TBATPB nitrobenzene solution and 0.1 mol dm^{-3} LiCl aqueous solution. The solid lines are the theoretical curves based on the Gouy-Chapman theory (the right-hand sides of Eq. (14)). The dashed lines are the theoretical curves shifted by -3×10^{-12} mol cm^{-2} for correction of (Eq. (10))

distribution using the Gouy-Chapman theory. Gavachi et al. [15] envisaged the existence of a compact layer at the surface of contact between the two immiscible phases similar to the inner layer existing at a metal/solution interface. One may expect that water and organic (oil) molecules are oriented, or more or less structured in this region which we will call hereafter the *inner layer*. Then the overall potential difference across the O and W phases has three components (Fig. 4):

$$\Delta\phi_O^W = \Delta\phi_{W2}^W + \Delta\phi_{O2}^{W2} - \Delta\phi_{O2}^O, \tag{13}$$

where $\Delta\phi_{W2}^W = \phi^W - \phi_2^W$ is the potential difference between the bulk of the W phase (ϕ^W) and the outer-Helmholtz plane in the W phase (ϕ_2^W), $\Delta\phi_{O2}^O = \phi^O - \phi_2^O$ that between the bulk of the O phase (ϕ^O) and the outer-Helmholtz plane in the O phase (ϕ_2^O), and $\Delta\phi_{O2}^{W2} = \phi_2^W - \phi_2^O$ that between the two outer-Helmholtz planes.

The relative surface excesses of ionic components can be determined from $-(\partial\gamma/\partial\mu_i)_{T,p,E,\mu_{j+i}}$ of the electrocapillary curves. Thus Kakiuchi and Senda [18] calculated $\Gamma_{Li/W}$ and $\Gamma_{Cl/W}$ (since $q^W = \Gamma_{Li/W} - \Gamma_{Cl/W}$) and also $\Gamma_{TBA/NB}$ and $\Gamma_{TPB/NB}$ (since $q^W = \Gamma_{TPB/NB} - \Gamma_{TBA/NB}$) as functions of q^W for a 0.1 mol 0.1 mol dm^{-3}

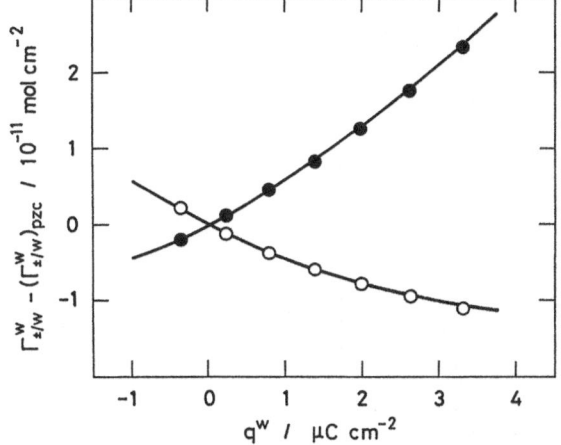

Fig. 6. Relative surface excesses of Li^+ ($\bullet$) and Cl^- ($\circ$) ions obtained from the differential capacity curves of the interface between 0.1 mol dm^{-3} TBATPB(NB) and 0.1 mol dm^{-3} LiCl(W). The solid lines are the theoretical curves calculated by the right-hand sides of Eq. (14)

TBATPB(NB)/0.1 mol dm^{-3} LiCl(W) interface. It has been shown [18] that the $\Gamma_{j/W}$ (and $\Gamma_{j/NB}$) vs. q^W plots are well explained by the surface excess of j ion ($j = Li^+$ etc.) in the diffuse double layer, $\Gamma^d_{j(W)}$ (and $\Gamma^d_{j(NB)}$) which can be calculated by the Gouy-Chapman theory by assuming the absence of charged components in the inner layer, as given, e.g., for Li^+ and Cl^- ions by:

$$\Gamma_{Li/W} - \Gamma_{Li/W, pzc} = \Gamma^d_{Li(W)} = \frac{A^W}{F}\left(\frac{q^W}{2A^W} + \sqrt{\left(\frac{q^W}{2A^W}\right)^2 + 1} - 1\right), \tag{14a}$$

$$\Gamma_{Cl/W} - \Gamma_{Cl/W, pzc} = \Gamma^d_{Cl(W)} = \frac{A^W}{F}\left(-\frac{q^W}{2A^W} + \sqrt{\left(\frac{q^W}{2A^W}\right)^2 + 1} - 1\right), \tag{14b}$$

where $\Gamma_{j/W, pzc}$ is the $\Gamma_{j/W}$ value at *pzc* and $A^W = (2RT\varepsilon^W c^W_{LiCl})$, ε^W and c^W_{LiCl} being the permittivity of water and the concentration of LiCl(W), respectively. Alternatively, the relative surface excesses can be determined by double integration of $(\partial c_{dl}/\partial \mu_i)_{T, p, E, \mu_{j \neq i}}$ of the differential capacity (c_{dl}) curves. Figure 6 shows the $\Gamma_{j/W} - \Gamma_{j/W, pzc}$ vs. q^W ($j = Li^+$ and Cl^-) plots for a 0.1 mol dm^{-3} TBATPB(NB)/0.1 mol dm^{-3} LiCl(W) interface obtained by Osakai et al. [45] using their differential capacity data. The solid lines are the theoretical curves calculated by Eq. (14a, b). Thus we may conclude that the specific adsorption of ions in the inner layer does not exist or is negligibly small at the TBATPB(NB)/LiCl(W) interface. This conclusion was further supported by the γ_{pzc} vs. $\ln a^{\pm W}_{LiCl}$ and γ_{pzc} vs. $\ln a^{\pm O}_{TBATPB}$ plots [18], the Essin-Markov plots [18], and the Grahame-Soderberg [49] plots [45]. Samec et al. [25–27, 50] and Reid et al. [38] compared the observed differential capacity curves with the theoretical diffuse double layer capacity curves calculated by the Gouy-Chapman theory and found a reasonable fit of the former to the latter.

In the situation that the net charge in the inner layer is zero, that is, in the absence of specific ion adsorption the values of $\Delta\phi^W_{W2}$ and $\Delta\phi^O_{O2}$ can be calculated by the Gouy-Chapman theory for given values of q^W by

$$\Delta\phi^W_{W2} = \frac{2RT}{F}\sinh^{-1}\frac{q^W}{2A^W} \tag{15}$$

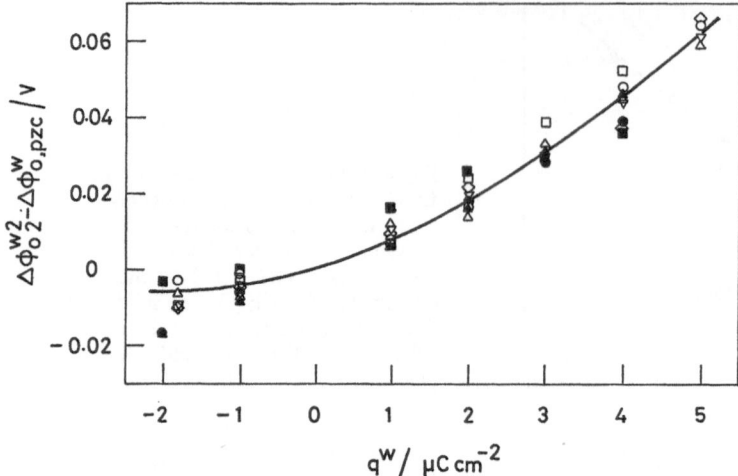

Fig. 7. Inner layer potential difference across the two outer Helmholtz planes as a function of the surface charge density in aqueous phase for the interface between TBATPB(NB) and LiCl(W) at various electrolyte concentrations. The concentration of LiCl was 0.05 (□), 0.1 (△), 0.2 (◇), 0.5 (▽), and 1.0 (○) mol dm^{-3} when the concentration of TBATPB was 0.1 mol dm^{-3} and the concentration of TBATPB was 0.05 (■), 0.1 (●), and 0.17 (▲) mol dm^{-3} when the concentration of LiCl was 0.1 mol dm^{-3}

and a similar equation with A^O in place of A^W for $\Delta\phi_{O2}^{}$. Then the $\Delta\phi_{O2}^{W2}$ can be calculated by Eq. (13) Figure 7 shows $\Delta\phi_{O2}^{W2}$ referred to its value at *pzc*, $\Delta\phi_{O,pzc}^{W}$, for the TBATPB(NB)/LiCl(W) interface at various electrolyte concentrations. The experimental points are scattered on a single curve with small deviations, indicating that the structure of the inner layer remains virtually unchanged with changes of the electrolyte concentration for fixed values of surface charge density. This result is consistent with the previous conclusion of the absence of specific ion adsorption in the inner layer.

The slope of the curve in Fig. 7 gives the reciprocal of the inner layer capacity c_i, which is defined by $c_i = (\partial q^W / \partial \phi_{O2}^{W2})$. Also, the differential capacity c_{dl} is given by a series combination of c_i and the two diffuse double layer capacities c_{2W} and c_{2O} which are defined by $c_{2W} = \partial q^W / \partial \Delta\phi_{W2}^W$ and $c_{2O} = -\partial q^W / \partial \Delta\phi_{O2}^O$:

$$\frac{1}{c_{dl}} = \frac{1}{c_{2W}} + \frac{1}{c_{2O}} + \frac{1}{c_i}. \tag{16}$$

The c_{2W} and c_{2O} values can be calculated by the Gouy-Chapman theory assuming the absence of specific ion adsorption in the inner layer. Then c_i can be determined from experimental values of c_{dl} using Eq. (16). The result is shown in Fig. 8 [45], indicating again the validity of the assumption that the inner layer capacity is solely a function of the surface charge density and not of the electrolyte concentrations. The inner layer capacity decreases with increasing positive surface charge density in the aqueous solution side. The decrease of the inner layer capacity with increasing q^W near the *pzc* is known also for the inner layer capacity at the mercury/aqueous solution interface, though the inner layer capacity values at the TBATPB(NB)/LiCl(W) interface are very large compared with those at the mercury/aqueous solution interface.

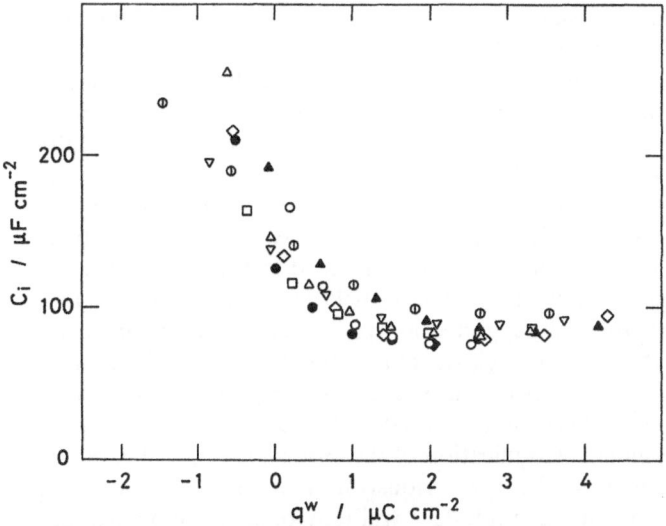

Fig. 8. Inner layer capacity as a function of the surface charge density in aqueous phase for the interface between TBATPB(NB) and LiCl(W) at various electrolyte concentrations. The concentration of LiCl was 0.02 (○), 0.05 (△), 0.1 (□), 0.2 (◇), 0.5 (▽), and 1.0 (⊕) mol dm^{-3} when the concentration of TBATPB was 0.1 mol dm^{-3} and the concentration of TBATPB was 0.05 (●) and 0.17 (▲) mol dm^{-3} when the concentration of LiCl was 0.1 mol dm^{-3}

The magnitude of $\Delta\phi_{O2}^{W2}$ can be evaluated if the value of $\Delta\phi_{O,pzc}^{W}$ is known (Fig. 7). Taking $\Delta\phi_{TBA}^{\phi} = -0.248$ V, $\Delta\phi_{O,pzc}^{W}$ was evaluated to be 0.02 V [18], independent of the electrolyte concentrations. Recently, Samec et al. [27], on the basis of electrochemical data, [25, 27] have suggested that $\Delta\phi_{TBA}^{\phi}$ should be shifted by -27 mV, then $\Delta\phi_{O,pzc}^{W}$ should read -0.007 V. On the other hand, Samec et al. [27] evaluated its value to be 0.015 V on the assumption that the differential capacity minimum corresponds to the *pzc*. Since for the TBATPB(NB)/LiCl(W) interface the capacity minimum potential appears at about 30 mV more positive potential than the *pzc* (Fig. 2), Samec et al.'s value should be corrected accordingly (probably a shift of -30 mV). Other investigators [16, 24] also reported relatively small values of $\Delta\phi_{O,pzc}^{W}$. These investigators also assumed that the variation of $\Delta\phi_{O2}^{W2}$ with q^{W} should be negligibly small. However, this is not consistent with the experimental results shown in Fig. 7.

The general trend of reported values of $\Delta\phi_{O,pzc}^{W}$, though differing from investigator to investigator in magnitudes, is that the potential drop across the inner layer at *pzc* is small, much smaller than the surface potential at mercury/solution interface. Such a small $\Delta\phi_{O,pzc}^{W}$ value may imply the compensation of the surface potential due to water dipoles with that of oppositely oriented nitrobenzene dipoles and/or the less oriented structure of the inner layer compared with that at the mercury/aqueous solution interface.

It has been suggested [18] on the basis of the negative values of $\Gamma_{Li/W,pzc}$ and. $\Gamma_{TPB/NB,pzc}$ that were found to be -3×10^{-12} (see Fig. 5) and -5×10^{-12} mol cm^{-2} (see Fig. 4 in [18]), respectively, that the inner layer consists of a laminate of a monolayer of water molecules and that of nitrobenzene molecules, the thickness being the closest-approach distance of solvated ions at the surface of contact of immiscible phases. On the other hand, the evaluation of the surface excess of water at the interface between an organic solvent (nitrobenzene, 1,2-dichloroethane, or hexane) and water containing electrolytes at relatively high concentrations implied that [21] the inner layer provides only a fraction of the monolayer of the solvent molecules. This led the authors of Ref. [21] to the conclusion that the concept of an ion-free layer at the oil/water interface is doubtful, while the continuous change in the compositin from one phase to the other is a more realistic picture. The observed c_i values of about $100\,\mu\text{F}\cdot\text{cm}^{-2}$ or more (see Fig. 8) are too large to be explained by the simple model of the inner layer consisting of a laminate of continuous dielectrics of nitrobenzene and water of monomolecular thickness, as suggested above, but could be explained by assuming a smaller value for the effective thickness of the inner layer. Consideration of a diffuse rather than sharp boundary between the space-charge region and the solvent layer is suggested [27]. The above discussion is based on the Gouy-Chapman theory which rests on a very simple model. Modification of the Gouy-Chapman theory has been discussed [51]. If the diffuse layer theory is seriously in error, then these models of the double layer structure of the oil/water interface will require drastic modification.

Electrocapillary Curves of Nonpolarized Oil/Water Interfaces

A comprehensive thermodynamic theory of electrocapillarity of the nonpolarized oil/water interface has been discussed elsewhere [30]. Here we discuss the following two cases IIb and IIIb, which correspond to the above systems IIa and IIIa, respectively;

$$\alpha_1 \begin{array}{|c|c|} b\ \text{mol dm}^{-3} & c\ \text{mol dm}^{-3} \\ \text{RTPB (NB)} & \text{RCl (W)} \end{array} \alpha_2, \tag{IIb}$$

$$\alpha_1 \begin{array}{|c|c} b\ \text{mol dm}^{-3} & c\ \text{mol dm}^{-3}\ \text{RCl (W)} \\ \text{RTPB (NB)} & d\ \text{mol dm}^{-3}\ \text{NaCl (W)} \end{array} \alpha_2. \tag{IIIb}$$

Here RTPB and RCl (R=TMA, TEA, etc.) represent tetraalkylammonium, tetraphenylborate, and tetralkylammonium chloride (R=tetramethylammonium, tetraethylammonium, etc.), respectively. When the R^+ ion is transferable across the interface while TPB$^-$, Cl$^-$, and Na$^+$ ions are practically not transferable across the interface, the electrocapillary equation for the nonpolarized interface of case IIb at constant temperature and pressure is given by:

$$-(\mathrm{d}\gamma)_{T,p} = \Gamma_{\text{Cl/W}}\,\mathrm{d}\mu_{\text{RCl}}^{\text{W}} + \Gamma_{\text{TPB/NB}}\,\mathrm{d}\mu_{\text{RTPB}}^{\text{NB}}, \tag{17}$$

where $\Gamma_{\text{Cl/W}}$ and $\Gamma_{\text{TPB/NB}}$ are the relative surface excesses of Cl$^-$ and TPB$^-$ ions and are defined by:

$$\Gamma_{\text{Cl/W}} = \Gamma_{\text{Cl}} - \frac{\chi_{\text{RCl}}^{\text{W}}}{\chi_{\text{W}}^{\text{W}}}\,\Gamma_{\text{W}}, \tag{18a}$$

$$\Gamma_{\text{TPB/NB}} = \Gamma_{\text{TPB}} - \frac{\chi_{\text{RTPB}}^{\text{NB}}}{\chi_{\text{NB}}^{\text{NB}}}\,\Gamma_{\text{NB}}, \tag{18b}$$

with

$$\Gamma_{Cl} + \Gamma_{TPB} = \Gamma_R. \tag{19}$$

When the reference electrodes α_1 and α_2 are reversible to TPB$^-$(NB) and Cl$^-$(W) ions, respectively, we have $FdE_{O^-}^W = d\mu_{RCl}^W + d\mu_{RTPB}^{NB}$ at constant temperature and pressure. Therefore Eq. (17) can be rewritten as follows:

$$-(d\gamma)_{T,p} = -F\Gamma_{Cl/W}dE_{O^-}^{W-} + (\Gamma_{Cl/W} + \Gamma_{TPB/NB})d\mu_{RTPB}^{NB}, \tag{20a}$$

or

$$-(d\gamma)_{T,p} = (\Gamma_{Cl/W} + \Gamma_{TPB/NB})d\mu_{RCl}^W + F\Gamma_{TPB/NB}dE_{O^-}^{W-}. \tag{20b}$$

These equations clearly show that the the slope of the electrocapillary curve of nonpolarized interface does not give the "surface charge density" but the relative surface excess of ionic components, as defined by Eq. (18) for case IIb. In other words, the electrocapillary maximum potential does not correspond to the "potential of zero charge". An approach to investigate the surface charge density and the double layer structure may be predicted as follows. When the values of the second terms of the right-hand sides of Eq. (18) (that is, the Γ_W and Γ_{NB} values), are known or estimated on reasonable argument, Γ_{Cl} and Γ_{TPB} (so that Γ_R by Eq. (19)) can be found from the slope of the electrocapillary curve. In this system Γ_{Cl} and Γ_{TPB} are the sole anionic components in the W and NB sides of the interface, respectively, whereas the sole cationic component Γ_R may be divided formally into the component in the W side Γ_R^W and that in the NB side Γ_R^{NB}, that is $\Gamma_R = \Gamma_R^W + \Gamma_R^{NB}$. Then the surface charge density in the W side, q^W, can be defined by $q^W = \Gamma_R^W - \Gamma_{Cl}^W = \Gamma_{TPB}^{NB} - \Gamma_R^{NB}$. Assuming the absence of specific ion adsorption in the inner layer, we have [18] approximately $\Gamma_{Cl/W} = \Gamma_{Cl(W)}^d - (\chi_{RCl}^W/\chi_W^W)\Gamma_W$ and $\Gamma_{TPB/NB} = \Gamma_{TPB(NB)}^d - (\chi_{RTPB}^{NB}/\chi_{NB}^{NB})\Gamma_{NB}^i$, Γ_W^i, and Γ_{NB}^i being the surface ammounts of W and NB in the inner layer per unit interfacial area. Accordingly, when the values of Γ_W^i and Γ_{NB}^i are known or estimated on reasonable argument, $\Gamma_{Cl(W)}^d$ and $\Gamma_{TPB(NB)}^d$ can be calculated from the slope of the electrocapillary curves. Then the q^W value can be calculated either from $\Gamma_{Cl(W)}^d$ by Eq. (14b) or from $\Gamma_{TPB(NB)}^d$ by the equation similar to Eq. (14b) for the NB side. The two q^W values should coinside with each other. Also, $\Gamma_{R(W)}^d$ can be calculated from q^W by the equation similar to Eq. (14a) for R^+ ion in W and $\Gamma_{R(NB)}^d$ by the equation similar to Eq. (14a) for R^+ ion in NB. The relation $\Gamma_{Cl(W)}^d + \Gamma_{TPB(NB)}^d = \Gamma_{R(W)}^d + \Gamma_{R(NB)}^d$ should be satisfied. Furthermore the $\Delta\phi_{W2}^W$ value can be calculated from q^W by Eq. (15) and the $\Delta\phi_{O2}^O$ value by an equation similar to Eq. (15) for the NB side, and finally the $\Delta\phi_{O2}^{W2}$ can be determined by Eq. (13).

The electrocapillary equation for case IIIb at constant temperature and pressure is given by:

$$-(d\gamma)_{T,p} = \Gamma_{(Cl-Na)/W}d\mu_{RCl}^W + \Gamma_{TPB/NB}d\mu_{RTPB}^{NB} + \Gamma_{Na/W}d\mu_{NaCl}^W, \tag{21}$$

where

$$\Gamma_{(Cl-Na)/W} = \Gamma_{Cl} - \Gamma_{Na} - \frac{\chi_{RCl}^W}{\chi_W^W}\Gamma_W, \tag{22a}$$

$$\Gamma_{Na/W} = \Gamma_{Na} - \frac{\chi_{NaCl}^W}{\chi_W^W}\Gamma_W, \tag{22b}$$

$$\Gamma_{TPB/NB} = \Gamma_{TPB} - \frac{\chi_{RTPB}^{NB}}{\chi_{NB}^{NB}}\Gamma_{NB}, \tag{22c}$$

with

$$\Gamma_{Na} + \Gamma_R = \Gamma_{Cl} + \Gamma_{TPB}. \tag{23}$$

The structure of the electric double layer of the interface of case IIIb should be discussed in the same way as above for case IIb. In the special case when $c \ll d$ and $c \ll b$ in case IIIb, we may suppose that (a) $\Gamma_R = \Gamma_R^O + \Gamma_R^W \simeq \Gamma_R^O$ and (b) $\Gamma_{(Cl-Na)/W} \simeq \Gamma_{Cl} - \Gamma_{Na}$ since χ_{RCl}^W / χ_W^W is very small, then $-(\partial\gamma/\partial\mu_{RCl}^W)_{T, p, \mu_i} = \Gamma_{Na} - \Gamma_{Cl} = \Gamma_{TPB} - \Gamma_R^O = q^W$. Therefore we might expect that the electrocapillary curve obtained with this special case of IIIb is similar to that obtained with case Ib. This consideration probably explains why the electrocapillary curve obtained with a case of type IIIb [15] appeared very much like those obtained with case Ib [17, 18]. The structure of electric double layer of nonpolarized oil/water interfaces was discussed by several investigators [15, 16, 19] on the basis of electrocapillary curve measurements. However, their argument was seemingly based on the assumption that Γ_W^i and Γ_{NB}^i are negligible in Eq. (18) and so on, which is not generally justified.

References

1. Gustalla, J.: Proc. Second International Congress of Surface Activity III, Butterworths, London, p. 112, 1957
2. Blank, M., Feig, S.: Science *141*, 1173 (1963)
3. Blank, M.: J. Colloid Interface Sci. *22*, 51 (1966)
4. Watanabe, A., Matsumoto, M., Tamai, H., Goto, R.: Kolloid Z. Z. Polym. *220*, 152 (1967), *221*, 47 (1967), *228*, 58 (1968)
5. Watanabe, A., Tamai, H.: Hyomen *13*, 67 (1975)
6. Dupeyrat, M., Michel, J.: J. Colloid Interface Sci. *29*, 605 (1969)
7. Dupeyrat, M., Nakache, E.: ibid. *73*, 332 (1980)
8. Joos, P., Van Bockstaele, M.: J. Phys. Chem. *80*, 1573 (1976)
9. Verburgh, Y., Joos, P.: J. Colloid Interface Sci. *74*, 384 (1980)
10. Gavach, C.: Experimentia Suppl. *18*, 321 (1971)
11. Gavach, C., d'Epenoux, B.: C. R. Acad. Sci. Paris *272*, 872C (1971)
12. d'Epenoux, B., Gavach, C.: J. Colloid Interface Sci. *56*, 138 (1976)
13. Spurny, J.V.: Colloid Polymer Sci. *255*, 902 (1977)
14. Koryta, J., Vanysek, P., Brezina, M.: J. Electroanal. Chem. Interfacial Electrochem. *75*, 211 (1977)
15. Gavach, C., Seta, P., d'Epenoux, B.: ibid. *83*, 225 (1977)
16. Gros, M., Gromb, S., Gavach, C.: ibid. *89*, 29 (1978)
17. Kakiuchi, T., Senda, M.: Bull. Chem. Soc. Jpn. *56*, 1322 (1983)
18. Kakiuchi, T., Senda, M.: ibid. *56*, 1735 (1983)
19. Reid, J.D., Melroy, O.R., Buck, R.P.: J. Electroanal. Chem. Interfacial Electrochem. *147*, 71 (1983)
20. Girault, H.H.J., Schiffrin, D.J., Smith, B.D.V.: ibid. *137*, 207 (1982)
21. Girault, H.H.J., Schiffrin, D.J.: ibid. *150*, 43 (1983)
22. Girault, H.H.J., Schiffrin, D.J.: ibid. *161*, 415 (1984)
23. Girault, H.H.J., Schiffrin, D.J.: ibid. *179*, 277 (1984)
24. Girault, H.H.J., Schiffrin, D.J., Smith, B.D.V.: (private communication)
25. Samec, Z., Marecek, V., Homolka, D.: J. Electroanal. Chem. Interfacial Electrochem. *126*, 121 (1981)
26. Marecek, V., Samec, Z.: ibid. *149*, 189 (1983)
27. Samec, Z., Marecek, V., Homolka, D.: Faraday Discussion Chem. Soc. 77, 10 (1984)
28. Osakai, T., Kakutani, T., Kakiuchi, T., Senda, M.: Rev. Polarogr. (Kyoto) *27*, 51 (1981)
29. Osakai, T., Kakutani, T., Senda, M.: Bull. Chem. Soc. Jpn. *57*, 370 (1984)

30. Kakiuchi, T., Senda, M.: ibid. *56*, 2912 (1983)
31. Koryta, J.: Electrochimica Acta *24*, 293 (1979)
32. Koryta, J.: ibid. *29*, 445 (1984)
33. Koryta, J.: Ion-selective Electrode Rev. (Ed. Thomas, J.E.) *5*, 131 (1981)
34. Koryta, J.: Adv. Electrochem. Electrochem. Eng. (Ed. Gerischer, H., Tobias, C.) *12*, 113 (1981)
35. Senda, M., Kakutani, T.: Hyomen *18*, 535 (1980)
36. Senda, M., Kakutani, T., Osakai, T.: Denki Kagaku *49*, 322 (1981)
37. Vanysek, P., Buck, R.P.: J. Electroanal. Chem. Interfacial Electrochem. *163*, 1 (1984)
38. Reid, J.D., Melroy, O.R., Bronner, W.E., Hughes, H.C., Vanysek, P., Buck, R.P.: Paper presented at the NATO A. S. I. Conference, July 2–13, 1984, Viano do Castelo, Portugal, 1984
39. Karpfen, F.M., Randles, J.E.B.: Trans. Faraday Soc. *49*, 823 (1953)
40. Minc, S., Koczorowski, Z.: Electrochim. Acta *8*, 575 (1963)
41. Hung, L.Q.: J. Electroanal. Chem. Interfacial Electrochem. *115*, 159 (1980)
42. Hung, L.Q.: ibid. *149*, 1 (1983)
43. Senda, M.: Rev. Polarogr. (Kyoto) *30*, 19 (1984)
44. Kakutani, T., Ohkouchi, T., Osakai, T., Kakiuchi, T., Senda, M.: Anal. Sci. (in press) *1*, 219 (1985); abstract, Rev. Polarogr. (Kyoto) *30*, 46 (1984)
45. Osakai, T., Kakutani, T., Kakiuchi, T., Senda, M.: (unpublished results)
46. Senda, M., Kakiuchi, T.: Rev. Polarography *27*, 49 (1981)
47. Verwey, E.J.W.: Colloid Science, Vol. 1 (Ed. Kruyt, H.R.), Elsevier Publ. Co., Amsterdam, p. 137, 1952
    Verwey, E.J.W., Niessen, K.F.: Phil. Mag. (7), *28*, 435 (1935)
48. van den Tempel, M.: Rec. Trav. Chim. Pays-Bas *72*, 419 (1953)
49. Grahame, D.C., Soderberg, B.A.: J. Chem. Phys. *22*, 449 (1947)
50. Hajkova, P., Homolka, D., Marecek, V., Samec, Z.: J. Electroanal. Chem. Interfacial Electrochem. *151*, 227 (1983)
51. Reeves, R.: Comprehensive Treatise of Electrochemistry (Ed. Bockris, J.O'M., Conway, B.E., Yeager, F.) Plenum Press, New York, Vol. 1, Chapt. 3, 1980

Study of the Electrical Double Layer at the Interface Between Two Immiscible Electrolyte Solutions by Impedance Measurements

Zdeněk Samec and Vladimir Mareček

In the first theoretical approach to the interface between two immiscible electrolyte solutions (ITIES) by Verwey and Niessen [1], the interface was represented by a diffuse double layer, i.e. one phase containing an excess of the positive space charge and the other phase an equal excess of the negative space charge. The space charge regions have been treated by the Gouy-Chapman theory [2, 3]. Gavach et al. [4] have suggested that the two space charge regions are separated by an inner or compact layer of oriented solvent molecules – a modified Verwey-Niessen (MVN) model. The inner layer can be either free of ions or the ions are located in the inner layer as specifically adsorbed particles. Boguslavsky et al. [5, 6] and Joos and Vanden Bogaert [7] neglected the diffuse double layer and assumed an ionic monolayer or bilayer at the ITIES formed by the specifically adsorbed ions.

The electrical double layer arising at the ITIES has been studied by measuring the surface tension [4, 7–16, 25] or the impedance [17–26] mainly of water/nitrobenzene [4, 7–15, 17, 19–24] and water/1,2-dichlorethane [12, 16, 18, 25, 26] systems. This contribution reviews the principles and the results of the impedance measurements, in particular those based on the AC impedance or galvanostatic pulse techniques, which have been used most frequently for the study of the double layer at the ITIES. The quantity, which can be inferred from the impedance measurements, and which is related to the double-layer structure, is the interfacial capacitance. We shall discuss first the thermodynamic background for the capacitance of the electrical double layer at the ITIES.

Thermodynamic Background

The typical galvanic cell, the potential of which is controlled or measured in impedance measurements, can be represented by:

$$M|M^+X^-|R^+X^-\|S^+Y^-|S^+X^-|M^+X^-|M', \qquad (I)$$
$$\quad\;(s)\qquad(w)\qquad(o)\qquad(w')\qquad(s)$$

where M and M' are two pieces of the same metal (e.g. silver) between which the potential difference $E = \varphi(M) - \varphi(M')$ is measured, and $M|M^+X^-|R^+X^-$ and $M'|M^+X^-|S^+X^-$ are the reference electrodes reversible to an anion X^- (e.g. chloride) in the aqueous phases w and w', respectively. The phases w' and o contain a common cation S^+, the partition of which safeguards the constant potential difference across the w'/o interface, cf. the analysis of $\Delta_o^{w'}\varphi$ in Ref. [27].

The Interface Structure and Electrochemical Processes
at the Boundary Between Two Immiscible Liquids
Editor: V. E. Kazarinov
© Springer-Verlag Berlin, Heidelberg 1987

The most general thermodynamic treatment of the electrical double layer at the ITIES was given by Kakiuchi and Senda [28]. Here we shall follow a simpler but instructive analysis by Girault and Schifrin [16].

Assuming the ion pairs R^+X^-, R^+Y^-, S^+Y^-, and S^+X^- can in general be formed in each phase, the thermodynamics of the interfacial region is described by the Gibbs equation [16]:

$$-d\gamma_{(T,\,p\,=\,const)} = \sum_{i=1}^{4} m_i d\tilde{\mu}_i + \sum_{j=1}^{4} m_j d\mu_j + m_w d\mu_w + m_o d\mu_o, \tag{1}$$

where the first and second sum on the right-hand side are the contributions from the ions and ion pairs, respectively; m_k is the excess interfacial concentration of the species k:

$$m_k = n_k^{(\sigma)}/A = (n_k - n_k^{(w)} - n_k^{(\sigma)})A, \tag{2}$$

where $n_k^{(\sigma)}$ is the number of moles of k in the interfacial region σ; $n_k^{(w,o)}$ are the k numbers of moles of the same species, which would be present in the homogeneous phase w or o, if that phase continues up to the hypothetical plane of the contact. By using the Gibbs-Duhem equation for each phase:

$$\sum_{i=1}^{4} n_i^{(w)} d\tilde{\mu}_i + \sum_{j=1}^{4} n_j^{(w)} d\mu_j + n_w^{(w)} d\mu_w + n_o^{(w)} d\mu_o = 0 \tag{3}$$

$$\sum_{i=1}^{4} n_i^{(o)} d\tilde{\mu}_i + \sum_{j=1}^{4} n_j^{(o)} d\mu_j + n_w^{(o)} d\mu_w + n_o^{(o)} d\mu_o = 0 \tag{4}$$

it is possible to eliminate two variables in Eq. (1), i.e.

$$-d\gamma_{(T,\,p\,=\,const)} = \sum_{i=1}^{4} \Gamma_i d\tilde{\mu}_i + \sum_{j=1}^{4} \Gamma_j d\mu_j. \tag{5}$$

Owing to the low mutual miscibility of the two solvents, $n_w^{(w)} n_o^{(o)} \gg n_w^{(o)} n_o^{(w)}$, $n_w^{(o)} n_{BA}^{(w)}$, $n_o^{(w)} n_{BA}^{(o)}$ and the relative surface excesses of the ions Γ_i or the ion pairs Γ_j are given by the equations:

$$\Gamma_k = m_k - m_w(n_k^{(w)}/n_w^{(w)}) - m_o(n_k^{(o)}/n_o^{(o)}). \tag{6}$$

By taking into account the electroneutrality condition:

$$\Gamma_{R^+} + \Gamma_{S^+} = \Gamma_{X^-} + \Gamma_{Y^-} \tag{7}$$

and the relationships $d\tilde{\mu}_X^{(w)} = -Fd\varphi(M)$, $d\tilde{\mu}_X^{(w')} = -Fd\varphi(M')$ and $d\tilde{\mu}_{S^+}^{(o)} = d\tilde{\mu}_{S^+}^{(w')}$ we get finally the electrocapillary equation:

$$-d\gamma_{(T,\,p\,=\,const)} = Q(dE_- - F^{-1} d\mu_{SX}^{(w')}) + \Gamma'_{R^+} d\mu_{RX} + \Gamma'_{Y^-} d\mu_{SY}. \tag{8}$$

Here Q is the thermodynamic surface excess charge density:

$$Q = -(\partial\gamma_{T,\,p}/\partial E_-)_{T,\,p,\,u} = F(\Gamma_{R^+} - \Gamma_{X^-} + \Gamma_{R^+Y^-} - \Gamma_{S^+X^-}) \tag{9}$$

and Γ'_{R^+} and Γ'_{Y^-} are the thermodynamic surface excesses:

$$\Gamma'_{R^+} = -(\partial\gamma/\partial\mu_{RX})_{T,\,p,\,E_-,\,\mu_{SY},\,\mu_{SX}} = \Gamma_{R^+} + \Gamma_{R^+X^-} + \Gamma_{R^+Y^-}, \tag{10}$$

$$\Gamma'_{Y^-} = -(\partial\gamma/\partial\mu_{SY})_{T,\,p,\,E_-,\,\mu_{RX},\,\mu_{SX}} = \Gamma_{Y^-} + \Gamma_{S^+Y^-} + \Gamma_{R^+Y^-}. \tag{11}$$

The coefficients (9) to (11) can be inferred from surface tension measurements. Moreover, the differential capacitance C can be measured, which is defined as:

$$C = (\partial Q/\partial E_-)_{T,p,\mu} = -(\partial^2 \gamma/\partial E_-)_{T,p,\mu}. \tag{12}$$

Obviously, the physical meaning of the capacitance can be complex, inasmuch as it involves the contributions from both the free charge and the charge bound in the ion pairs. If there is no ion transfer across the interface, $n_{R^+}^o = n_{X^-}^o = n_{S^+}^w = n_{Y^-}^w = 0$, the system behaves as an ideally polarizable interface in a sense that the interfacial state can be controlled by the charge supplied from outside. In particular, in the absence of the ion pairs, the capacitance is given by:

$$C = (\partial q/\partial E_-)_{T,p,u}, \tag{13}$$

where the surface charge density q is

$$q = m_{R^+} - m_{X^-} = m_{Y^-} - m_{S^+}. \tag{14}$$

AC Impedance Measurements

In the classical impedance measurements a small periodic voltage or current signal is applied to the interface and the electrical current or voltage response, respectively, is measured, so that the complex impedance $\tilde{Z}$ can be evaluated from the complex voltage $\tilde{U}$ and the complex current $\tilde{I}$ as:

$$\tilde{Z} = \tilde{U}/\tilde{I}. \tag{15}$$

It is suitable to perform the measurements in the potential range in which the ion transfer is negligible and the boundary behaves as an ideally polarizable interface.

Figure 1 shows a cyclic voltammogram of the water/nitrobenzene interface in the presence of hydrophillic ions in water (Na$^+$, Br$^-$) and hydrophobic ions in nitrobenzene (tetrabutylammonium$^+$, tetraphenylborate$^-$). In the potential range 0.13–0.45 V the current is controlled mainly by charging the interface and the system

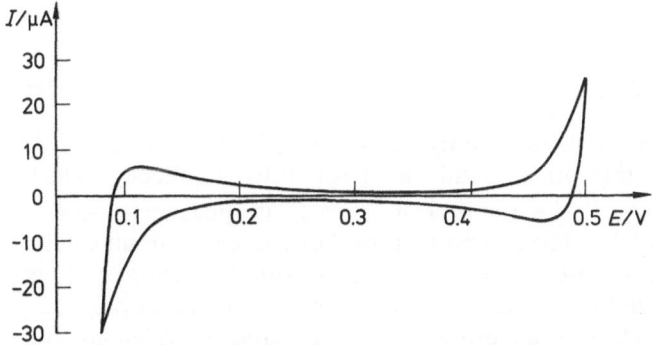

Fig. 1. Cyclic voltammogram of 0.01 M NaBr in water and 0.01 M tetrabutylammonium tetraphenylborate in nitrobenzene. Scan rate 0.1 V s^{-1}. Ohmic potential drop compensation adjusted to 1.35 kΩ [21]

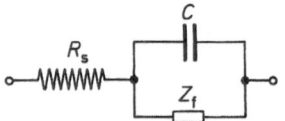

Fig. 2. Equivalent circuit for ITIES: C is the capacitance of the interface, Z_f is the faradaic impedance, R_s the solution resistance between the tips of the Luggin capillaries

has the property required, while at more positive or more negative potentials the interfacial transfer of the ions present in the system prevails. Such an interface can be represented by the electrical equivalent circuit shown in Fig. 2, which consists of a parallel combination of the interfacial capacitance C and the faradaic impedance Z_f, with the solution resistance R_s between the tips of the Luggin capillaries in series [17].

In the case of a fast, diffusion controlled, ion-transfer process:

$$X^z(\text{w}) \rightleftarrows X^z(\text{o}), \tag{II}$$

where z is the charge number of the ion X, the impedance Z_f becomes the Warburg impedance z_w:

$$Z_f = Z_W = (1-j)\sigma\omega^{-1/2}, \tag{16}$$

where $j = \sqrt{-1}$, ω is the angular frequency and the characteristic parameter σ of the faradaic process is:

$$\sigma = (4RT/z^2F^2Ac^0\sqrt{2D})\cosh^2[zF(E - E_{1/2}^{\text{rev}})/2RT], \tag{17}$$

where A is the interfacial area, c^0 the bulk ionic concentration, D the diffusion coefficient, and E and $E_{1/2}^{\text{rev}}$ are the potential and the reversible polarographic half-wave potential, respectively. The real Z' and the imaginary Z'' components of the complex impedance become:

$$Z' = |Z|\cos\delta = R_s + Z_C X[(X+1)^2 + 1]^{-1}, \tag{18}$$

$$Z'' = |Z|\sin\delta = Z_C X(X+1)[(X+1)^2 + 1]^{-1}, \tag{19}$$

where δ is the phase shift between the applied and measured signal. The capacitance Z_C is given by:

$$Z_C = (\omega C)^{-1} \tag{20}$$

and the parameter X is given by:

$$X = (Z_W/Z_C)\sqrt{2} = 2\sigma C\omega^{1/2}. \tag{21}$$

Apparently, if $X \gg 1$ (high frequency limit), $Z' = R_s + Z_C X^{-1} = R_s + (2C^2\sigma)^{-1}$. $\omega^{-3/2}$ and $Z'' = Z_C$. On the other hand, if $X \ll 1$ (low frequency limit), $Z' = R_s + Z_W\sqrt{2}$ and $Z'' = Z' - R_s$, i.e. the plot of Z'' vs. Z' (impedance plot) is a straight line with an angle of 45°. The components of the equivalent circuit can be resolved using the following procedure. First, the solution resistance R_s is evaluated as the high frequency limit of Z', e.g. by the extrapolation of the plot of Z' vs. $\omega^{-3/2}$ to $\omega^{-3/2} = 0$ [18]. Second, using R_s it is possible to calculate the parameter X and the capacitance Z_C from Eqs. (18) and (19).

For the experimental determination of the impedance of a measured system the alternating current bridge is the most accurate method. This method, utilizing a

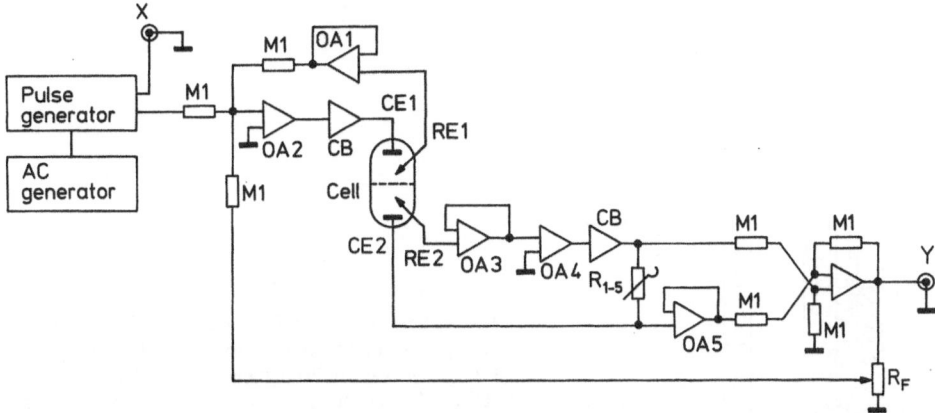

Fig. 3. Block scheme of the four-electrode potentiostat with a positive feedback for the ohmic drop compensation [30]

four-electrode potentiostat for *DC* polarization of the water/organic solvent interface, was described by Figaszewski [29].

As a balance detector an oscilloscope with a tuned amplifier can be used. This approach, however, is rather time consuming and tedious as compared with an *AC* impedance technique, based on alternating current voltammetry (polarography). In the latter case, a sinusoidal voltage of small amplitude (typically about 5 mV peak-to-peak) and of frequency ranging from units of cycles per second to tens of kilocycles per second is superposed on the slow triangular voltage pulse and fed to an potentiostat [17], cf. Fig. 3. The alternating current flowing through the interface is analysed as a function of the *DC*-potential *E* by means of phase-sensitive detection (lock-in-amplifier). The in-phase $I(0°)$ and quadrature $I(90°)$ components of the current signal are recorded on an $X - Y$ recorder as a function of E and are transformed into the impedance Z and the phase shift according to the relationships:

$$|Z| = V_0/(I(0°)^2 + I(90°)^2)^{1/2}, \tag{22}$$

$$\delta = \tan^{-1}(I(90°)/I(0°)). \tag{23}$$

The scheme of a four-electrode cell [31] suitable for impedance measurements at the water/nitrobenzene or 1,2-dichlorethane interfaces, is shown in Fig. 4. The lower-density aqueous phase occupies the bottom of the cell, with the organic solvent phase above it. The stable location of a flat and reproducible interface in the round hole cut in the glass barries *B* is ensured by the low surface tension between water and glass and by the incompressibility of the liquid. Prior to use, the inner space of the cell is made hydrophilic by washing with a concentrated solution of sodium hydroxide, doubly distilled water and by drying at a temperature above 100°C in a dry-box. The cell, in which the location of the phases is inverted, i.e. the higher-density organic solvent occupies the bottom of the cell, can be used as well [17]. However, in order to establish a flat interface in this cell, it is necessary to "hydrophobize" the inner space which is in contact with the organic solvent either

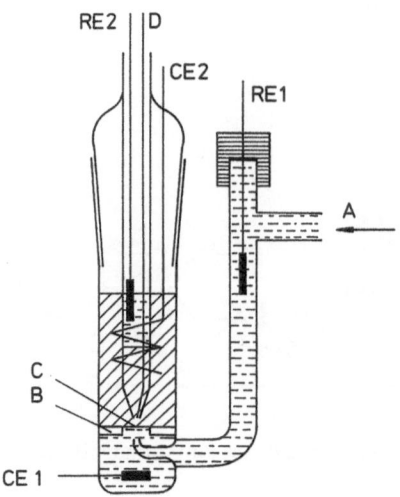

Fig. 4. Scheme of the four-electrode cells with two platinum counter electrodes (CE 1 and CE 2) and two reference electrodes (RE 1 and RE 2): A – connection to a microsyringe for the adjustment of the interface, B – glass barrier with a round hole, C – liquid/liquid interface, D – insulated copper wire [31]

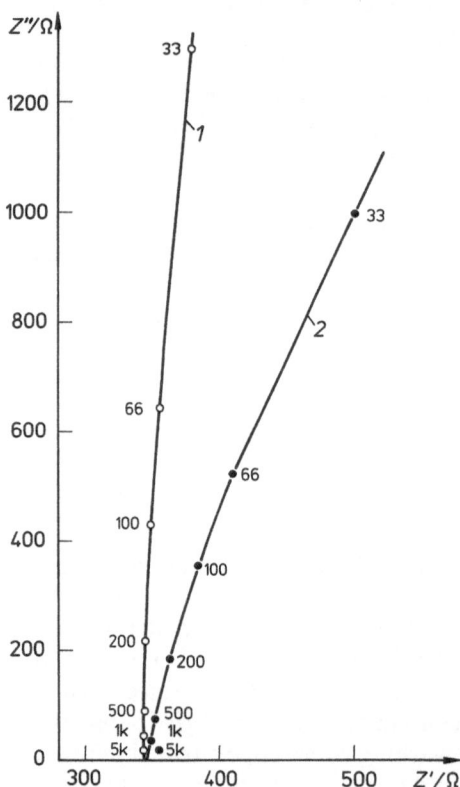

Fig. 5. Impedance plot for the water/nitrobenzene interface at (*1*) $E = 250$ mV and (*2*) $E = 400$ mV. Composition of the aqueous phase: 0.05 M NaBr, nitrobenzene phase 0.05 M tetrabutylammonium tetraphenylborate. Interfacial area: 27.4 mm^2 [31]

Table 1. Analysis of the impedance plots for the system of 0.05 M NaBr in water and 0.05 M tetrabutylammonium tetraphenylborate in nitrobenzene at $E = 250$ mV(A) and $E = 400$ mV(B). The values of the capacitance are given in parentheses; evaluated from $C^{-1} = Z''\omega$ [31]

frequency (Hz)	A^a $X\omega^{-1/2}$ (rad$^{-1/2}$ s$^{1/2}$)	C (Fm^{-2})	B^b $X\omega^{-1/2}$ (rad$^{-1/2}$ s$^{1/2}$)	C (Fm^{-2})
11	2.2	0.159 (0.177)	–	–
33	2.7	0.165 (0.165)	0.39	0.182 (0.219)
66	2.8	0.168 (0.171)	0.36	0.181 (0.208)
100	3.4	0.166 (0.165)	0.36	0.181 (0.203)
200	6.1	0.167 (0.167)	0.31	0.181 (0.199)
500	–	– (0.163)	0.25	0.181 (0.196)
1000	–	– (0.162)	–	– (0.200)
5000	–	– (0.095)	–	– (0.076)

[a] Solution resistance 345 Ω
[b] Solution resistance 348 Ω

by treating it with dimethyldichlorosilane [33] or by inserting a piece of Teflon tubing [34].

The impedance plots for the water/nitrobenzene interface are illustrated in Fig. 5. At frequencies approaching 5 kHz, the correction was made for the phase shift of the measured signal caused by the potentiostat itself. Sometimes a semicircular arc appears on the impedance plot frequencies and low electrolyte concentrations [23, 35], but such a behaviour is probably due to the unsuitable cell construction or the potentiostat failure. The results of the analysis of the impedance plots shown in Fig. 5 are summarized in Table 1. With the increasing frequency ω, the parameter X increases and its evaluation from the equation:

$$Z''(Z' - R_s)^{-1} = 1 + X, \tag{24}$$

becomes less reliable because $Z' \approx R_s$. On the other hand, if $X \gg 1$, the equation:

$$Z'' \approx Z_C = (\omega C)^{-1}, \tag{25}$$

holds with sufficient accuracy and can be used for the evaluation of the capacitance C, cf. the data in parentheses in column A of Table 1. Nevertheless, the use of Eq. (25) leads always to an overestimation of the interfacial capacitance as seen from the data in parentheses in column B of Table 1.

Galvanostatic Pulse Technique

In general, the current I flowing through the polarizable interface is the sum of the double layer charging current I_c and of the faradaic current I_f:

$$I = I_c + I_f = C(dE/dt) + I_f, \tag{26}$$

where C is the double-layer capacitance and dE/dt is the rate of the change of the potential drop across the interface. When a current step $\delta I = I_0 = $ const. is imposed on the interface, the charging current I_c decreases with time, while the faradaic current I_f increases. These current/time relationships were described theoretically for a simple electron transfer reaction at a metal electrode assuming a small decline of the system from the equilibrium [36]. Obviously these relationships also apply to the simple ion transfer reaction at the ITIES (II). For a small variation $\delta E \ll RT/zF$ of the potential difference E from its equilibrium value E_r:

$$E_r = E_{1/2}^{\text{rev}} + (RT/zF)\ln[D_X(\text{o})^{1/2} c_X^*(\text{o})/D_X(\text{w})^{1/2} c_X(\text{w})], \tag{27}$$

I_c decreases with time according to [20, 21]:

$$\frac{I_c}{I_0} = \frac{RTC\{1 + \exp[(zF/RT)(E_r - E_{1/2}^{\text{rev}})]\}}{(zF)^2 (\pi D_X(\text{o})\Delta t)^{1/2} c_X^*(\text{o})}, \tag{28}$$

where Δt is the time elapsed from the pulse initiation ($t = t_0$), $D_X(\text{o})$ and $D_X(\text{w})$ are the diffusion coefficients and $c_X^*(\text{o})$ and $c_X^*(\text{w})$ are the concentrations at the interface of the ion X^z in the phases o and w, respectively, and $E_{1/2}^{\text{rev}}$ is given by:

$$E_{1/2}^{\text{rev}} = E_X^0 + (RT/zF)\ln[D_X(\text{w})/D_X(\text{o})]^{1/2} + \Delta E', \tag{29}$$

where E_X^0 is the standard potential difference and $\Delta E'$ is the activity coefficient term. Obviously, at very short times $(I_c/I_0) \to 1$ and:

$$I_0 = C(dE/dt)_{\Delta t \to 0}. \tag{30}$$

Consequently, the interface behaves as a resistor R_s and capacitor C in series, cf. Fig. 2. Under these conditions the ohmic potential drop $\delta E_0 = I_0 R_s$ appears as a step on the galvanostatic transient at the beginning of the pulse ($t = t_0$) and the slope of this transient at $t > t_0$ is controlled exclusively by the capacitance C, Eq. (30). The potential range over which Eq. (30) holds with sufficient accuracy ($I_c/I_0 > 0.99$) in the course of a pulse duration $\Delta t = 5$ ms can be estimated from Eqs. (27) and (28). For the electrolyte concentrations 0.01 mole dm^{-3} this estimate shows [20, 21] that Eq. (30) is applicable when E_r (vs. TBA$^+$) falls between ca. 150 and 450 mV or for the galvanic potential difference $\Delta_o^w \varphi$ between ca. -150 and 150 mV.

Figure 6 shows the block scheme of the microcomputer-measuring system for the fast performance galvanostatic pulse technique [31]. The experimental interface can be divided into three parts. Two of them are the pulse generator and the data acqustion system operated by a 20 kB microcomputer. The third one is a four-electrode galvanostat/potentiostat, which is the only analogue component of the system. The pulses were generated at the output of a 12-bit digital-to-analogue converter (DAC 12

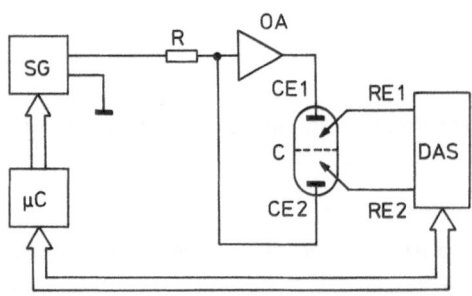

Fig. 6. Block scheme of the galvanostat with the microcomputer control: OA – operational amplifier, R – 1 kΩ resistor, SG – pulse generator, DAS – data aquisition system, μC-microcomputer

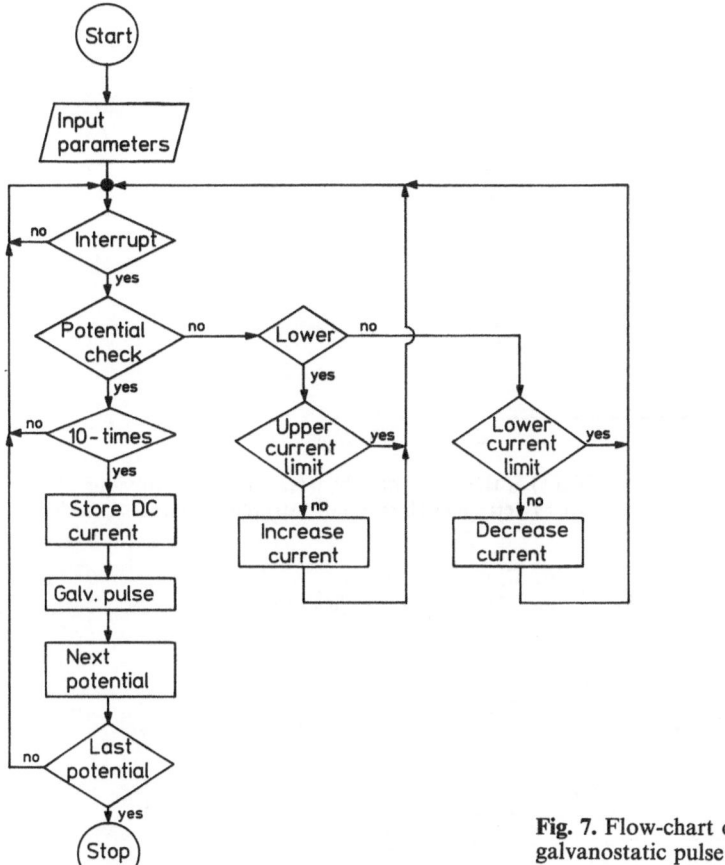

Fig. 7. Flow-chart of the program for the galvanostatic pulse measurements [31]

QZ, Analog Devices) combined with a 8-bit digital-to-analogue converter (WNC 041, Tesla, Czechoslovakia). The output potential range was ± 2048 mV with 0.05 mV resolution corresponding to a maximum current of 2.048 mA in steps ± 50 nA in the galvanostatic mode. The measured signal was sampled with a minimum sampling time of 100 µs by means of a 12-bit data acquisition system (SDM 853, Burr Brown) and the data were stored in the microcomputer memory. Figure 7 shows the flow-chart of the programs for the galvanostatic pulse measurements. Input parameters were: initial potential, spacing of the potential values, potential range, magnitude of the current step and maximum value of the pre-polarization current. In the measurements, the interface was usually pre-polarized by a current $|I| < 12$ µA. The potential difference E across the galvanic cell (I) was read every 3 ms in an interrupt mode and compared with the pre-programmed value starting at the initial potential. When the pre-programmed potential value was reached and remained constant (± 10 mV) for at least 30 ms, the current pulse of amplitude $I_0 = \pm 25$, 50, 75 or 100 µA and of duration $\Delta t = 2.5$ ms was generated across the cell. The potential difference E was sampled every 0.1 ms for 2.5 ms before the current pulse was applied and for the next 2.5 ms of pulse duration. The whole galvanostatic experiment, which comprised the measurement and storage of

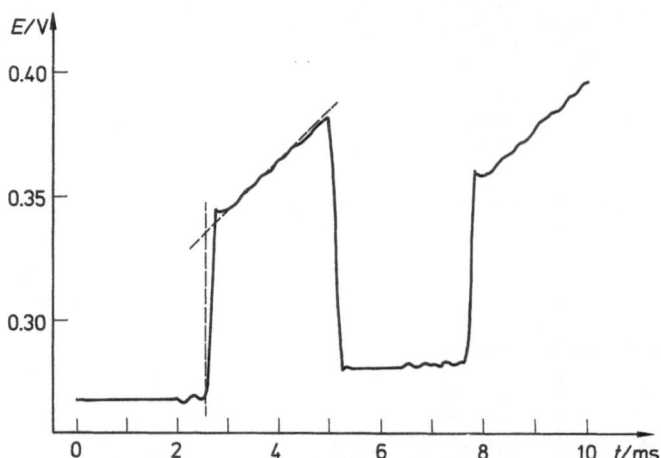

Fig. 8. Galvanostatic transient ($I_0 = 50\ \mu A$) at the water/nitrobenzene interface. Aqueous phase: 0.05 M LiCl, nitrobenzene phase: 0.01 M tetraphenylarsonium tetraphenylborate. Interfacial area: 22 mm² [32]

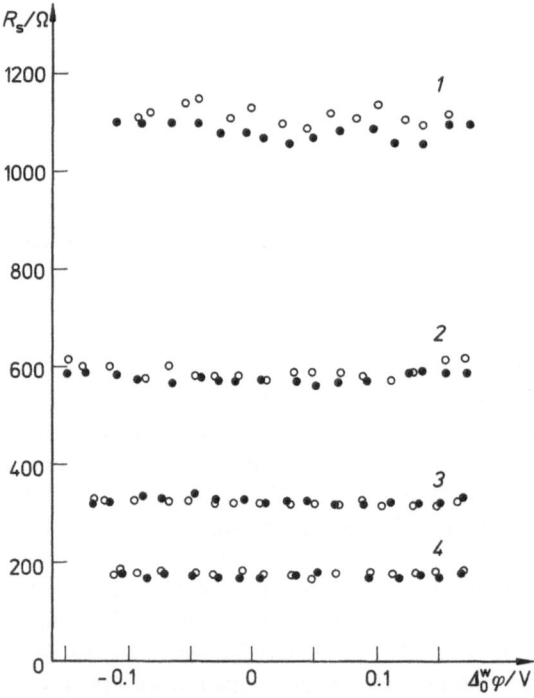

Fig. 9. Solution resistance R_s between the tips of the Luggin capillaries from the galvanostatic transient at $I < 0$ (●) or $I > 0$ (○). Aqueous phase: (1) 0.005 M, (2) 0.01 M, (3) 0.025 M, and (4) 0.05 M MgSO₄. Nitrobenzene phase: (1) 0.01 M, (2) 0.02 M, (3) 0.05 M, and (4) 0.10 M tetrabutylammonium tetraphenylborate. Interfacial area 22 mm² [32]

about 30 galvanostatic transients at different potentials, was completed in about 20 seconds.

An example of two consecutive galvanostatic transients is shown in Fig. 8. The solution resistance R_s, which was evaluated from the step on the galvanostatic transient at $\Delta t = 0$, is plotted vs. potential E in Fig. 9. As one can expect, the solution resistance is

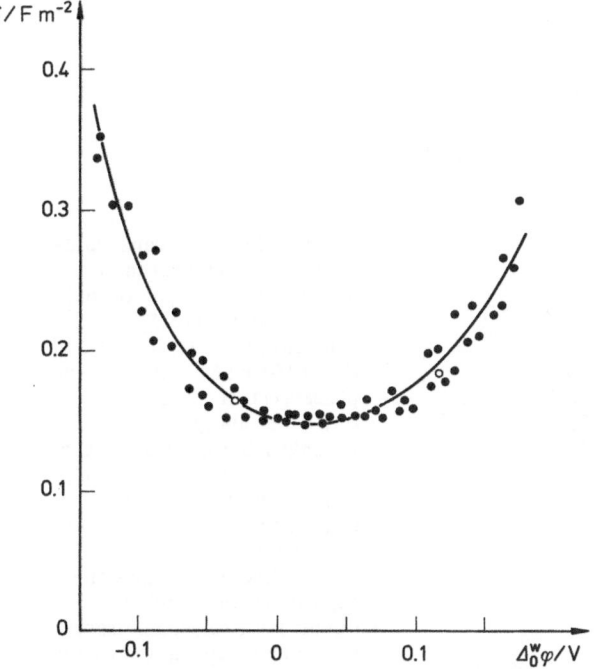

Fig. 10. Capacitance of the water/nitrobenzene interface vs. potential difference $\Delta_o^w\varphi$ from galvanostatic (●) or AC impedance (○) measurements. Composition of the aqueous phase: 0.05 M NaBr, nitrobenzene phase: 0.05 M tetrabutylammonium tetraphenylborate [31]

independent of both the direction of the current pulse and the potential E, and decreases with increasing electrolyte concentration.

In Fig. 10 the capacitance data obtained by the fast performance galvanostatic pulse technique are compared with those inferred from the AC impedance measurements, cf. Table 1. The agreement is very good.

Capacitance Data

Figure 11 shows the plots of the double layer capacitance as a function of the potential difference $\Delta_o^w\varphi$ for the system of NaBr in water and tetrabutylammonium tetraphenylborate (TBATPB) in nitrobenzene.

The conversion of the potential E into the potential difference $\Delta_o^w\varphi$ was accomplished by using tetramethylammonium (TMA$^+$) cation as the reference ion in situ [21], i.e. after the AC or galvanostatic measurements were completed, TMA$^+$ was dissolved in the nitrobenzene or the aqueous phase and the cyclic voltammogram was measured. Since the transfer of TMA$^+$ across the water/nitrobenzene interface is fast and reversible, it is possible to infer the standard potential $E_{TMA^+}^o$ of TMA$^+$ ion transfer from the cyclic voltammogram by using Eq. (29). The ratio $D_{TMA^+}(w)/D_{TMA^+}(o) = 2.64$ was estimated [21] from the limiting ionic conductivities, and the activity coefficients term $\Delta E'$ was evaluated from the extended Debye-Hückel equation. The latter correction amounts to several millivolts depending on the electrolyte concentration. Finally, E is converted into $\Delta_o^w\varphi$ using the standard potential $E_{TMA^+}^o$ determined by cyclic voltammetry and the standard Galvani potential differ-

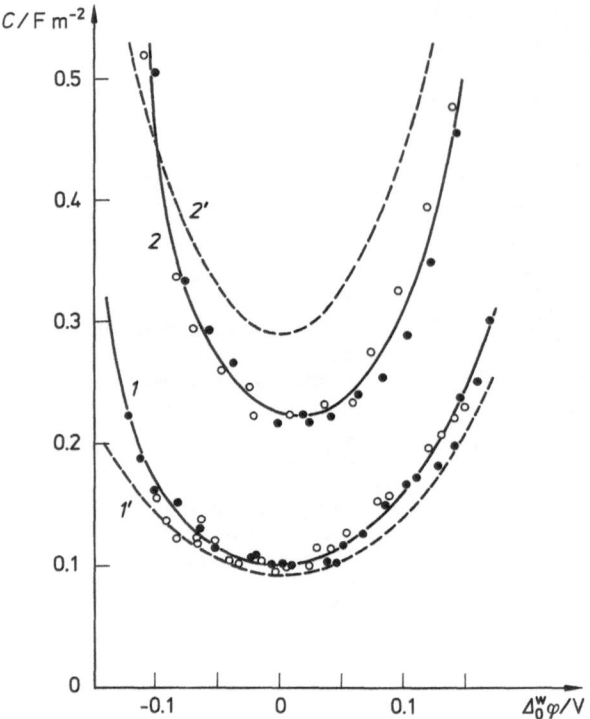

Fig. 11. Plot of the capacitance C of the water/nitrobenzene interface as a function of the potential difference $\Delta_0^w\varphi$. Aqueous phase: (*1*) 0.01 M and (*2*) 0.10 M NaBr. Nitrobenzene phase: (*1*) 0.01 M and (*2*) 0.10 M tetrabutylammonium tetraphenylborate. Capacitance data were evaluated from the slope of the galvanostatic transient for the positive (○) or negative (●) current step. Dashed lines (*1'*, *2'*) show the capacitance of the diffuse double layer calculated using the Gouy-Chapman theory for $\Delta_0^w\varphi_i = 0$ V [32]

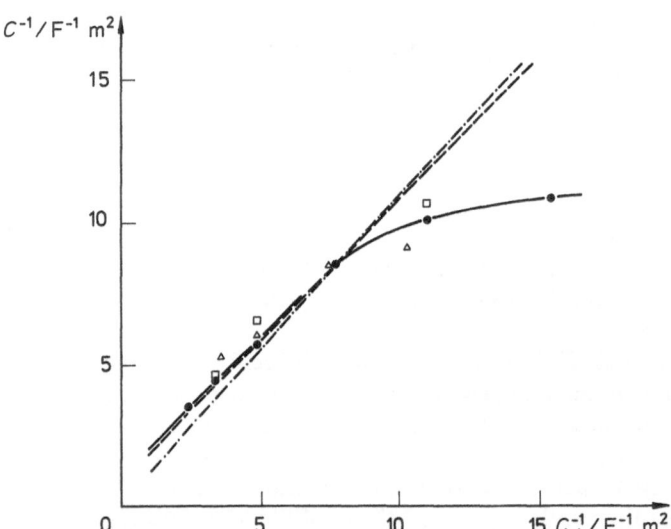

Fig. 12. Inverse capacitance C^{-1} of the interface between aqueous NaBr (●) LiCl (□) or MgSO$_4$ (△) and a solution of tetrabutylammonium tetraphenylborate in nitrobenzene as a function of C_d^{-1} calculated from Eq. (31). The dashed-and-dotted line calculated from Eq. (41) [32]

ence $\Delta_o^w \varphi_{TMA^+}^o = 0.035$ V [37]. A similar procedure can be adopted for the water/1,2-dichlorethane interface [37].

In case that the concentrations of NaBr ($c^0(w)$) and TBATPB ($c^0(o)$), are equal, the double layer capacitance decreases with decreasing electrolyte concentration in the whole potential range unless this concentration is lower than about 0.02 mol dm^{-3}. For $c^0(w) = c^0(o) < 0.02$ M, the capacitance becomes independent of the concentration. This is illustrated in Fig. 12, where the inverse capacitance C^{-1} at $\Delta_o^w \varphi = 0$ is plotted as a function of the inverse capacitance of the diffuse double layer. The latter capacitance was evaluated using the linearized Gouy Chapman theory from the equation:

$$C_d^{-1} = (\varepsilon_0 \varepsilon^w \varkappa^w)^{-1} + (\varepsilon_0 \varepsilon^o \varkappa^o)^{-1}, \tag{31}$$

where ε is the relative dielectric permittivity, ε_0 is the permittivity of vacuum, $\varkappa$ is the Debye screening length:

$$\varkappa^{-1} = \left(\frac{F^2}{\varepsilon \varepsilon_0 RT} \sum_i z_i^2 c_i \right)^{-1/2}, \tag{32}$$

z_i is the charge number of an ion, c_i the ionic concentration, and the superscript w and o refer to water and organic phase, respectively. In case of MgSO$_4$ aqueous solutions, the ionic concentrations were evaluated as $c_i = (1 - \alpha) c^0$, where α is the degree of the ion association [32]. At higher electrolyte concentrations the plot of C^{-1} vs. C_d^{-1} seems to satisfy the equation:

$$C^{-1} = C_d^{-1} + C_i^{-1}, \tag{33}$$

with the concentration-independent part of the inverse capacitance $C_i^{-1} \approx 1.2 \, m^2 F^{-1}$.

Zero-Charge Potential Difference

From the surface tension measurements of the interface between the aqueous solution of LiCl and the nitrobenzene solution of TBATPB, the zero-charge potential difference was estimated as $\Delta_o^w \varphi_{pzc} \approx 0.020$ V [14] on the basis of the standard potential difference $\Delta_o^w \varphi_{TBA^+}^o = -0.248$ V [37] for the reference tetrabutylammonium cation. If the corrected value $\Delta_o^w \varphi_{TBA^+}^o = -0.275$ V [21] is used instead, the zero-charge potential difference becomes $\Delta_o^w \varphi_{pzc} = -0.007$ V, which is obviously very close to zero. The surface tension data for the system of NaBr in water and tetraalkylammonium tetraphenylborate in nitrobenzene also indicate that $\Delta_o^w \varphi_{pzc}$ is zero [11, 13].

The value of $\Delta_o^w \varphi_{pzc}$ can also be inferred from the capacitance data. For the MVN model [4] the Galvani potential difference can be written as:

$$\Delta_o^w \varphi = \Delta_o^w \varphi_i + \varphi_2^o - \varphi_2^w, \tag{34}$$

where $\Delta_o^w \varphi_i$ is the potential difference across the inner layer; φ_2^o and φ_2^w are the potential differences across the diffuse layers in the organic and aqueous phase, respectively. In the absence of the specific adsorption, the double-layer capacitance $C = dq^w/d\Delta_o^w \varphi$ is given by Eq. (33), where $C_i = dq^w/d\Delta_o^w \varphi_i$ is the inner-layer capacitance. The capacitance of the diffuse double layer can be evaluated using the Gouy-Chapman theory for a $z:z$

Table 2. The potential difference $\Delta_o^w \varphi_0$ corresponding to the minimum of the double-layer capacitance at the water/nitrobenzene interface in the presence of tetrabutylammonium tetraphenylborate in nitrobenzene and NaBr(I), MgSO$_4$(II) or LiCl(III) in water. Temperature 25°C [32]

	$\Delta_o^w \varphi_0$/mV		
$c^0(o)/M^a$	I	II	III
0.005	11	–	–
0.010	5	3	18 (15)[b]
0.020	4	– 8	– (25)
0.050	0	15	24 (29)
0.100	11	16	50 (45)

[a] For NaBr and LiCl electrolytes $c^0(w) = c^0(o)$, for MgSO$_4$ $c^0(w) = c^0(o)/2$
[b] From the capacitance data [17] using $\Delta_o^w \varphi_{TBA^+}^0 = -0.275$ V [21]

electrolyte [17]:

$$C_d^{-1} = C_{2-w}^{-1} + C_{2-o}^{-1}, \tag{35}$$

$$C_{2-o} = -\partial q^o/\partial \varphi_2^o = (zFA^o/RT)\cosh(zF\varphi_2^o/2RT), \tag{36}$$

$$C_{2-w} = -\partial q^w/\partial \varphi_2^w = (zFA^w/RT)\cosh(zF\varphi_2^w/2RT), \tag{37}$$

$$A^{o(w)} = (2RT\varepsilon^{o(w)}\varepsilon_0 c^0)^{1/2}, \tag{38}$$

where q^w and q^o are the surface charge density on the aqueous and organic solvent side of the interface, respectively. From the plot shown in Fig. 12 we can estimate $C_i \approx 0.8\,Fm^{-2}$ and, obviously at low electrolyte concentrations $C_i \gg C_d$ and, consequently, $C \approx C_d$. In such a case, the capacitance should have its minimum just at the zero surface charge.

The potential difference $\Delta_o^w \varphi_0$ corresponding to the minimum of the double-layer capacitance was found by a least square fit of the capacitance plot around the minimum to a third order polynomial [32]. The values of $\Delta_o^w \varphi_0$ are summarized in Table 2, and it is seen that they are rather close to zero. A slight shift of $\Delta_o^w \varphi_0$ towards the more positive potential differences, which is pronounced most for the aqueous LiCl solutions, can be ascribed to the variation of the inner-layer capacitance C_i with the surface charge density. On this basis the conclusion was made, that for all the systems studied, the zero-charge potential difference is $\Delta_o^w \varphi_{pzc} \approx 0$ V [32].

Inner-Layer Potential Difference and Capacitance

The dashed lines in Fig. 11 show the capacitance of the diffuse double-layer C_d calculated using the Gouy-Chapman theory from Eq. (35) for $\Delta_o^w \varphi_i = \text{const} = 0$. As it has been noticed [21], two trends are apparent on comparing the experimental capacitances with those calculated using the GC theory. First, at low surface charge

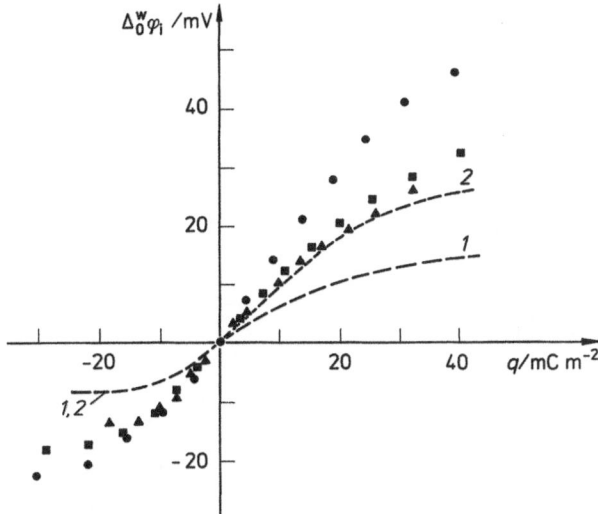

Fig. 13. Potential difference across the inner layer $\Delta_o^w\varphi_i$ at the water/nitrobenzene interface as a function of the surface charge density q^w on the aqueous side of the interface evaluated from the experimental data using non-iterative HNC results for the diffuse double layer ($\sigma_i^w = \sigma_i^o/2 = 0.425$ nm, $\varepsilon^w = 78.4$, $\varepsilon^o = 34.8$) at concentrations of NaBr in water and tetrabutylammonium tetraphenylborate in nitrobenzene: (▲) 0.02 M, (■) 0.05, and (●) 0.10 M. The dashed lines are from experimental data using the Gouy-Chapman theory for the diffuse double layer at concentrations (*1*) 0.05 M and (*2*) 0.10 M [32]

densities and higher electrolyte concentrations there is a drop in the experimental capacitance from the GC value, which increases with increasing electrolyte concentration. Such a drop can be due to the finite capacitance C_i of an inner layer at the water/nitrobenzene interface. For the system of NaBr in water and TBATPB in nitrobenzene, the inner-layer potential difference $\Delta_o^w\varphi_i$ and capacitance C_i were evaluated [21] using the GC theory (Eqs. (33–38)) from the surface charge density q obtained by integrating the plot of capacitance against the potential difference:

$$q^w = -q^o = \int_{\Delta_o^w\varphi_{pzc}}^{\Delta_o^w\varphi} C d\Delta_o^w\varphi. \tag{39}$$

The dashed lines in Fig. 13 show the inner-layer potential difference $\Delta_o^w\varphi_i$ as a function of the surface charge density. It is apparent that at a constant value of q, $\Delta_o^w\varphi_i$ is not independent of the electrolyte concentration and the same is true for the inner-layer capacitance C_i [21]. The second trend is obvious at higher surface charge densities, where the experimental capacitance tends to rise above the Gouy-Chapman value, which would correspond to the negative inner-layer capacitance C_i.

A recent theoretical analysis of the electrical double layer [39] has implied that the Gouy-Chapman theory overestimates the potential drop across the space charge region and underestimates the corresponding capacitance. This error becomes more pronounced as the ionic diameter and the charge number increase or the solvent dielectric permittivity decreases [39]. Since this may be the case for the ITIES, potential

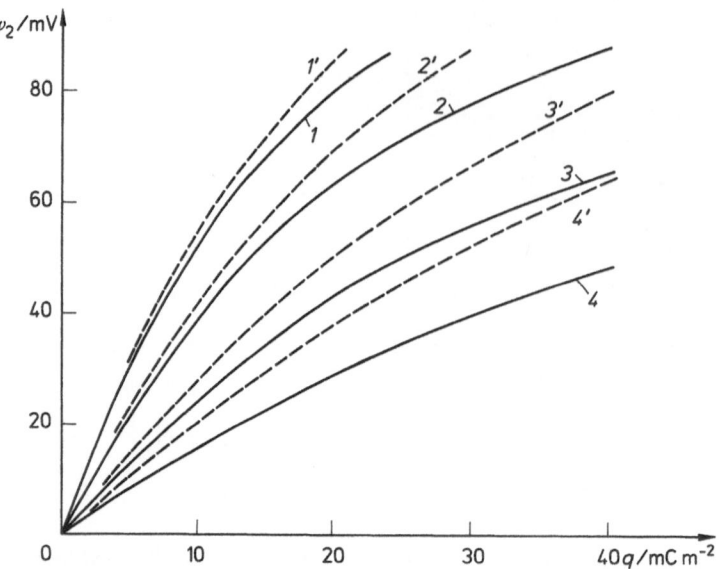

Fig. 14. Potential difference between the plane of closest approach and the bulk solution evaluated using the non-iterative HNC procedure [40] for the primitive electrolyte as a function of the surface charge density q at electrolyte concentration: (1, 1') 0.01 M, (2, 2') 0.02 M, (3, 3') 0.05 M and (4, 4') 0.10 M. The parameters: $z = 1$, $T = 298$ K, $\sigma_i^o = 0.850$ nm, $\varepsilon^o = 34.8$. The dashed lines: Gouy-Chapman theory [32]

drops across the space charge regions were evaluated [32] using the non-iterative procedure [40], which is based on the HNC equation for the primittive model (PM) of the electrolyte solution. The size effect is illustrated in Fig. 14 for the space charge region in the organic solvent phase. Using these results, the potential drop across the diffuse double layer $\varphi_2^o - \varphi_2^w$ and the capacitance C_d were found for a given surface charge density q, which was obtained by integrating the experimental capacitance plot, and from Eqs. (33) and (34) the inner-layer capacitance C_i and the potential drop $\Delta_o^w \varphi_i$, respectively, were evaluated [32]. Figure 11 shows the plots of $\Delta_o^w \varphi_i$ vs. q at various electrolyte concentrations for the system of NaBr in water and TBATPB in nitrobenzene. It is apparent that upon taking into account the ion size, the value of the inner-layer potential difference is roughly twice as large and the effect of the electrolyte concentration is reduced considerably as compared with the analysis based on the GC theory. Analogously, at a constant surface charge, the inner-layer capacitance C_i becomes almost independent of the electrolyte concentration. As shown in Fig. 15 the plot of C_i vs. q has a single minimum, the position of which depends on the nature of the electrolyte. Thus, upon taking into acount the size effect, the modified Verwey-Niessen model [4] seems to describe well the structure of the water/nitrobenzene interface. However, the inner-layer potential difference $\Delta_o^w \varphi_i$ is rather small and the corresponding inner-layer capacitance C_i is rather large as compared with the metal/electrolyte interface. This points to some penetration of the ions into the inner layer [21, 32] as a result of which the inner-layer potential difference is in part reduced. The ion penetration does not necessarily mean a specific ionic adsorption, but rather a less

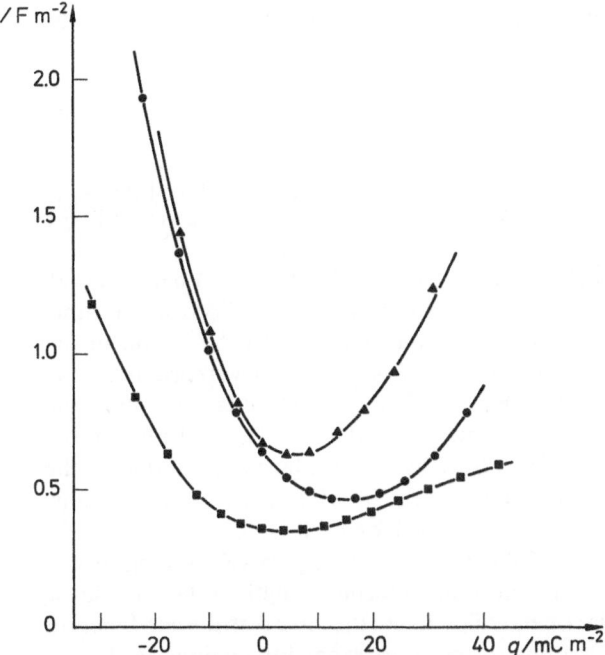

Fig. 15. Capacitance of the inner layer C_i at the water/nitrobenzene interface as a function of the surface charge density q^w on the aqueous side of the interface evaluated from the experimental data using the non-iterative HNC results for the diffuse double layer ($\sigma_i^w = \sigma_i^o/2 = 0.425$ nm, $\varepsilon^w = 78.4$, $\varepsilon^o = 34.8$). Aqueous phase: (●) 0.1 M NaBr, (▲) 0.1 M LiCl, (■) 0.05 M MgSO$_4$. Nitrobenzene phase: 0.1 M tetrabutylammonium tetraphenylborate (●, ▲, ■) [32]

restricted ionic movement close to the plane of the contact compared with the metal/electrolyte interface. Since the ions can enter the inner layer, its properties may depend on the nature of the electrolytes present in the system [32], cf. Fig. 15.

Close to the zero-charge potential difference the effect of the ion penetration on the interfacial capacitance can be estimated by solving the linearized Poisson-Boltzmann equations in all three regions of the MVN model [41]. For the sake of simplicity, it was assumed [32] that only the ions from the organic solvent phase enter the inner layer, so that their concentrations differ from zero at $x_2^w < x < \infty$, while they turn to zero at $x < x_2^w$, the x-axis being oriented towards the organic solvent phase. In that case, the double-layer capacitance C is [32]:

$$C = \varkappa^w \varepsilon^w \varepsilon_0 \left[\frac{(1+s)\exp(\varkappa^i \delta) + (1-s)\exp(-\varkappa^i \delta)}{(1+r)(1+s)\exp(\varkappa^i \delta) + (1-r)(1-s)\exp(-\varkappa^i \delta)} \right], \tag{40}$$

where $\delta = x_2^o - x_2^w$, $s = \varepsilon^i \varkappa^i / \varepsilon^o \varkappa^o$ and $r = \varepsilon^w \varkappa^w / \varepsilon^i \varkappa^i$. Equation (40) can be formally written as:

$$C^{-1} = C_d^{-1} + C_i^{-1} + \Delta, \tag{41}$$

with C_d^{-1} being given by Eq. (31) and the inverse capacitance of the inner layer C_i^{-1} given by:

$$C_i^{-1} = (\varepsilon^i \varepsilon_0)^{-1} \delta, \tag{42}$$

where δ is actually the inner-layer thickness. The parameter Δ, which accounts for the ion penetration, is then given by:

$$\Delta = (\varepsilon^i \varepsilon_0 \varkappa^i)^{-1} \left[\frac{(1-s^2)\tanh(\varkappa^i\delta)}{1+s\tanh(\varkappa^i\delta)} - \varkappa^i\delta \right]. \tag{43}$$

In the above equations, the superscript i refers to the inner-layer region. Apparently, without the ionic penetration $\varkappa^i = 0$, $s = 0$, $\Delta = 0$ and $C^{-1} = C_d^{-1} + C_i^{-1}$. For $\varkappa^i \neq 0$, the parameter Δ is negative and the inverse capacitance is reduced.

Using Eq. (41), the inverse capacitance was evaluated [32] for $\delta = 1.6$ mm, $\varepsilon^i = 25$ and $\varkappa^i = (\varepsilon^0/\varepsilon^i)\varkappa^0$, which corresponds to an extension of the Boltzmann ionic distribution from the region of the dielectric permittivity ε^0 into the region of the permittivity ε^i. At higher electrolyte concentrations, the theory reproduces the experimental data well, cf. the dot-and-dashed line in Fig. 12, but it fails to follow the drop in the inverse capacitance at low concentrations. If the screening length in the inner layer is estimated as f. $\varkappa^i = (\varepsilon^0/\varepsilon^i)^{1/2}\varkappa^0$ with f being the adjustable parameter, the theoretical fit for the experimental capacitance of the interface between the aqueous NaBr and nitrobenzene TBATPB solutions is reached for $f = 0.41, 0.72, 1.00, 1.05, 1.25$ or 1.5 at $c^0(w) = c^0(o) = 0.005, 0.01, 0.02, 0.05, 0.1$ or $0.2\,M$, respectively. Qualitatively, this would correspond to a slower variation of the screening length in the inner layer than in the space-charge region. In fact, a modification of the ionic distribution function close to the interface can be expected due to the short-range interactions [42].

The properties of the water/nitrobenzene interface at low electrolyte concentrations are somewhat surprising, since one would expect that they can be explained in terms of the Gouy-Chapman theory. On the contrary, the experimental capacitance is much higher than the GC value. It is noteworthy that this effect is even more pronounced for the water/1,2-dichlorethane interface [38]. In spite of the fact that the inner-layer structure at the water/1,2-dichlorethane interface is much less clear due to the extensive ion association in the organic solvent phase, the drop from the GC prediction seems to increase with the decreasing dielectric permittivity of the organic solvent.

References

1. Verwey, E.J.W., Niessen, K.F.: Philos. Mag. 28, 435–446 (1939)
2. Gouy, G.: C. R. Acad. Sci. 149, 654 (1910)
3. Chapman, D.L.: Philos. Mag. 25, 475 (1913)
4. Gavach, G., Seta, P., d'Epenoux, B.: J. Electroanal. Chem. 83, 225–235 (1977)
5. Krylov, V.S., Myamlin, V.A., Boguslavsky, L.I., Manvelyan, M.A.: Elektrokhimiya 13, 834–840 (1977)
6. Gugeshashvili, M.I., Manvelyan, M.A., Boguslavsky, L.I.: ibid. 10, 819–822 (1974)
7. Joos, P., Vanden Bogaert, R.: J. Colloid. Interface Sci. 56, 206–212 (1976)
8. Kahleweit, M., Strehlow, H.: Z. Elektrochem. 58, 658 (1954)
9. Boguslavsky, L.I., Frumkin, A.N., Gugeshashvili, M.I.: Elektrokhimiya 12, 856–860 (1976)
10. Seta, P., d'Epenoux, B., Gavach, C.: J. Electroanal. Chem. 95, 191–199 (1979)
11. Reid, J.D., Melroy, O.R., Buck, R.P.: ibid. 147, 71–82 (1983)
12. Girault, H.H., Schiffrin, D.I.: ibid. 150, 43–49 (1983)
13. Gros, M., Gromb, S., Gavach, C.: ibid. 89, 29–36 (1978)
14. Kakiuchi, T., Senda, M.: Bull. Chem. Soc. Jpn. 56, 1753–1760 (1983)
15. Kakiuchi, T., Senda, M.: ibid. 56, 1322–1326 (1983)

16. Girault, H.H., Schiffrin, D.J.: J. Electroanal. Chem. *170*, 127–141 (1984)
17. Samec, Z., Mareček, V., Homolka, D.: ibid. *126*, 121–129 (1981)
18. Hájková, P., Homolka, D., Mareček, V., Samec, Z.: ibid. *151*, 277–282 (1983)
19. Homolka, D., Hájková, P., Mareček, V., Samec, Z.: ibid. *159*, 233–238 (1983)
20. Mareček, V., Samec, Z.: ibid. *149*, 185–192 (1983)
21. Samec, Z., Mareček, V., Homolka, D.: Faraday Discuss. Chem. Soc. 77, 197–208 (1984)
22. Reid, J.D., Vanýsek, P., Buck, R.P.: J. Electroanal. Chem. *161*, 1–15 (1984)
23. Reid, J.D., Vanýsek, P., Buck, R.P.: ibid. *170*, 109–125 (1984)
24. Osakai, T., Kakutani, T., Senda, M.: Bull. Chem. Soc. Jpn. *57*, 370–376 (1984)
25. Girault, H.H.J., Schiffrin, D.I.: J. Electroanal. Chem. *161*, 415–417 (1984)
26. Geblewicz, G., Figaszewski, Z., Koczorowski, Z.: ibid. *177*, 1–12 (1984)
27. Le Hung, Q.: ibid. *115*, 159–174 (1980)
28. Kakiuchi, T., Senda, M.: Bull. Chem. Soc. Jpn. *56*, 2912–2918 (1983)
29. Figaszewski, Z.: J. Electroanal. Chem. *139*, 309–315 (1982)
30. Samec, Z., Mareček, V., Weber, J.: ibid. *100*, 841–852 (1979)
31. Mareček, V., Samec, Z.: ibid. *185*, 263–271 (1985)
32. Samec, Z., Mareček, V., Homolka, D.: ibid. *187*, 31–51 (1985)
33. Kakutani, T., Osakai, T., Senda, M.: Bull. Chem. Soc. Jpn. *56*, 991–996 (1983)
34. Melroy, O.R., Bronner, W.E., Buck, R.P.: J. Electrochem. Soc. *130*, 373–380 (1983)
35. Silva, F., Moura, C.: J. Electroanal. Chem. *177*, 317–323 (1984)
36. Berzins, T., Delahay, P.: J. Am. Chem. Soc. 77, 6448–6453 (1955)
37. Koryta, J., Vanýsek, P., Březina, M.: J. Electroanal. Chem. 75, 211–228 (1977)
38. Samec, Mareček, V., Holub, K., Hájková, P.: J. Electroanal. Chem. (in press)
39. Henderson, D., Blum, L., Lozada-Casson, M.: J. Electroanal. Chem. *150*, 291–303 (1983)
40. Henderson, D., Blum, L.: ibid. *111*, 217–222 (1980)
41. Vorotyntsev, M.A.: Kornyshev, A.A.: Elektrokhimiya *20*, 3 (1984)
42. Outhwaite, C.W., Bhuigan, L.B., Levine, S.: J. Chem. Soc. Faraday Trans. II 76, 1388–1408 (1980)

Redox and Photochemical Reactions at the Interface Between Immiscible Liquids

L. I. Boguslavsky and A. G. Volkov

I. Introduction

One of the important specific features of the interface between immiscible liquids – ionic conductors – is the fact that, for the reaction of electron transfer across the interface to occur, electron donors and acceptors must be present in the contacting phases. Artificial systems of immiscible liquids in which redox processes occur are as yet only a subject of investigation [1–3]. In nature, however, the redox reactions occurring in the membrane/electrolyte system are quite widespread [4]. For instance, the energy conversion processes in the respiratory chain of mitochondria or energy storage in photosynthesis represent a chain of redox transformations taking place on a mitochondrial or photosynthetic membrane. From an electrochemical point of view, the membrane on which redox transformations take place represents for the electron a thin insulating layer consisting of two monolayers of lipids oriented with their hydrophobic tails towards each other. Electron donors or acceptors can be immersed in this lipid matrix, as well as the enzymes catalyzing the redox transformations [1]. As a rule, the attempts to investigate experimentally the redox transformations involving protein enzyme systems are faced with difficulties since an artificial bilayer is not everywhere capable of withstanding mechanically the adhesion to its surface of high-molecular enzyme systems. Nevertheless, on the bilayers it is possible to effect redox transformations involving both relatively low-molecular compounds and membrane protein complexes [5–8]. If, however, we investigate only one interface to exclude charge transfer through the bulk of the membrane, then with a proper choice of components it is possible not only to observe redox reactions involving relatively low-molecular compounds, but also to carry out processes with enzymes isolated from native objects serving as catalysts [9–14]. Regardless of the type of reactions carried out, a common feature of two-phase systems is the possibility to vary widely (within several orders) the equilibrium constant value of chemical and electron-exchange reactions by changing the nature of the organic phase and the ratio between the aqueous and organic phase volumes, and hence to shift the equilibrium of the thermodynamically unfavorable reactions in the direction of the desired product. At the same time a catalyst or enzyme can be localized in an optimum manner, i.e. be situated in a surrounding favorable to their functioning. For enzymes this feature is of particular importance since it can prevent its inactivation by the organic solvent, or denaturation due to an irreversible transformation, as is often the case at the water/air interface.

The Interface Structure and Electrochemical Processes
at the Boundary Between Two Immiscible Liquids
Editor: V. E. Kazarinov
© Springer-Verlag Berlin, Heidelberg 1987

Broadly, the redox reactions at the interface between immiscible liquids may be divided into two types. The first type includes spontaneous processes. It is precisely this type of redox transformation that was investigated in the vast majority of works on bioenergetics [4, 15] as well as in model systems of membranes [16, 17] and at oil/water interfaces [18–20]. The redox transformations of the second type were realized at the interface between immiscible electrolytes by applying a controlled voltage to this interface [21]. In the present study, the greatest attention will be given to spontaneous processes.

The occurrence of a redox reaction in the oil/water system is primarily established analytically by identification of the reaction products, often contained in different phases. Other indirect data, however, are also used, such as changes in the interfacial tension at the interface between immiscible liquids [22] or in the Volta potential [22–29]. A change in the transmembrane potential may also be an indication of a redox reaction at the membrane/electrolyte interface [6, 7, 26–28].

The reactions between the components adsorbed at the interface in the monolayer belong to a special type that may be conditionally considered along with heterogeneous reactions. It is possible, however, that some of the truly heterophase processes catalyzed by enzymes or inorganic catalysts at the interface include the reactions in monolayers.

II. Redox Reactions in Monolayers

Upon illumination of a mixed monolayer of chlorophyll with ferridoxine in the presence of ascorbate, Brody and Owens [22] detected a slight change in the isotherm, which in their opinion is an indication of a reaction between these components. Upon illumination of mixed films of chlorophyll and oxidized cytochrome c, a reduction in the mean area per molecule by about 18% is observed. On the other hand, illumination of films of chlorophyll with reduced cytochrome increases the area per molecule significantly (by about 10%). The difficulty of interpreting very small effects lies in the fact that in the dark the area per molecule increases both for the films with reduced cytochrome c and for the mixture containing reduced cytochrome. At any rate, for the case with reduced cytochrome c, Brody and Owens believe that they observed an electron-transfer reaction stimulated by light:

$$CHl + (Cyt c)_{ox} \overset{h\nu}{\rightleftarrows} (CHl)_{ox} + (Cyt c)_{red}. \tag{1}$$

Investigations of mixed monolayers of chlorophyll with plastocyanine revealed some slight reversible changes of the Volta potential. The magnitude of the effect depended on the illumination time and possibly points to the presence of an electron-exchange reaction between chlorophyll and plastocyanine:

$$CHl + (PC) \rightleftarrows CHl^- + (PC)^+. \tag{2}$$

III. Redox Processes in the Oil/Water System when Donor and Acceptor are Contained in Different Phases

1. Evidence for the Occurrence of the Process

Long before the processes at the interface between immiscible liquids began to be studied systematically, Bell [29] observed an oxidation reaction of benzoyl-*o*-toluidine by permanganate to benzoylanthranilic acid:

$$
\underset{\text{CH}_3}{\overset{\text{NHCOPh}}{\bigcirc}} \quad \xrightarrow{\text{KMnO}_4} \quad \underset{\text{COOH}}{\overset{\text{NHCOPh}}{\bigcirc}} \tag{3}
$$

The reaction occurred at the water/benzene interface and its kinetics did not depend on the rate of stirring of each of the phases.

Lately, this entirely novel type of heterogeneous process, the so-called "Phase transfer catalysis" [2], has been widely used by chemists for preparative purposes. Here belong a number of other reactions for which the fact of the occurrence of an oxidative transformation has been established but the factors affecting the process rate have not been examined [30]. Some progress in this direction was made in the investigation of a simple redox reaction between hydrophobic porphyrin* dissolved only in octane and a hydrophilic donor – sodium dithionite – dissolved only in water [31, 32].

2. Influence of Specific Adsorption of Halogen Ions on the Reduction of Hydrophobic Porphyrin [31, 32]

In the absence of specific adsorption at the interface between immiscible electrolytes the potential drop occurs mainly in two diffuse layers contained in different phases. True enough, in the octane/water system there are practically no ions in the hydrocarbon phase in a concentration sufficient to screen the adsorption potential at the interface. Due to this fact, in investigating the processes in the electric double layer, it is possible to use the Volta potential measurements for determining the surface potential. The surface potential ϕ_s is known to consist of two terms:

$$
\varphi_s = \varphi_d + \varphi_G. \tag{4}
$$

The first term of this sum is the potential due to the dipole moment of the adsorbed particles which yields a potential determined by the Helmholtz formula:

$$
\varphi_d = \frac{4\pi n_s \mu_{\text{eff}}}{\varepsilon}, \tag{5}
$$

where n_s is the concentration of adsorbed particles, μ_{eff} is the vertical component of the effective dipole moment, and ε the static permitivity of the medium surrounding the dipole. The second term in Eq. (1) is associated with the presence of a diffuse layer of ions and is determined by the well-known Gouy-Chapman equation. For charged

---

* The iron complex 2,7,12,17-tetramethyl-3,13-octadecyl-8,18-bi(carbometaxyethyl)porphyrin (FeP).

monolayers:

$$\frac{d\varphi_s}{d \lg c} = \frac{d\varphi_G}{d \lg c} = 58 \text{ mV} \tag{6}$$

assuming the area per molecule and the vertical component of the effective dipole moment of adsorbed particles to be unchanged [33].

It is known that iron complexes of porphyrins are salts in which the metal complexes act as cations. When dissolved in dry hydrocarbon, these salts are undissociated. At the interface with water or in "wet" octane, where water is, presumably, present as a microemulsion, hydrolysis takes place:

$$[FeP]^+Cl^- + H_2O \rightleftarrows [FeP]OH + HCl. \tag{7}$$

The reaction product, however, can undergo dimerization to form a complex:

$$2[FeP]OH \rightleftarrows FeP\text{--}O\text{--}FeP + H_2O. \tag{8}$$

Comparison of the adsorption spectra of FeP in dry and wet octane showed that hydrolysis transforms the FeP^+ spectrum into FePOH or FeP–O–FeP spectra. Acidification of the solution with hydrochloric acid shifts the reaction equilibrium to the left and again the initial form is obtained. In order to determine in which form porphyrins exist at the interface – as FePOH or as FeP–O–FeP – Volta potentials were measured during porphyrin adsorption at the interface and the influence of the ionic force of the subsolution on the surface potential at equal aqueous solution pH was investigated. At alkaline pH the potential shift did not change with increasing supporting electrolyte concentration. This means that φ_d makes the main contribution to the surface potential. In a more acid region at pH 6.5, the electric double layer is made up of positively charged FeP^+ complexes on the octane side and anions on the side of water. The slope $d\varphi_s/d \lg c_{KCl}$ is 55–60 mV per tenfold change of the salt concentration in the range from 10^{-4} M to 10^{-1} M KCl.

In order to find out whether the anions in the aqueous phase contribute to the dipole potential shift, Volta potentials were measured in the system:

$$\text{Au}\left|\text{air}\right|\begin{array}{c}\text{octane hydrophobic}\\ \text{metalloporphyrin}\end{array}\left|\begin{array}{c}\text{water}\\ \text{sodium dithionite}\end{array}\right|\begin{array}{c}\text{reference}\\ \text{electrode}\end{array} \tag{9}$$

when the salts KCl, KBr, KI are added. If it is assumed that no specific adsorption of halogens occurs at the octane/water interface, then φ_s should not depend on the ion nature. It proved, however, that φ_s shifts into the range of negative values in the series Cl^-, Br^-, I^- due to specific adsorption of halogens at the interface. This result was obtained qualitatively also for the system water/dichloroethane containing LiCl solutions in the aqueous phase and TBA TPB in dichloroethane [34]. When FeP, in concentrations by several orders lower than that of the supporting electrolyte, is added to this system, the point of zero charge shifts in the negative direction, which points to adsorption of metalloporphyrin in the compact part of the electric double layer. The shift of the zero-charge potential depends on the porphyrin concentration and the nature of the metal-forming part of the complex.

When dithionite is added to water, FeP undergoes reduction and the Volta potential shifts in the negative direction. A more thorough investigation of the FeP

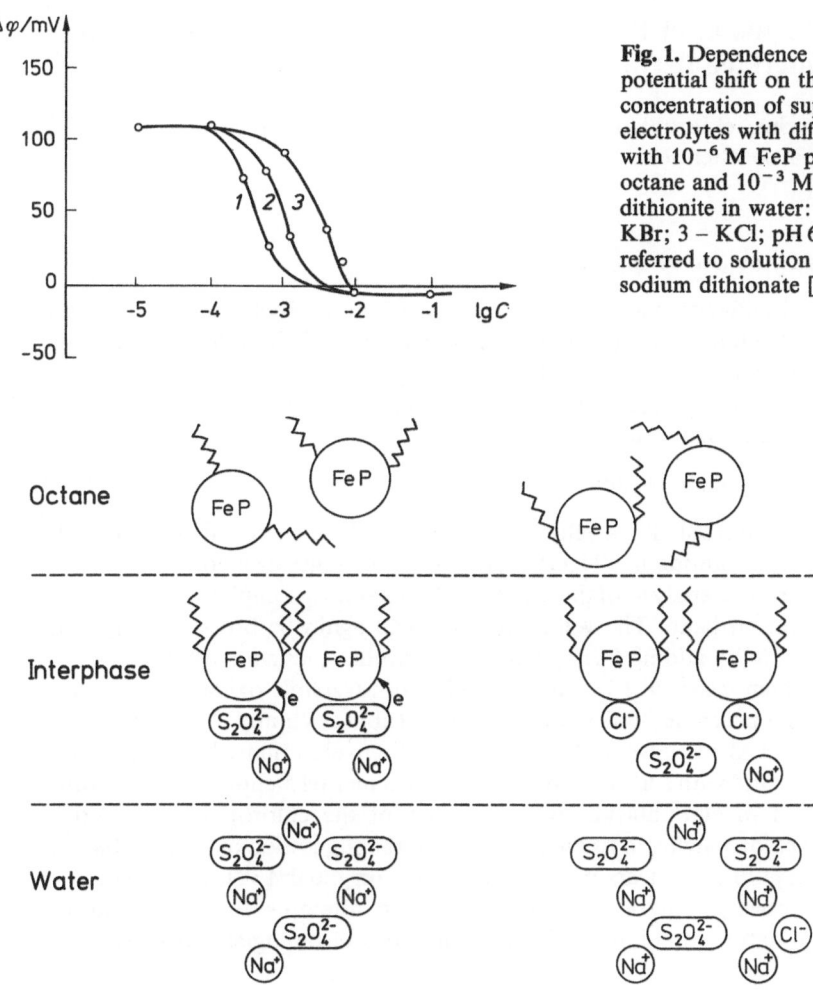

Fig. 1. Dependence of the potential shift on the concentration of supporting electrolytes with different halogens with 10^{-6} M FeP present in octane and 10^{-3} M sodium dithionite in water: 1 – KI; 2 – KBr; 3 – KCl; pH 6.5. Potential referred to solution without sodium dithionate [31]

Fig. 2. Schematic diagram of the adsorption layer structure at the octane/water interface: **a** in the absence of halogens; **b** in the presence of specifically adsorbed ions [32]

reduction reaction showed it to occur only at a low supporting electrolyte concentration. The inhibition of the porphyrin reduction reaction depended not only on the ionic force of the aqueous solution, but also on the nature of the supporting electrolyte anion. I^- anions, which are known to have a higher adsorbtivity, inhibit the reaction at lower concentration than Br^- anions do, the latter being more effective than Cl^- anions (Fig. 1). To transfer the electron to the adsorbed porphyrin, the dithionite anion must approach it quite closely occupying the place on the aqueous phase side. The cessation of the porphyrin reduction with increasing halide concentration implies that halogen anions displace dithionite anions from the adsorption layer. This process is shown schematically in Fig. 2.

IV. Metalcomplexes of Porphyrins – Catalysts of Redox Reactions at the Interface Between Immiscible Liquids

Among porphyrins, chlorophyll – a derivative of chlorin (dihydroporphyrin-7,8) attracts the greatest attention.

1. Redox Reactions Involving Chlorophyll

Timiryazev has already suggested that participation of chlorophyll in photosynthesis is due to its reversible redox transformations. Excited chlorophyll molecules can act as electron acceptors and donors [35, 36].

Rabinovich and Weiss [35, 36] investigated a homogeneous dark redox reaction of chlorophyll with ferric chloride:

$$CHL + Fe^{3+} \rightleftarrows CHL + Fe^{2+} , \tag{10}$$

where CHL is oxychlorophyll – a yellow oxide form of the pigment. When ferric salts are added to a CHL solution in CH_3OH, the solution changes its color from green to yellow. A spectroscopic analysis of the products shows almost complete disappearance of the red adsorption band. The yellow solution can again be brought back to its previous green state by adding ferric chloride or another reducing agent.

Krasnovsky demonstrated [37] the possibility of reversible oxidation of chlorophyll with benzoyl peroxide in alcoholic solution. Rabinovich and Weiss found that chlorophyll can act as a catalyst in the dark. They discovered that when ethylchlorophyllide is left to stand with a ferric chloride solution in methyl alchohol, ferric chloride is slowly reduced in amounts far exceeding that of the chlorophyll in solution. Investigation of chlorophyll and its analogues led to the suggestion of the possible role of porphyrins as photosensitizers in the origin of life on Earth [38]. Investigation of heterogeneous processes marks a new step in the attempts to understand the physicochemical nature of chlorophyll transformations in biological systems.

2. Adsorption of Chlorophyll at the Oil/Water Interface

An asymmetric chlorophyll molecule consists of a hydrophilic "head" formed by four pyrrole rings arranged around magnesium and a long "tail" – a hydrophobic chain (phytol residue). The hydrophilic head seems to be turned into water and the hydrophobic tail attaches chlorophyll to the nonaqueous phase.

A number of studies are concerned with the adsorption of chlorophyll in monolayers at the water/nitrogen and water/heptane interfaces [39–41]. Comparison of the areas per chlorophyll molecule for the nitrogen/water and heptane/water interface shows that at the interface with a hydrocarbon, the molecules are less inclined toward the interface. Upon illumination of the interface, the area per chlorophyll molecule changes slightly [41].

During chlorophyll adsorption at the octane/water interface, the interfacial tension changes markedly. This fact was utilized in determining the surface excess of chlorophyll by the Gibbs equation. From the slope of the initial section of the isotherm, and independently of the calculations by Frumkin's isotherm

$$Bc = \frac{\theta}{1-\theta} e^{-2a\theta} \tag{11}$$

the adsorption equilibrium constant $B = 9 \times 10^5$ l/mole and the attraction constant $a = 0.4$ were calculated. The sign and value of the constant point to a strong attractive interaction between chlorophyll molecules. This effect is in agreement with the well-known capacity of chlorophyll molecules for aggregation. Frumkin's adsorption isotherm describes very well the adsorption of chlorophyll at the octane/water interface at the coverages 0.9. With the use of the equation:

$$B = (1/55.5) \exp\left(-\frac{\Delta G^0}{RT}\right) \tag{12}$$

the free adsorption energy of chlorophyll at the octane/water interface was calculated as $\Delta G = -12.5$ kcal/mole. From the data in Refs. [42, 43] one can envisage the arrangement of chlorophyll molecules at the interface. With the number of particles per 1 cm^2 of interface at $\theta = 1$ known, it is possible to calculate the value of the projection of the area per molecule, which for chlorophyll was 53 Å^2. Comparison of the projection of the area per molecule with the area of the chlorophyll molecule itself permitted the scope of the plane of the chlorophyll molecule at the interface to be estimated. The slope is about 60°, which is close to the value obtained by optical methods both for chloroplast membranes and for chlorophyll incorporated into BLM [44].

3. Redox Reactions Catalyzed by Chlorophyll in the Oil/Water System

If a hydrophobic electron acceptor is added to a nonaqueous phase and a reducing agent to water, then the Volta potential measured in the chain:

$$\text{Au} \left| \text{air} \left| \begin{array}{c} \text{octane} \\ \text{electron acceptor} \\ \text{chlorophyll} \end{array} \right| \begin{array}{c} \text{water} \\ \text{NADH} \end{array} \right| \begin{array}{c} \text{reference} \\ \text{electrode} \end{array} \tag{13}$$

shifts in the negative direction due to the occurrence of an electron-exchange reaction at the interface catalyzed by chlorophyll:

$$(\text{NADH})^w + (\text{A})^o \xrightleftharpoons{\text{CHL}} (\text{NAD}^+)^w + (\text{H}^+)^o + (\text{A}^{2-})^o. \tag{14}$$

Here A is one of the electron acceptors used in the investigation of the catalytic properties of metalloporphyrins (e.g., 2N-methylamino-1,4-naphthoquinone). When the steady state is reached, the charge flows from water to octane and from octane to water controlled by the rate of the electron-exchange reaction at the interface; these flows are equal and therefore the total charge flow across the interface is zero. In the case of a reaction that alters the concentration of particles in the nonaqueous phase, the interfacial potential shift must change by a quantity that can be negative or positive depending on the sign of injected charges. Such an effect can be explained by the fact that during transfer of charged particles by a catalyst, the sign of the part of the electric double layer that lies in octane changes. The magnitude of the effect in question does

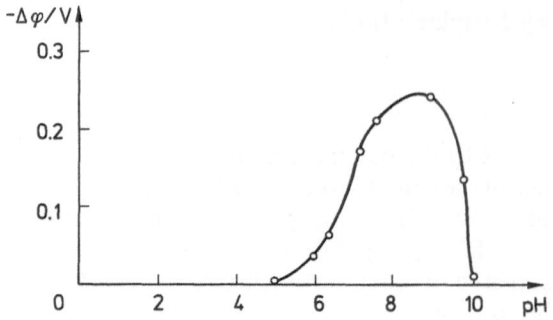

Fig. 3. Dependence of the potential shift at the octane/water interface on the pH of an aqueous solution. Medium: 5 g/ml chlorophyll, 20 mM tris-HCl, 5×10^{-4} M NADH, 10^{-5} M MANQ [45]

not depend on how the redox reaction was started: whether by the substrate NADH, chlorophyll or by the electron acceptor.

The dependence of the potential shift at the octane/water interface on the aqueous solution pH is shown in Fig. 3 [45]. The injection of electrons from water into octane proved to have an optimum in the pH range 6.5–9.1 and to be inhibited in acid and alkaline regions. The inhibition of the electron transfer reaction in the acid region may be due to chlorophyll losing magnesium to give pheophytin. The inhibition in the alkaline region may result from the fact that at pH = 10 the carbocyclic ring is ruptured, and salt of tribasic acids of chlorophyllins are formed since simultaneously ester links undergo hydrolysis [46].

4. Formation of the Boundary Layer Enriched in Protons

The observed Volta potential does not change at a sufficiently large buffer capacity of the solution (from 5 to 100 ml tris-HCl). This indicates that the change of the potential shift at the octane/water interface results from electron transfer across the interface and not from the pH change in the boundary layer. As shown in Refs. [8, 47], however, during a redox reaction on BLM incrusted with chlorophyll, a layer adjacent to the

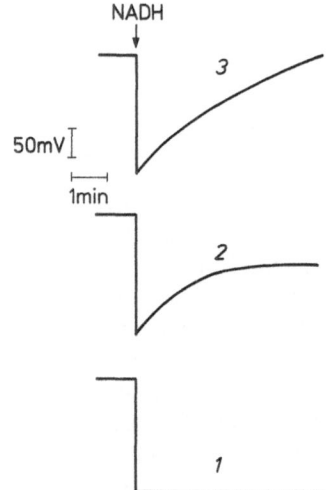

Fig. 4. Dependence of the Volta potential change in chain (15) on incubation time of the reaction mixture. Composition of medium: 10^{-5} M chlorophyll, 2×10^{-4} M NADH; 10^{-4} M vitamin K_3 and tris-HCl: *1* – 10 mM; *2* – 1 mM; *3* – 0 mM; pH 6.5 [46]

membrane is formed with a proton concentration different from that in the bulk [47, 48, 49]. Therefore, in Refs. [48, 49] the possibility of recording the formation of boundary layers in the octane/water system was looked into. As it follows from Eq. (14), a charge-transfer reaction must involve proton ejection into the aqueous phase. Figure 4 [46] shows the dependence of the Volta potential on the incubation time of the reaction mixture at different tris-HCl buffer concentrations. The aqueous phase pH was so chosen that acidification of the boundary layer was to result in inactivation of chlorophyll due to its pheophytinization. Indeed, as can be seen in Fig. 4, at low tris-HCl concentrations $\Delta\varphi$ decreases at $pH \leq 6.5$. In the buffered solution this effect is absent.

The value of the Volta potential being investigated does not depend on the manner the redox reaction is stated: by the substrate NADH in the aqueous phase, by chlorophyll at the interface, or by the electron acceptor in the nonaqueous phase. No changes in Volta potentials were observed when one of the components of the reaction under investigation was excluded.

5. Redox Reactions Catalyzed by Other Metalloporphyrins

The prosthetic groups of many membrane enzyme systems contain metalloporphyrins. Therefore, it is not surprising that an attempt has been made to reveal and study the catalytic effect of porphyrins during redox processes at the interface between immiscible liquids. Complexes of different metals with different porphyrins* exhibited catalytic activity in the oil/water system [50]. First of all, let us consider the transformations of metal complexes of porphyrins in the octane/water system. In wet octane, as a result of the hydrolysis and dimerization reactions, FeEP changes to a μ-complex and this results in a change in the absorption spectrum (Fig. 5). With a strong acidification of the aqueous phase, the equilibrium of Eqs. (12) and (13) shifts to the left and at pH 1 the positions of the maxima on the absorption spectra of ethioporphyrin in dry and wet octane coincide.

---

* EP – ethioporphyrin II, CP – iron complex of tetramethyl ether of coproporphyrin III, NADH – nicotinamine adenine dinucleotide, reduced.

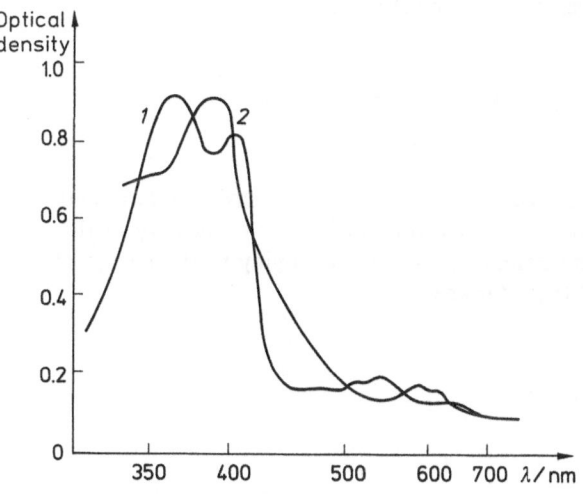

Fig. 5. Absorption spectra of FeEP in octane: I – in dry octane; 2 – in wet octane; pH 7.5 [50]

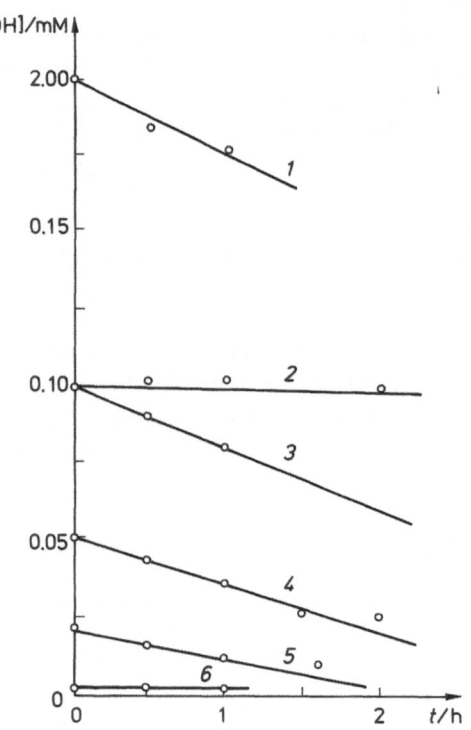

Fig. 6. Dependence of NADH concentration on incubation time of an octane/water system at different initial NADH concentrations (indicated) in argon atmosphere – 1–6. Medium: 10^{-6} M FeEP; 10^{-4} M vitamin K_3; 20 mM tris-HCl; pH 7.5. Control without FeEP-2 [50]

When FeEP is introduced into octane and NADH into water, FeEP does not undergo reduction but the absorption maximum of FeEP in the core region shifts towards the long-wavelength region by 3 nm, which testifies to the introduction of FeEP and NADH molecules. In the full system containing the electron donor NADH, the catalyst FeEP as well as the electron acceptor Vitamin K_3 or 1,4-naphthaquinone, it proved possible to record spectrally the oxidation of NADH (Fig. 5) and the reduction of the electron acceptor catalyzed by metalloporphyrins. The NADH concentration in the full system decrease with time (Fig. 6).

V. Evidence for the Heterogeneity of Redox Reactions Catalyzed by Metalcomplexes of Porphyrins

While, in general, the occurrence of redox reactions can be easily determined, additional evidence is required in order to prove the possible heterogeneity of this process. Therefore, this section will investigate some facts testifying to the nature of redox reactions occurring directly at the interface.

1. Adsorption of Catalyst at the Interface

The adsorption of catalyst at the interface indicates that a reaction must occur there. As shown by the Volta potential measurements in the chain:

$$\text{Au} \left| \text{air} \left| \begin{array}{c} \text{octane} \\ \text{catalyst} \end{array} \right| \text{water} \left| \begin{array}{c} \text{water} \\ \text{KCl, sat.} \end{array} \right| \text{Hg}_2\text{Cl}_2, \text{Hg} \right. \tag{15}$$

FeP is adsorbed at the octane/water interface and produces a positive potential shift with respect to water as large as 0.46 V (Fig. 7). The adsorption of surface-active catalysts, such as coproporphyrin (CP) [51, 52], at the octane/water interface has been studied. From the interfacial tension data, with the use of the Gibbs adsorption equation:

$$\Gamma = -\frac{1}{RT} \frac{d\sigma}{d \ln c} \tag{16}$$

the adsorption isotherm was plotted (Fig. 8). From the slope of the initial section of the isotherm the adsorption equilibrium constant was estimated as $B = 1.2 \times 10^5$ l/mole; the concavity of this section indicates an attractive nature of the interaction of adsorbed particles. Knowing the number of CP particles per 1 cm^2 of interface at the limiting coverage $\Gamma = \Gamma_{max}$, it is possible to calculate the limiting area per one CP

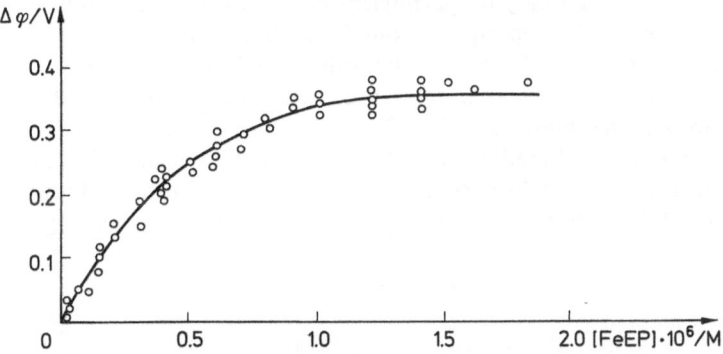

Fig. 7. Dependence of the Volta potential measured in chain (8) on FeEP concentration. The figure shows the results of 10 experiments obtained on different FeEP preparations. Medium: octane/water, pH 6.5 [50]

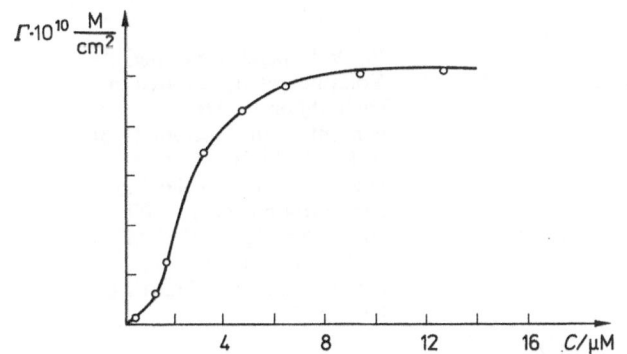

Fig. 8. Adsorption isotherm of CP at the octane/water interface [51]

molecule (32.4 Å²). Comparison of the limiting area per molecule with the area of the CP molecule itself permits the slope angle of the plane of the CP molecule to the octane/water interface to be estimated. With total surface coverage by CP molecules, this angle is equal to about 65°, which is very close to the slope angles of other porphyrins at the membrane/water interface in monolayers and at the water/air interface, as determined by optical methods [39, 53].

2. Cause of the Potential Shift and its Proportionality to the Concentration of Adsorbed Catalyst

First of all, it should be pointed out that the presence of an electron-exchange reaction was also recorded from a change of the interfacial potential shift [9, 45–60]. It can be seen in Fig. 9 [50] that when NADH is added into the cell containing Vitamin K_3 and FeEP, a potential shift arises, reaching -0.45 V with respect to the solution without NADH. The change of the potential shift depends on the NADH concentration, and the limiting potential value remains constant with increasing buffer capacity of solution from 5 to 100 mM. This proves conclusively that the Volta potential change is caused by the electron-exchange reaction at the interface rather than by the pH change in the boundary layer.

A similar effect of the appearance of a negative potential shift is also observed when FeEP is replaced by $Co^{3+}EP$ or by metalless porphyrins. This means that a redox reaction at the interface is catalyzed only by porphyrins having as a central atom a transition metal which is capable of accepting electrons. The influence of the nature of the central atom in the porphyrin on the Volta potential value and the reaction rate was also observed for other redox reactions at water/octane, water/decane, water/chloroform and water/dichloroethane interfaces [57–59].

If in the presence of CP and 2N-methylamino-1,4-naphthaquinone (MANQ) in octane, a reducing agent such as NADH or ascorbate is added to the aqueous phase; then a negative potential shift is observed caused by the catalytic electron-exchange reaction (Fig. 10) [51].

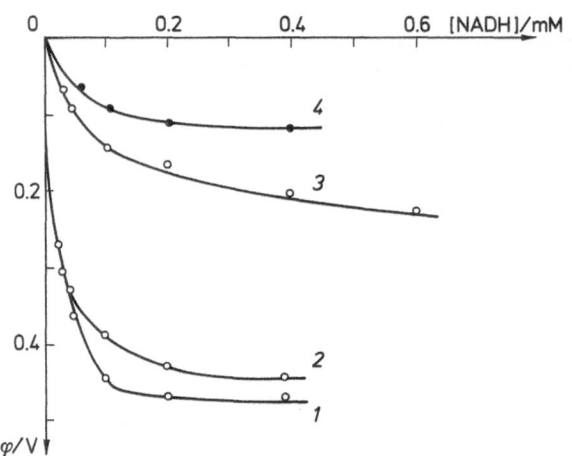

Fig. 9. Dependence of the Volta potential measured in chain (8) on NADH concentration. Medium: (1–3) – 10^{-6} M FeEP; 10^{-4} M vitamin K_3 and tris-HCl in concentrations of: *1* – 100; *2* – 20, *3* – 5 mM; *4* – 10^{-6} M Co^{3+}; 10^{-4} M 1,4-naphthoquinone, 20 mM tris-HCl; ph 7.7 [50]

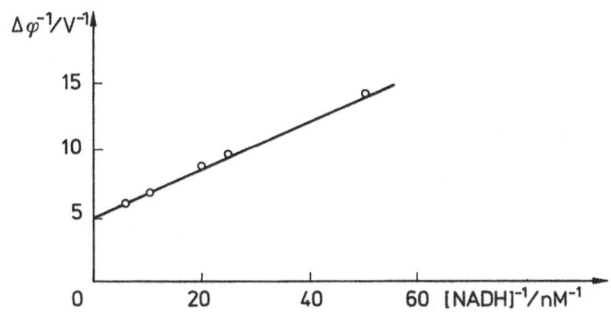

Fig. 10. Dependence of the Volta potential measured in chain (8) on NADH concentration in reciprocal coordinates. Medium: 10^{-4} M 2N-methylamino-1,4-naphthoquinone, 7×10^{-6} M CP, 10^{-2} M tris-HCl; pH 7.3. Potential referred to solution without NADH [52]

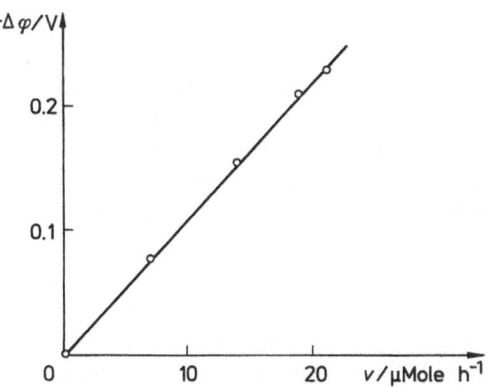

Fig. 11. Dependence of the Volta potential measured in chain (8) on the NADH consumption rate in full system during the first incubation hour. Medium: 10^{-6} M FeEP; 10^{-4} M vitamin K_3; 20 mM tris-HCl [54]

The dependence of surface potential and reaction rate on the NADH concentration are similar (Fig. 11), and the potential shift in the given system is proportional to the oxidation rate of NADH. The obtained result agrees well with the theoretical model of a heterogeneous catalytic reaction proposed earlier [23–25], according to which the potential shift at the interface between immiscible liquids relative to the supporting electrolyte is directly proportional to the rate of the heterogeneous electron-exchange reaction.

Thus, if an electron acceptor and a metal complex of porphyrin are introduced into the hydrophobic phase and a substrate capable of donating electrons is added to the aqueous phase, the potential shift in this system recorded by means of a vibrating electrode results from accumulation in the electric double layer of particles that have captured an electron from the hydrophilic donor.

The phenomenological theory of catalytic charge transfer across the interface between two immiscible liquids [23–25] implies that in the absence of side reactions, the Volta potential shift in chain (13) in the case of catalytic charge transfer across the interface is expressed by the formula:

$$\Delta\varphi = \frac{zFk_2[K]_0[S]_0}{Ck_3(K_m+[S]_0)} , \qquad (17)$$

where z is the particle charge, F the Faraday number, C the integral capacity of electric double layer, k_2 the rate constant of the catalytic reaction, k_3 the rate constant of charge ejection into water, $[K]_0$ the initial surface concentration of catalyst, $[S]_0$ the initial

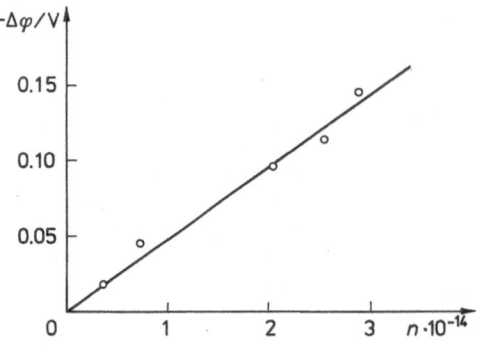

Fig. 12. Dependence of the Volta potential measured in chain (8) on the number of CP molecules adsorbed at octane/water interface. Medium: 10^{-3} M NADH; 10^{-4} M 2N-methylamino-1,4-naphthoqiunone 10^{-2} M tris-HCl; pH 7.3 [50]

concentration of substrate, and K_m the Michaelis constant. This formula permits two types of verification, which are given below. As can be seen from Fig. 12, the change in the potential shift due to the electron-exchange reaction is proportional to the number of catalyst molecules at the interface, which agrees with Eq. (17) and indicates the presence of a reaction at the interface. Furthermore, Eq. (17) can be written as follows:

$$\frac{1}{\Delta\varphi} = \frac{K_m}{\Delta\varphi_{max}} \frac{1}{[S]_0} + \frac{1}{\Delta\varphi_{max}} \tag{18}$$

whence from the experimental dependence (Fig. 12) by extrapolation of the experimental curve, the Michaelis constant is determined. As can be seen from Fig. 12, for the given reaction at pH 7.3, $K_m = 3 \times 10^{-4}$ M.

Along with relatively simple cases of reactions involving catalysts whose concentration can be assessed from the adsorption isotherm, reactions are known to occur in the oil/water system in which the membrane enzyme system acts as a catalyst of a heterogeneous process if it adheres to the interface [9, 10, 12, 19].

VI. Enzyme Complexes of the Mitochondrial Respiratory Chain in the Oil/Water Interface

Isolated oligoenzyne complexes of the respiratory chain of mitochondria – cytochrome oxidase, succinate-cytochrome c reductase, and NADH-CoQ reductase – catalyze the transfer of charges between water and octane that can be recorded from the change in the potential shift at the octane/water phase separation boundary by the vibrating plate method. A necessary condition for the appearance of this effect has proved [18, 62] to be the presence of the corresponding enzymes in the oxidation substrates in the aqueous phase and also the presence of a charge acceptor in the octane phase [10, 18, 59].

Aerobic oxidation of cytochrome c is accomplished by cytochrome c oxidase representing the terminal step of the respiratory chain of mitochondria. Cytochrome c oxidase can be considered as being the acceptor and donor of 4 electrons; but for the present, the mechanism of molecular oxygen reduction remains a controversial issue [63]. The experiments with cytochrome c oxidase in the octane/water system

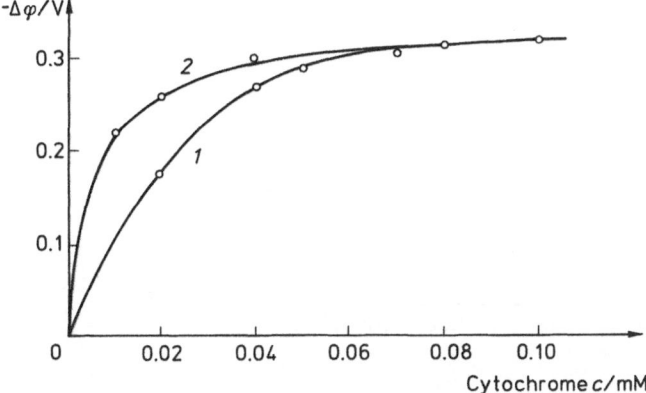

Fig. 13. Dependence of the potential shift at the octane/water interface on cytochrome c concentration. Incubation medium: 1 – 0.05 M tris-HCl (pH 7.4), 1 mM ascorbate; 20 g/ml cytochromoxidase protein and 10^{-5} M vitamin K_3; 2 – the same and 3 mM tetramethyl-phenylenediamine [18]

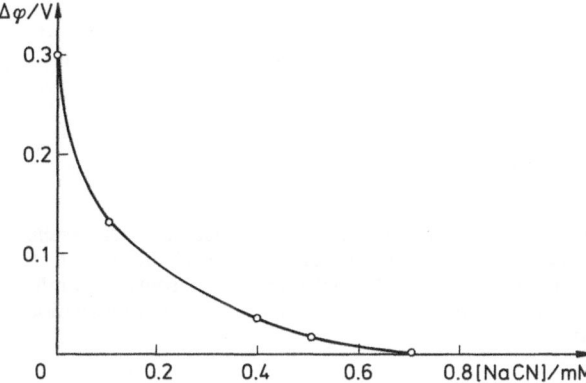

Fig. 14. Dependence of potential shift at the octane/water interface on NaCN concentration. Incubation medium: 0.05 M tris-HCl (pH 7.4); 1 mM ascorbate; 20 g/ml cytochromoxidase protein; 10^{-5} M vitamin K_3; 5×10^{-5} M cytochrome c [18]

showed that, when an enzyme suspension is added to the octane/water system, part of the particles adhere to the interface, and it is precisely this fact that in the incubation mixture of cytochrome c oxidase complex, cytochrome c and naphthaquinone is responsible for the negative Volta potential shift upon addition of adsorbate. The dependence of $\Delta \varphi$ on the concentration of the substrate of cytochrome c oxidase – cytochrome c is given in Fig. 13. It can be seen that the semimaximal effect is achieved at a cytochrome c concentration of 16 mM. This concentration decreases to 5 mM in the presence of the lipophilic electron carrier tetramethyl-N-phenylenediamine. A decrease in the cytochrome c concentration at which the curve becomes saturated was also observed in special experiments measuring the activity of cytochrome oxidase. A negative potential shift arising in the full system gradually decreased to zero when inhibitors of cytochrome c oxidase – cyanide and polylysin – were added in increasing concentrations (Fig. 14). No change of Volta potential was observed when enzyme was added to an aqueous phase containing small amounts of the cationic detergent

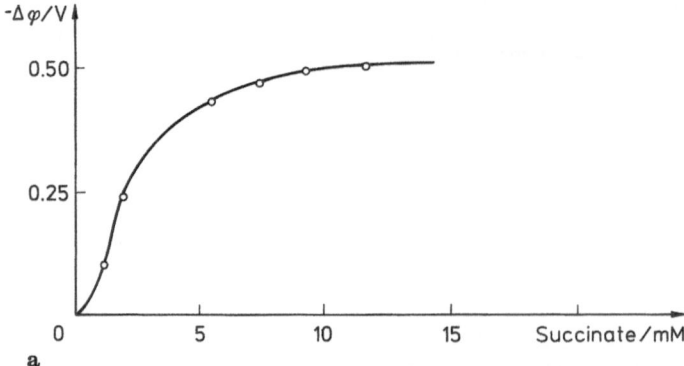

a

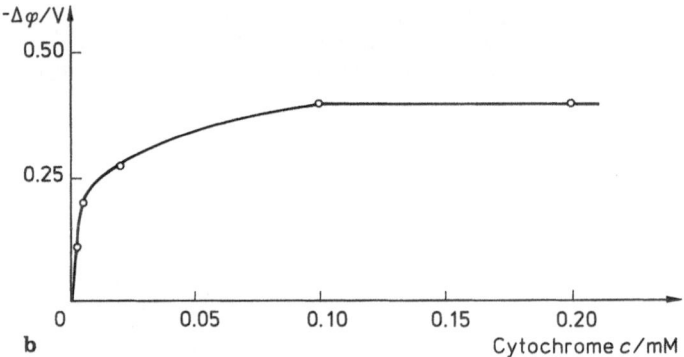

b

Fig. 15. Dependence of the displacement of the Volta potential in the octane/water system catalyzed by succinate-cytochrome c reductase on the concentration of succinate (**a**) and of cytochrome c (**b**). Incubation medium: 0.05 M tris-HCl (pH 7.4) 0.5 mM cytochrome c, 0.1 mM MNQ, and 0.4 mg of succinate-cytochrome c reductase protein per ml (for b the incubation medium also contained 4 mM succinate)

cetyltrimethylammonium bromide. This agent covered the interface, inhibiting a subsequent interaction of enzyme complexes with this surface. It is significant that, when the detergent was added after the enzyme interaction with the interface, it was ineffective.

Figure 15 shows the results of analogous experiments performed with the isolated succinate–cytochrome c reductase complex of mitochondria. It can be seen from Fig. 15 that inclusion of the cytochrome c reductase reaction with rising concentrations of succinate in the aqueous phase is accompanied by a displacement of the potential jump at the water/octane boundary of separation with an increase in the negative charge of the octane phase. Half-saturation of the system with the oxidation substrate is achieved at a concentration of succinate of approximately 2 mM. As in the experiments with cytochrome oxidase, the transfer of negative charges into the octane phase by cytochrome c reductase is achieved only when MNQ is present in the system. Nevertheless, the presence in the aqueous phase of a natural electron acceptor – cytochrome c – was necessary to achieve the effect.

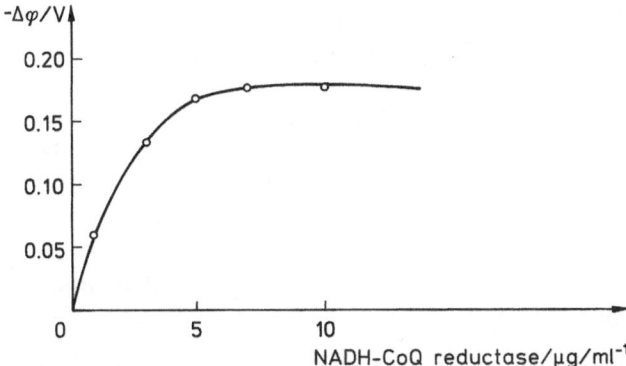

Fig. 16. Transfer of negative charges from water into octane catalyzed by NADH-CoQ reductase as a function of the concentration of enzymes. Incubation medium: 20 mM tris-HCl (pH 7.4), 0.2 mM NADH and 0.01 mM MNQ [18].

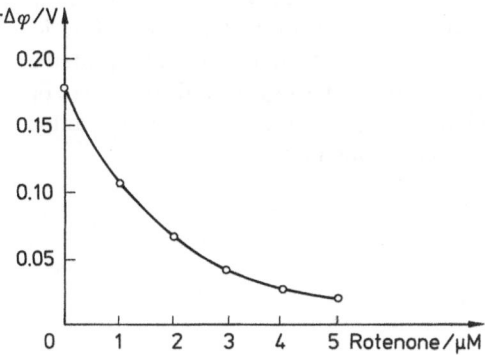

Fig. 17. Suppression by rotenone of the transfer of negative charges from water into octane catalyzed by NADH-CoQ reductase. For the incubation medium, see caption to Fig. 16. Concentration of NADH-CoQ reductase: 25 µg of protein per ml [18]

The half-saturation of the system with cytochrome c was achieved at a concentration of about 10 M (Fig. 15a, b). In all cases, the effect was reversed and prevented by antimycin – an inhibitor of the succinate-cytochrome c reductase activity of this enzyme complex – but was not suppressed by cyanide. In concluding this series of experiments, we investigated the isolated NADH-CoQ reductase complex of the respiratory chain of mitochondria. As can be seen from the results given in Fig. 16, an increase in the concentration of the enzyme complex was accompanied by an increase in the negative charge of the octane phase.

The presence of NADH in the aqueous phase and of MNQ in the system proved to be necessary for the effect to be achieved. The semimaximum effect was reached at a concentration of the protein of the enzyme complex of 1.5 µg/ml (Fig. 16); the semimaximum saturation of the system with the substrate of NADH reaction was observed at a concentration of 25 µM.

It was shown by special measurement that the reduction of MNQ by this enzyme complex in an aqueous medium is not suppressed by rotenone, while the electrogenic function of NADH-MNQ reductase in the octane/water system proved to be sensitive to rotenone.

As can be seen from Fig. 17, the semimaximum suppression of the process of charging the octane phase was reached at a rotenone concentration of 2 µM.

VII. Redox Reactions in the Oil/Water System Accompanied by Protonation of Acceptor in the Nonaqueous Phase

This section is concerned with more complex processes in which the redox reactions catalyzed by enzymes at the oil/water interface are accompanied by the capture of one of the reaction products – protons – by acceptor contained in the oil phase and at the interface.

1. Enzyme-Catalyzed Redox Reactions Accompanied by Capture of Proton by Acceptor in the Nonaqueous Phase

As in the preceding experiments (Figs. 15–17), in the presence of the proton acceptor 2,4-dinitrophenol (DNP) instead of MNQ, the positive shift of the Volta potential is characterized by saturation curves with the semimaximum values reached at a concentration of DNP of 70 M (Fig. 18) and a concentration of NADH of 22 M (Fig. 19). In contrast to the system in which MNQ was used as the electron acceptor, the charging of the octane phase accompanied by the reduction of ferricyanide proved to be insensitive to rotenone. Rotenone also did not suppress the NADH-ferricyanide activity of the enzyme complex measured in an aqueous medium.

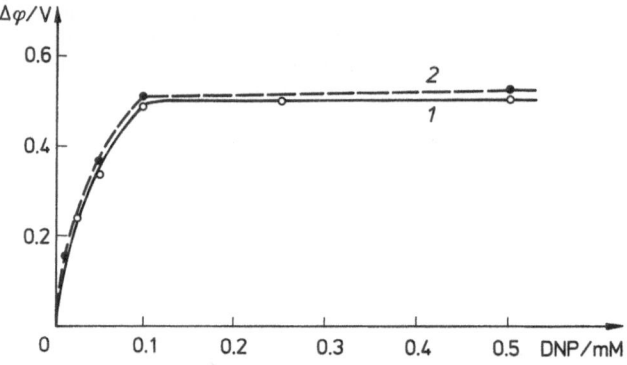

Fig. 18. Catalysis of the transfer of positive charges from water into octane by NADH-CoQ reductase (1) and by succinate-cytochrome c reductase (2) as a function of the concentration of DNP. Incubation medium: 1) 20 mM tris-HCl (pH 7.4); 0.2 mM NADH; and 25 μg of NADH-CoQ reductase protein per ml; 2) 20 mM tris-HCl (pH 4.7), 7 mM succinate, 0.2 mM cytochrome c, and 40 μg of succinate-cytochrome c reductase protein per ml. In both experiments, the medium contained 1 mM ferricyanide [18]

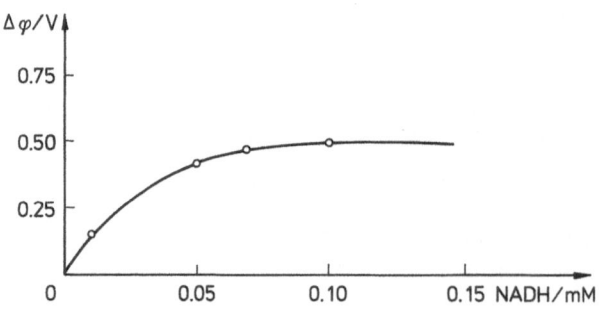

Fig. 19. Dependence of the transfer of positive charges from water into octane catalyzed by NADH-CoQ reductase on the concentration of NADH. The incubation medium (see caption to Fig. 18, curve 1) also contained 1 mM DNP [18]

Mersalyl – a reagent that blocks SH groups – proved to be an effective charge into the octane phase.

The capacity for catalyzing the transfer of positive charges into octane also proved to be characteristic of the isolated succinate-cytochrome c reductase complex when the reducing equivalents – succinate, cytochrome c, ferricyanide, and DMP – were present in a system of the corresponding donors and acceptors. The process proved to be sensitive to antimycin. It must be mentioned that the displacement of the potential could not be detected in the absence of ferricyanide, in spite of the presence in the aqueous phase of relatively high concentrations of cytochrome c, reaching 5×10^{-4} M. Nevertheless, cytochrome c proved to be a necessary component of the system for detecting the effect in the presence of ferricyanide.

Prerequisite to the transfer of charges through a boundary of separation proved are conditions ensuring the catalysis of the redox reactions characteristic of the oligoenzyme complexes used. In the case of cytochrome oxidase, the presence of a lipophilic electron acceptor – MNQ – proved to be necessary. The sensitivity to cyanide of the process of charging the octane phase by cytochrome oxidase showed that this acceptor takes up electrons either from cytochrome a_3, which is localized in the octane phase, or from ionized oxygen produced by the cytochrome oxidase. So far as concerns the succinate-cytochrome c reductase complex, in this case MNQ probably "removes" electrons from the cytochrome c localized at the phase boundary or in the octane phase. All attempts to affect the transfer of charge from an aqueous phase into octane in such systems as cytochrome c+ascorbate in the aqueous phase and MNQ in octane, cytochrome c+ascorbate+cytochrome c reductase in the aqueous phase and MNQ in octane, ferrocyanide in the aqueous phase and MNQ in octane proved to be unsuccessful. This fact and also the results given above of the action of specific inhibitors of the respiratory chain in the system investigated show that by no means any redox reaction can lead to a transfer of charges through a water/octane phase separation boundary. In particular, the reduction of MNQ by succinate-cytochrome c reductase in the presence of antimycin proved to be ineffective in this respect.

The transfer of charges into octane catalyzed by the NADH-CoQ reductase complex requires special consideration. When a lipophilic electron acceptor (MNQ) is present in the octane, this complex catalyzes the transfer of an electron into the octane by a route sensitive to rotenone. In addition to this, the possibility is created for the transfer of protons into the octane which appears as a suppression of the negative charge of the octane phase in the presence of a lipophilic proton acceptor – DNP. The latter process becomes the only possible one when cyanide is used as the electron acceptor in the aqueous phase and DNP as the proton acceptor in the octane phase.

The increase in the positive charge in the octane phase cannot be explained simply by a shift in the pH value of the aqueous phase at the boundary of separation during the enzymatic reaction, since the magnitude and the kinetic characteristics of the effect did not change with a rise in the concentration of *tris*-buffer in the aqueous phase from 3 to 50 mM. It may be assumed that the phenomenon of the transfer of H^+ ions into octane by NADH-CoQ reductase models the function of the first energy coupling site, since the oxidation of NADH on the inner surface of the mitochondrial membrane is accompanied by the generation of a trans-membrane gradient of H^+ ions. The fact that the reduction of ferricyanide and the transfer of protons into the octane phase is achieved during a reaction insensitive to rotenone confirms the idea that the first energy coupling site in the respiratory chain is localized before the respiratory transfer agent interacting with rotenone.

In relation to the succinate-cytochrome c reductase segment of the respiratory chain, which includes the second energy coupling site, it may be assumed that it contains at least two sections interacting with the nonpolar octane phase. One of them is localized at the site of the action of antimycin and probably consists of cytochrome c associated in a difinite manner with the cytochrome c_1 reductase complex and immersed in the octane phase to such an extent as to ensure the transfer of the electron to the acceptor in the octane. In addition to this, catalysis by the succinate-cytochrome c reductase reaction in the aqueous phase (final acceptor ferricyanide) is accompanied by the transfer of H^+ ions to a lipophilic proton acceptor (DNP) in the octane phase. This functional response can probably be considered as modelling the "proton pump" of the second energy coupling site catalyzing the transfer of H^+ ions through the mitochondrial membrane. It must be mentioned that the reduction of ferricyanide by this enzyme complex in an aqueous phase is a process insensitive to antimycin. The suppression by antimycin of the electrogenic function of the succinate-cytochrome c reductase complex in the octane/water system means that the transfer both of negative and of positive charges into the octane requires the transport of electrons through all the components of a second energy coupling site of the chain.

In recent years, a general procedure has been developed to transfer proteins directly in their active form from biological membranes into organic solvents as protein-lipid complexes. These complexes in the hydrophobic media resemble model membranes, i.e. monolayers, planar bilayers and liposomes [14]. In Ref. [14] the transfer of three enzymes of the inner mitochondrial membrane into organic solvents as protein-lipid complexes has been studied to understand better the extraction process. The enzymes studied were cytochrome c oxidase, ATPase and succinate dehydrogenase. These enzymes were transferred into hexane and diethyl ether in an active state. However, the extracted activities varied quantitatively, depending on the amount of protein of the starting preparation, the concentration of phospholipids and the cation employed. In all conditions, cytochrome c oxidase was extracted with the highest yield and specific activity, and it was actually enriched in the organic extract. The values for succinate dehydrogenase and ATPase were lower, but their specific activities were similar to those of the starting material. This indicates that some membrane proteins are preferentially extracted into organic solvents in a functional state. The enzymes, as protein-lipid complexes, are fairly stable in organic solvents; during one month of storage at $4°C$ in hexane, some enzymes loose less than 50% of their activity.

In Ref. [61] the properties of bacteriorhodopsin in *azolectin*/decane solution and of bilayer lipid membranes, formed from this mixture were studied. The dynamic light scattering revealed that bacteriorhodopsin in such a mixture is a suspension with particle sizes of about 2μ. Four types of BLM were investigated: (a) membranes formed according to the Mueller-Rudin technique; (b) membranes formed according to the Montal-Mueller technique from two monolayers obtained from this mixture at the water/air interface; (c) the same as (b), one monolayer having no bacteriorhodopsin; (d) membranes modified with liposomes containing bacteriorhodopsin in the presence of Ca^{+2} cations. The BLM of (a), (c), and (d) types generated a photocurrent.

An increase in the membrane conductivity induced by the uncoupler TTFB was accompanied by an enhancement of the stationary photocurrent for the (a) and (d)-

type BLM, but it did not produce any alteration of the photocurrent of the (c)-type BLM. This implies the presence of structures called "Third water" for (a)- and (d)-type BLM, and their absence in (c)-type BLM. In the (d)-type BLM, experimental data indicates that the photoresponses observed are due not only to bacteriorhodopsin in the proteoliposomes adhered to the BLM, as generally believed, but also to bacteriorhodopsin in the BLM itself. The transfer of bacteriorhodopsin to the BLM may result from the complete fusion of liposomes with the planar membrane. The existence of such a transfer is indicative of the possibility of obtaining a most convenient model system for bacteriorhodopsin studies: a planar biomolecular lipid membrane containing incorporated bacteriorhodopsin.

The light-induced proton translocation by bacteriorhodopsin at the planar interface of octane/water [10, 12, 19, 20] and in octane-water emulsions [64] has been studied. A retinotoxin thought to form a stable shift base with retinol in rhodopsin, inhibited the light-actived proton transport [64].

2. Chlorophyll-Catalyzed Redox Reaction Accompanied by the Capture of Proton by Acceptor in the Nonaqueous Phase

In the presence of a proton acceptor in the nonaqueous phase, such as 2,4-dinitro-phenol (DNP), a reducing agent – NADH – and an oxidant – $K_3Fe(CN)_6$ – in water, chlorophyll and metalloporphyrin catalyze the process which was recorded as a Volta potential change in the chain [48, 57, 65]:

$$\text{Au} \left| \text{air} \right| \text{DNP} \left| \begin{array}{c} \text{heptane} \\ \text{metallocomplex} \\ \text{of porphyrin} \end{array} \right| \begin{array}{c} \text{water} \\ K_3Fe(CN)_6 \\ \text{NADH} \end{array} \left| \begin{array}{c} \text{water} \\ \text{KCl, sat.} \end{array} \right| Hg_2Cl_2, Hg. \quad (19)$$

Figure 20 shows the change of the potential shift in the full system upon illumination as a function of the incident light intensity. The maximum potential shift is $+0.2$ V. As evidenced by the measurements, the potential shift arising in the system when DNP is added is equal to -0.2 V and results from the existence at the interface of the anionic form of a weak acid, RO^-. Therefore, the potential shift in the positive direction by $+0.2$ V indicates the disappearance of this shift due to the formation of the protonated form, ROH. Presumably, in the course of the reaction a proton was generated by the reducing agent NADH as it was oxidized by ferricyanide:

$$(\text{NADH})^w + 2\,Fe(CN)_6^{-3} + (RO)^o \xrightarrow{h\nu} NAD^+ + 2\,Fe(CN)_6^{-4} + ROH. \quad (20)$$

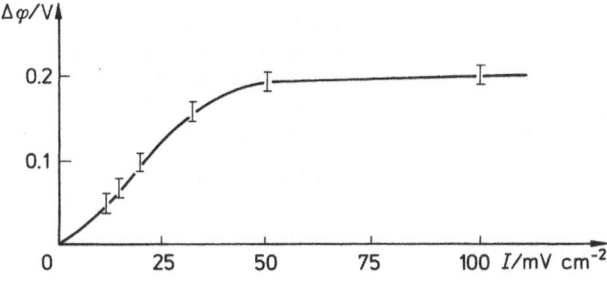

Fig. 20. Dependence of the Volta potential measured in chain (19) on incident light intensity. Medium: 2×10^{-4} M NADH, tris-HCl (pH 7.4), 2×10^{-2} M, 10 µg/ml chlorophyll, 10^{-3} M $K_3Fe(CN)_6$, 1 mM DNP [65]

A process of this type was studied in more detail in experiments involving water as a hydrogen donor in Eq. (20).

3. Photooxidation of Water Catalyzed by Chlorophyll Adsorbed at the Interface Between Two Immiscible Liquids

When an octane/water system containing chlorophyll a, DNP and $K_3Fe(CN)_6$ is illuminated with visible light, oxygen evolution takes place, accompanied by appearance of a positive photopotential [42, 43, 46, 48, 49, 65].

Introduction of molecular oxygen from water was catalyzed not only by chlorophyll, but also by other metal complexes of porphyrins, e.g. by an iron complex of tetramethyl ether of coproporphyrin III (CP) [66, 86]. Oxygen evolution was recorded by polarography with a dropping electrode [66–67] and Clark electrode [42–43]. It was proved by mass-spectrometry that the detected oxygen evolves from water [68]. Later it was found that the chlorophyll immobilized an electrode [69–71], and that the bilayer lipid membrane [72] and Aerosil [73] sensitize oxygen evolution in the water photooxidation reaction. Figure 21 shows the dependence of the oxygen evolution rate on the electron acceptor concentration in water measured with a dropping mercury electrode. It is clear from this figure that at acceptor concentrations of over 10^{-2} M/l there exists a correlation between the rate of the oxygen evolution and the change in the potential shift at the octane/water interface. In the absence of one of the reaction components in the system, i.e. either chlorophyll, electron acceptor in water, or proton acceptor in octane, no oxygen was found. Also without octane, neither photopotential generation in chain (20) nor oxygen evolution was observed.

Special check experiments with a Clark electrode [42–43] along with the mass-spectrometry data (obtained earlier) conclusively prove that the presence of oxygen resulting from the reaction can by no means be due to the termal effect of the illumination on the Clark electrode, as was suggested by Tien [27] in discussing the results of Ref. [72].

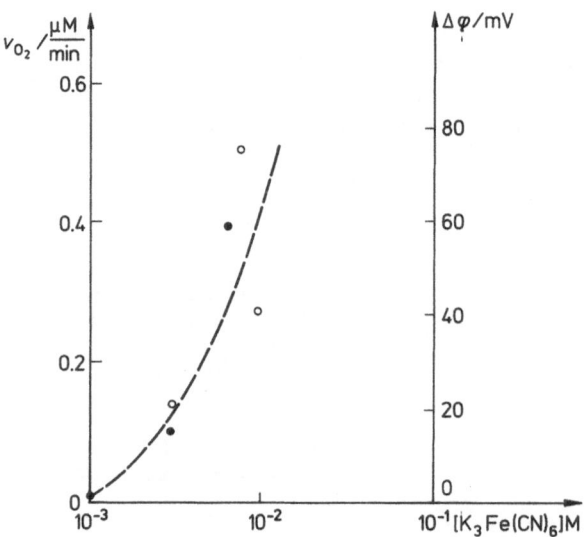

Fig. 21. Dependence of the oxygen evolution rate (●) and Volta potential (○) measured in chain (15) on $K_3Fe(CN)_6$ concentration upon illumination. Medium: 2×10^{-6} M chlorophyll, 2×10^{-2} M tris-HCl, pH 7.4 and 10^{-3} M DNP [43]

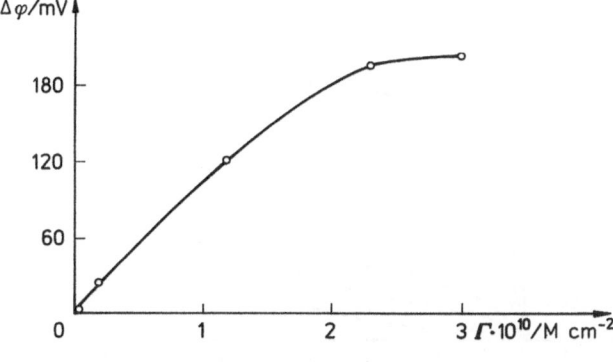

Fig. 22. Dependence of the Volta potential measured in chain (19) on adsorbed chlorophyll amount. Medium: 10^{-3} M DNP, 10^{-2} M $K_3Fe(CN)_6$, pH 7.4 [43]

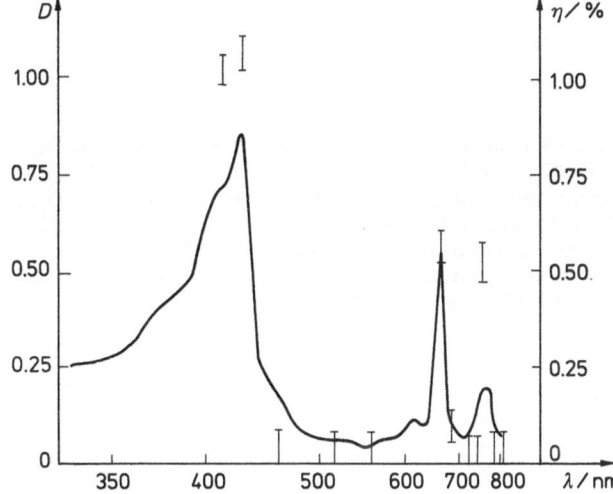

Fig. 23. Absorption spectrum of chlorophyll in octane and action spectrum of oxygen evolution reaction (dashed line). The oxygen evolution rate was determined in the medium: water/octane, 10^{-3} M PCP, 10^{-3} M NAD, 2×10^{-2} M tris-HCl, pH 7.5 [42, 43]

Figure 22 shows the dependence of the photopotential, arising upon illumination of the system, on the surface excess of chlorophyll at the octane/water interface. As can be seen from Fig. 22, the potential shift is proportional to the amount of chlorophyll adsorbed at the interface. With increasing surface concentration of chlorophyll, the dependence deviates from linear and tends to saturation. This fact reflects the interaction of adsorbed chlorophyll molecules with one another, which can result in a change in their orientation on the surface or in an increased size of aggregates with unchanged orientation of the molecules of each aggregate.

The most spectacular proof of the participation of chlorophyll in the water photooxidation reaction is the action spectrum measured from the oxygen evolution rate on a Clark electrode and showing the relation of the number of oxygen molecules to the number of incident light quanta (Fig. 23). Thus, the data obtained from Fig. 23 with allowance for the proportionality between the oxygen evolution rate and the photopotential (Fig. 21), as well as the data in Fig. 22, lead to the conclusion that the process is catalyzed by chlorophyll adsorbed at the interface.

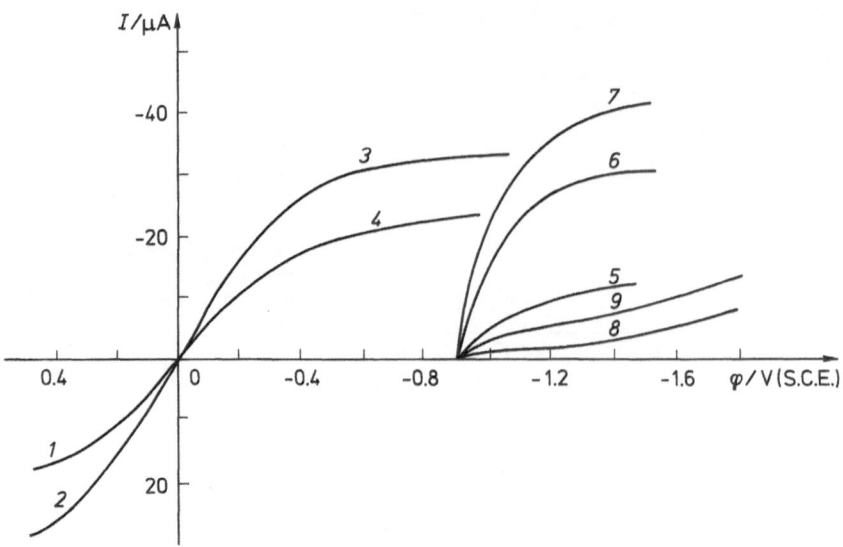

Fig. 24. Polarograms of Fe(CN)$_6^{4-}$ oxidation on an Hg electrode in the system: 20 mM tris-HCl, pH 7.9, 1 mM DNP, 10 mM K$_3$Fe(CN)$_6$, 10^{-6} M CP *1)* without illumination; *2)* 0.5 h illumination. Polarograms of O$_2$ reduction on an Hg electrode in the system: 10^{-6} M chlorophyll, 1 mM DNP, 20 mM tris-HCl, pH 7.7, 10 mM K$_3$Fe(CN)$_6$, *3)* without illumination, *4)* 0.5 h illumination; On an In–Ga electrode in the system 10^{-6} M chlorophyll, 1 mM DNP, 0.5 mM NADP, 20 mM tris-HCl, pH 7.9: *5)* without illumination. In the same system with NAD instead of NADP, *8)* without illumination, *9)* 1 h illumination [67]

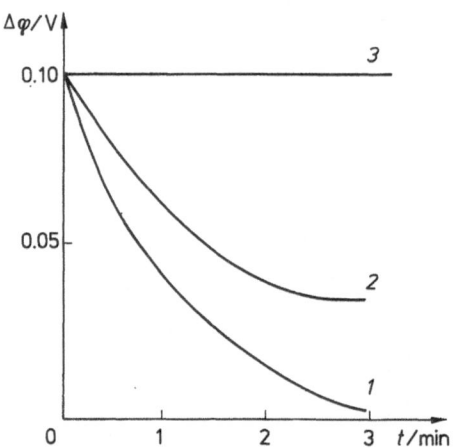

Fig. 25. Dependence of the Volta potential measured in chain (19) on incubation time. Medium: 1 mM DNP, 10^{-6} M chlorophyll, 20 mM K$_3$Fe(CN)$_6$ and tris-HCl: *1)* 0 mM; *2)* 10 mM; *3)* 20 mM, pH 5.9 [49]

    Formation of another reaction product – ferrocyanide – was recorded with the use of the dropping electrode [66] (Fig. 24). An anodic wave due to ferrocyanide oxidation arose after illumination under the conditions indicated and had a half-wave potential of +0.2 V with respect to the saturated calomel electrode. Comparison of the limiting currents of ferricyanide specially introduced into this system with those of the products resulting from Eq. (20) shows that the concentration of the latter is approximately

7×10^{-4} M. The necessity for stirring the solution in order to determine the true ferrocyanide concentration is an additional indication that $Fe(CN)_6^{4-}$ is formed not uniformly throughout the bulk solution but at the interface.

During the water photooxidation reaction by chlorophyll involving proton transfer across the interface, a boundary unstirred layer is formed, enriched in products. The dependence of the Volta potential on the pH of the incubation medium was measured. It proved that $\Delta\varphi$ has a maximum at pH 6–9 and is inhibited in the acid and alkaline regions due to the reasons considered in the preceding section. Figure 25 shows the dependence of $\Delta\varphi$ on the incubation time at different buffer concentrations measured at the initial pH 5.9. If during the reaction of Eq. (20) a boundary unstirred proton-enriched layer is formed, then with decreasing buffer capacity, the photopotential value decreases due to pheophytinization of chlorophyll [67].

4. Carotene – a Photosensitizer of the Water Photooxidation Reaction

In the wavelength region between 430 and 680 nm chlorophyll absorbs rather weakly (Fig. 26). Therefore, regarding chlorophyll only as a catalyst, in order to be able to observe the photosensitization of the water photooxidation reaction, one can introduce into the system a pigment which absorbs in this region. One of the major photosensitizers in photosynthesis is carotene [75–79]. It is known that a Van-der-Waals interaction exists between β-carotene absorbing light in the range of 400–520 nm and the tetrapyrrole cycle of chlorophyll resulting in the formation of a weak complex [76, 77]. Carotene not only protects chlorophyll from photooxidation but it is also capable of transferring the excitation energy to it. The carotinoids themselves, however, cannot achieve photosynthesis without chlorophyll [76, 77].

Figure 26 [78, 79] (curve a) shows an absorption spectrum of chlorophyll a in dry octane. The main absorption maxima lie at the wavelengths 370, 409, 412, 428, 505, 534, 577, 614, and 662 nm. Curve b of this figure represents the spectrum of β-carotene in dry octane (absorption maxima at 458 and 495 nm). It should be noted that the

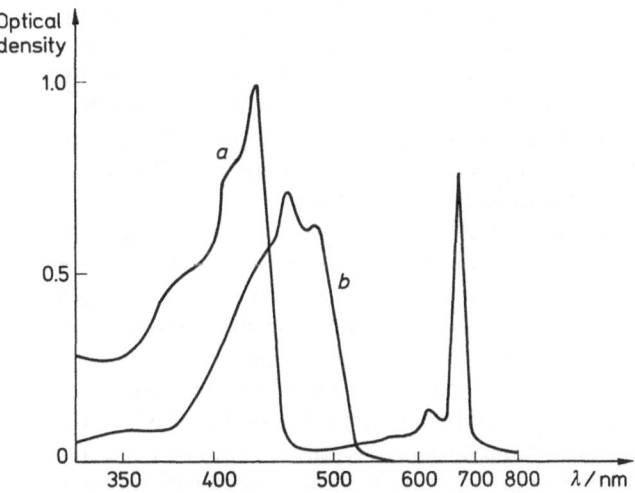

Fig. 26. Absorption spectrum of chlorophyll a (a) and β-carotene (b) in dry octane [78]

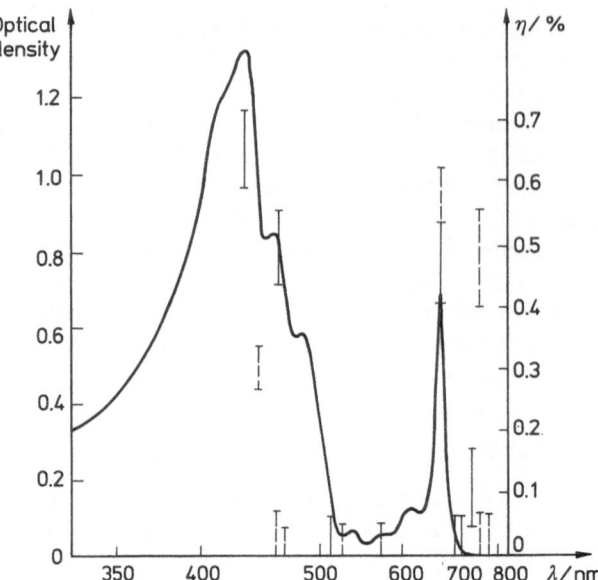

Fig. 27. Absorption spectrum of chlorophyll a and β-carotene mixture in octane; cuvette thickness 1 cm, pigment concentration 10^{-5} M [78]

absorption spectra of β-carotene in "wet" and dry octane coincide. This indicates that the hydrophobic carotene molecule interacts weakly with water and does not seem to form hydrates. When the solutions of β-carotene and chlorophyll in octane are mixed, their spectra are transferred to the spectrum presented in Fig. 27 (solid line). A complex of chlorophyll a with β-carotene absorbs light in the region 370–500 nm and 650–700 nm. The spectrum presented in Fig. 27 is not obtained by simple summation of the chlorophyll a and β-carotene spectra, which confirms the presence of complexing. The spectra shown in Figs. 26 and 27 refer to the octane phase bulk, but the complex is assumed to exist at the octane/water interface as well.

Figure 28 shows the dependences of the amount of evolved oxygen on the illumination for different order of introduction of ingredients. When the proton acceptor DNP, the electron acceptor NAD$^+$, and chlorophyll are introduced into the octane/water system, oxygen evolution takes place upon illumination. The action spectrum coincides with the chlorophyll absorption spectrum [42, 43]. At the wavelength 462 nm, illumination does not give rise to O_2 evolution (Fig. 28). When β-carotene was added, however, oxygen evolution was recorded with the use of the Clark electrode. When illumination is turned off, oxygen evolution stops completely, starts again when it is turned on again and proceeds at the same rate as before. The quantum yield of oxygen at the wavelength 462 nm is as high as 0.5% (as calculated per incident light). Oxygen evolution in the full system does not depend on the order of adding chlorophyll and β-carotene (see Fig. 28). In the absence of chlorophyll, however, no oxygen evolution takes place in the system in question (Fig. 28).

Figure 29 shows the dependence of the molecular oxygen evolution rate on the concentration of β-carotene in octane. With increasing β-carotene concentration, the oxygen evolution rate rises and reaches a limiting value at a carotene concentration of 10^{-5} M. It is not clear whether the dependence shown in Fig. 29 indicates that the

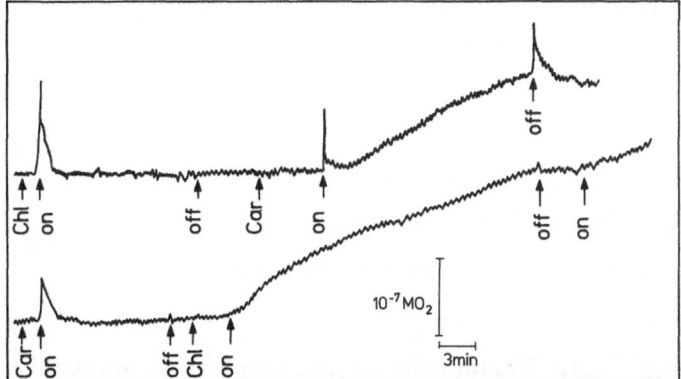

Fig. 28. Dependence of evolved oxygen amount on incubation time and order of adding reagents. Composition: water/octane/water, 10^{-3} M NAD, 2×10^{-2} M tris-HCl, pH 7.4, 10^{-3} M PCP, 4×10^{-6} M chlorophyll, and 10^{-5} M β-carotene were added to octane fraction. The interface was illuminated with monochromatic light at a wavelength of 462 nm ($\Delta\lambda = 8$ nm) [75]

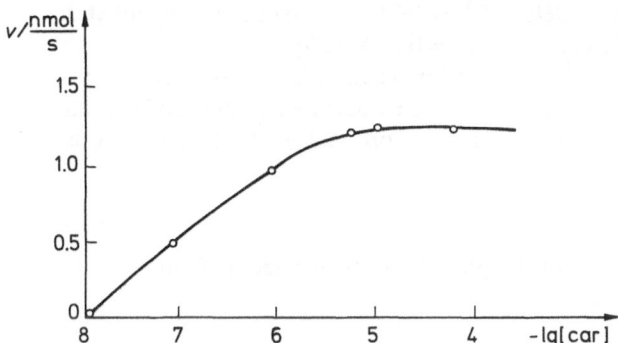

Fig. 29. Dependence of oxygen evolution rate on β-carotene concentration. Medium: water/octane, pH 7.4, β-carotene in concentrations as indicated in Fig. 28 [75]

reaction occurs at the interface or that there is simply an optimum relation between the amounts of chlorophyll and β-carotene. The deviation from linear of the dependence of the O_2-evolution rate with increasing β-carotene concentration points in favor of the former assumption. The fact that the oxygen evolution rate is independent of the order of introduction of β-carotene and chlorophyll into the system favors the latter (Fig. 28). It is unlikely that in the case of a competition between β-carotene and chlorophyll for the interface, the oxygen evolution rates for the two cases should be identical, as is already known for other systems. If, however, a catalytic complex of composition $CHL_mCar_n(H_2O)_x$ is formed, the sequence of its assembly can affect the rate of molecular oxygen evolution.

Figure 28 shows the action spectra of the oxygen photooxidation reaction by chlorophyll in the presence (and absence) of β-carotene. It is seen from the figure that β-carotene sensitized the water photooxidation reaction in the spectral region where light is absorbed by β-carotene rather than by chlorophyll. The presence of an absorption maximum at 742 nm indicates that chlorophyll participates in the reaction as a hydrated oligomer.

Formally, the water oxidation reaction by a water-soluble electron acceptor (Ox), catalyzed by a hydrated oligomer of chlorophyll and sensitized by β-carotene, can be written as follows:

$$\beta\text{-carotene} \rightleftarrows \beta\text{-carotene}^*,$$

$$CHl_m^*(H_2O)_x \curvearrowright CHl_m(H_2O)_x, \tag{21}$$

$$CHl_m^*(H_2O)_x + Ox \rightarrow CHl_m^+(H_2O)_x + Red,$$

$$4\,CHl_m^+(H_2O)_x + 2\,H_2O \rightarrow 4\,CHl_m(H_2O)_x + O_2 + 4\,H^+. \tag{22}$$

5. Possible Mechanism of Water Photooxidation Sensitized by Chlorophyll

Reaction Thermodynamics

Water can be oxidized to molecular oxygen by CHl^{2+} ($E^0 = 0.96$ V [85]), that can be formed only on an electrode at high anodic potentials, and by the cations of a hydrated oligomer of chlorophyll ($E^0 = 0.92$ V [70]), which is the most likely participant in water oxidation [80, 83–86]. The monomer CHl^+ ($E^0 = 0.76$ V) can oxidize water but this process does not proceed effectively since $E^0_{H_2O/O_2} = 0.81$ V [83].

In a most general form the specificity of the redox reaction in a two-phase system, where the starting substances and the reaction products obey the distribution law, can be represented as follows. Let the following reaction proceed in the heterogeneous system:

$$Red \rightleftarrows Ox + e^-. \tag{23}$$

The standard redox potentials in each of the phases can be written as follows:

$$nFE^0_{H_2O} = {}_{Red}\mu^0_{H_2O} - {}_{Ox}\mu^0_{H_2O} - {}_e\mu^0, \tag{24}$$

$$nFE^0_{oil} = {}_{Red}\mu^0_{oil} - {}_{Ox}\mu^0_{oil} - {}_e\mu^0. \tag{25}$$

Subtracting Eq. (24) from Eq. (25) yields the change in the standard potential at the interface:

$$nF\Delta E^0 = \left({}_{Red}\mu^0_{oil} - {}_{Red}\mu^0_{H_2O}\right) - \left({}_{Ox}\mu^0_{oil} - {}_{Ox}\mu^0_{H_2O}\right) \tag{26}$$

or

$$\Delta E^0 = \frac{RT}{nF} \ln \frac{B_{Red}}{B_{Ox}} \tag{27}$$

where n is the number of electrons and B_i is the distribution coefficient of the i ion:

$$RT \ln B_+ = {}_{Red}\mu^0_{oil} - {}_{Red}\mu^0_+, \tag{28}$$

$$RT \ln B_- = {}_{Ox}\mu^0_{oil} - {}_{Ox}\mu^0_-. \tag{29}$$

Equation (27) was first obtained by Volkov [81–83].

At present, more than 100 different mechanisms describing possible oxidation pathways of water to molecular oxygen have been documented. Possible thermodynamic pathways of water oxidation to molecular oxygen are illustrated in Fig. 30 [81].

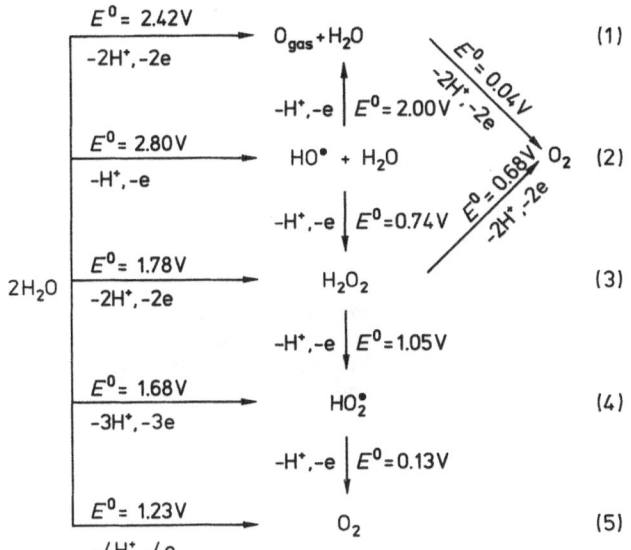

Fig. 30. Thermodynamic scheme of possible pathways of water oxidation to molecular oxygen [82]

The oxygen evolution reaction at the octane/water interface occurs upon illumination at the wavelength 670 nm, so that the thermodynamic possibility of the reactions following the mechanism of Eqs. (1–3) is excluded. For a three-electron reaction (4) a stronger oxidant than the cation radical of chlorophyll, bivalent cation of chlorophyll and the cations of a hydrated oligomer of chlorophyll is necessary. It should be noted that Eqs. (1–3) are thermodynamically possible if the intermediate particles formed are in the adsorbed state.

The original schemes of Kutyurin [84–85] and Renger [74] are one-electron mechanisms in which O_2 evolution proceeds via the formation of intermediate radicals. The radical $O^{\cdot}$, $HO^{\cdot}$, and $H_2O^{\cdot}$, however, are extremely reactive and can easily enter into side hydroxylation and oxidation reactions leading to destruction of chlorophyll and other reagents. Possibly, such processes can be realized in model systems where the quantum yield is low, the condition for four- or two-electron reactions are lacking and where the life-time of the sensitizer is negligible. So there remains only one possible pathway – the four-electron oxidation of water to oxygen [80–83, 86].

In Nature (on membranes of chloroplasts) water is oxidized by the four-electron mechanism and the life-time of chlorophyll can be as long as one day. In a model octane/water system, oxygen evolution occurs also during several hours. This is indirect evidence in favor of a many-electron reaction. Therefore we shall consider a four-electron mechanism of water oxidation sensitized by chlorophyll adsorbed at the oil/water interface proposed by Volkov [81]. It was shown above that the interface is the most likely site for the water photooxidation reaction. Thus it is assumed that water can be oxidized by a reaction complex adsorbed at the interface that consists of a hydrated oligometer of chlorophyll, a hydrophilic electron acceptor and hydrophobic proton acceptor [81, 82, 86]. The water in the reaction complex is linked coordinatively with the magnesium of one of the chlorophyll molecules, by hydrogen bonds with the carbonyl group of another chlorophyll molecule, and with the phenol anion also

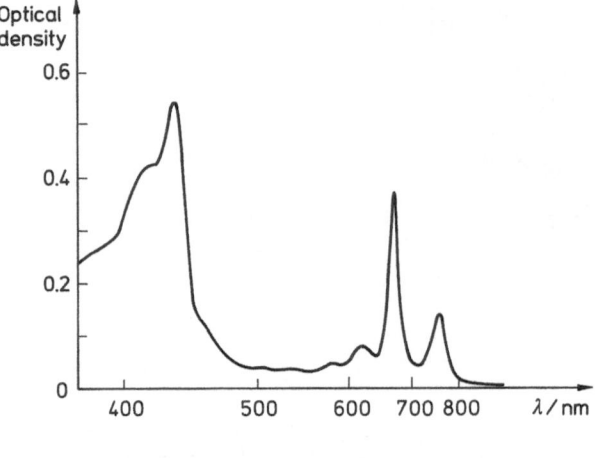

Fig. 31. Absorption spectrum of chlorophyll *a* in "wet" (equilibrated with water) octane [81]

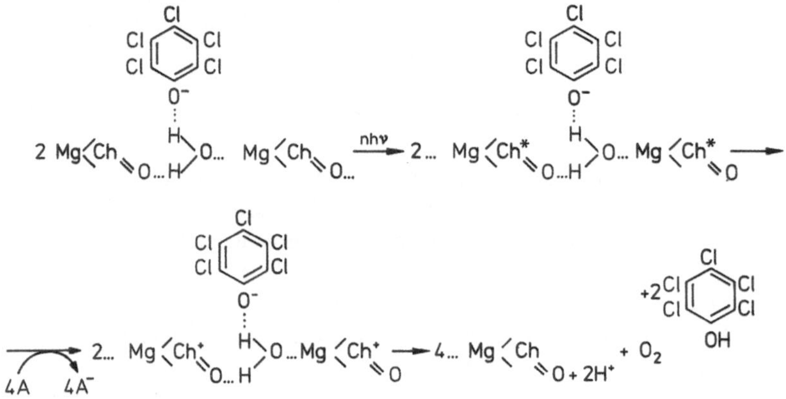

Fig. 32. Scheme of water photooxidation with hydrated oligomer of chlorophyll at the interface between two liquids [82]

adsorbed at the interface, which is necessary for adsorption binding since it accepts some of the protons liberated during the reaction. It was revealed by NMB that 2,2-dinitrophenol catalyzed the H^+ transport between water and oil, the characteristic time of this reaction being of the order of 0.1–1 ms.

A diluted solution of chlorophyll *a* in dry octane has the characteristic adsorption spectrum shown in Fig. 26. If octane is saturated with water, an additional maximum at 742 nm appears in this "wet" octane (Fig. 31) [86] that is due to formation of the hydrated oligomer $CHl_m(H_2O)_x$. When interacting with water, oligomers of chlorophyll undergo hydration, but this process occurs slowly since it involves breaking of old coordination bonds between chlorophyll molecules in dry octane $=C=O \ldots Mg$ and formation of new bonds $=C=O \ldots \overset{\text{H}}{\underset{|}{}} H-O \ldots Mg$. If the oxygen atoms of two water molecules in the oligomer are located close enough to each other between closely packed chlorophyll molecules, this situation will favor a many-electron process of water oxidation represented schematically in Fig. 32. According to this scheme, hydrated oligomer of chlorophyll $CHl_m(H_2O)_x$ adsorbed at the interface and closely packed so that overlapping of electronic clouds of porphyrin rings is possible, is excited under the action of light and this results in the formation of an oxidized form of pigment and a reduced form of acceptor.

It is quite possible, as was supposed by Kutyurin [84, 85], that during chlorophyll photooxidation a positive charge arises on one of the nitrogen atoms in the porphyrin ring and it becomes a strong electrophilic reagent. This leads to opening of the cyclic system of conjugated bonds in the tetrapyrrol ring, closing of the conjugated system at the keto group of the fifth ring, and to oxidation of the water molecule structurally bound with this group. The detachment of electrons from the oxygen of water leads to a redistribution of the electron density in the fifth ring, whereby the former system of cyclic bonds is restored and this hinders the reverse reaction of electron transfer. Some of the released protons are bound by proton acceptors residing in the interfacial region on the octane side. This results in generation of a photopotential. Some of the protons are ejected into water, and the pH in the weak buffer solution shifts into the acid region [67].

Water oxidation to oxygen seems to be a many-electron process [81–82, 86, 87] if the reaction proceeds effectively with a high quantum yield. For this purpose it is necessary to abstract four electrons from two water molecules forming part of the reaction complex. This can be effected by a tetramer or two dimers of chlorophyll bound into a multi-center complex adsorbed at the interface.

VII. Coupling of Reactions at the Interface Between Immiscible Liquids

Here, we shall consider two cases of the coupling of reactions, both of which occur at the interface. In the first case we have to deal with a reaction similar to the already-mentioned processes of of electron transfer across the oil/water interface in which 1,4-naphthaquinone, vitamin K_3 and $2 N$-methylamino-1,4-naphthoquinone were generally used as electron acceptors. These acceptors can, however, enter into side hydroxylation reactions. Therefore, to study the coupling of reactions, we chose as an acceptor tetramethylquinone (TMQ) which is free of this shortcoming. Under these conditions it proved possible to record the coupling of two redox reactions at the octane/water interface involving NADH, FeEP, and oxygen [56].

In the octane/water system containing FeEP, TMQ, and NADH, a NADH oxidation reaction catalyzed by porphyrin occurs at the interface. The consumption of NADH recorded spectroscopically at the wavelength 340 nm is proportional to the incubation time of the reaction mixture (Fig. 33). As is seen from Fig. 33, when check experiments are carried out in the octane/water system not containing one of the reaction components (TMQ, FeEP) in an inert atmosphere, NADH is practically not consumed. In the full system the NADH concentration decreases monotonically with time (Fig. 33, curve 4). In the presence of atmospheric oxygen the consumption of NADH is appreciable even in a system without TMQ (Fig. 33, curve 3). In the full system, however, NADH consumption in the presence of oxygen is much higher than in the control without one of the ingredients or in an inert atmosphere. Comparison of NADH consumption in air and in argon atmosphere shows that in the presence of

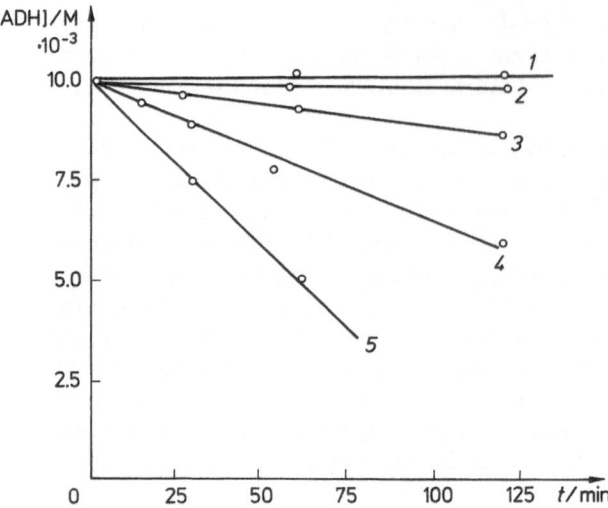

Fig. 33. Dependence of NADH concentration on incubation time in argon atmosphere (curves *1*, *2*, *4*) and air (curves *3*, *5*). Medium: 10^{-6} M FeEP, 2×10^{-5} M TMQ, 20 mM tris-HCl, pH 7.5, *2)–3)* control without TMQ and *1)* without TMQ [56]

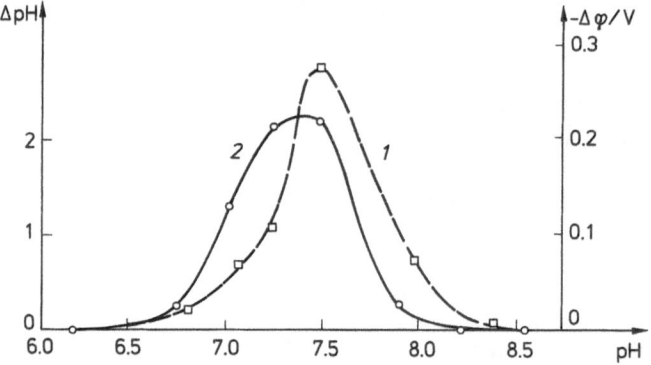

Fig. 34. Dependence of the potential shift (1) and change in pH of aqueous phase (2) on initial pH in aqueous solution. Medium: 10^{-6} M FeEP, 2×10^{-5} M NADH, 5 mM tris-HCl [56]

oxygen a reaction occurs at the interface that may be considered as being NADH oxidation by oxygen.

The oxygen reduction reaction is known to be accompanied by alkalization of the solution in the reaction zone. Measurement of the aqueous solution pH in the full system in air showed that in the presence of a reaction at the interface, the aqueous phase undergoes alkalization which depends on NADH concentration and initial pH of the system (Fig. 34). When TMQ is substituted by vitamin K_3 or 1,4-naphthoquinone, no change of pH in a weakly buffered solution (5 mM tris-HCl) takes place. This can be attributed to the different structure of the reaction complex at the interface.

Spectroscopic studies carried out in a system containing a complete set of all reaction ingredients revealed that in the course of a reaction in the inert gas atmosphere

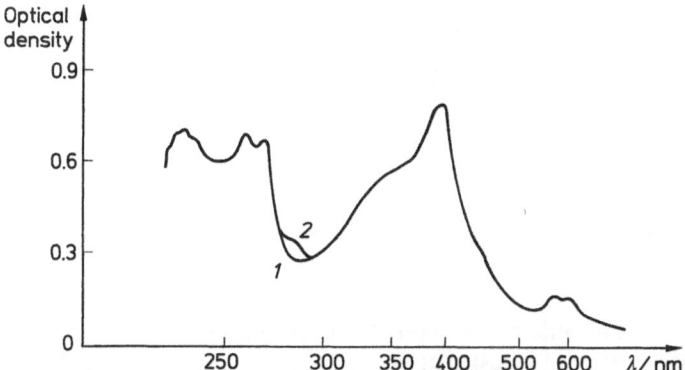

Fig. 35. Absorption spectrum of octane fraction after 5 h incubation in argon atmosphere. Medium: 10^{-6} M FeEP, 2×10^{-5} M TMQ, 2×10^{-4} M NADH, 20 mM tris-HCl, pH 7.5, *1)* before incubation with NADH, *2)* after incubation with NADH [56]

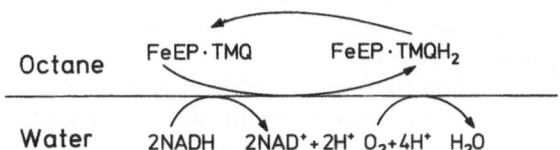

Fig. 36. Scheme of NADH oxidation and oxygen reduction [55]

(argon, nitrogen) a reduced form of acceptor is produced – $TMQH_2$ (Fig. 35); TMQ absorbs in the ultraviolet (260, 265 nm). The peak due to the reduced form lies at 285 nm. $TMQH_2$ can be again oxidized by intensive mixing with freshly prepared silver oxide. This fact was established spectroscopically. It should be stressed that in the range of NADH concentrations used, in the absence of FeEP in solution, TMQ is not directly reduced in appreciable amounts.

The TMQ reduction in the full system was also recorded chromotographically. For the purpose of control, TMQ and $TMQH_2$ were applied to chromatographic systems.

When the reaction is started, the potential shift at the interface changes. This change depends on the concentration of all ingredients and the aqueous solution pH. According to the spectral data, $TMQH_2$ can be obtained between pH 7 and 8. The greatest negative potential shift as well as the maximum alkalization of the solution occur in the same pH range. This indicates that the NADH oxidation reaction and oxygen reduction are coupled, presumably via $TMAH_2$. The process scheme is shown in Fig. 36.

The synthesis and hydrolysis of ATP at the membranes of mitochondria, chloroplasts and bacteria is coupled with the functioning of the electron transport chain (respiration or photosynthesis) via the membrane transport of protons:

$$ADP + P_i + nH^+ \rightleftarrows ATP + H_2O. \tag{30}$$

The molecular mechanism for the synthesis of ATP at the cost of the energy of proton re-solvation between the membrane and water matrix was proposed by Kornyshev and Volkov [88, 89].

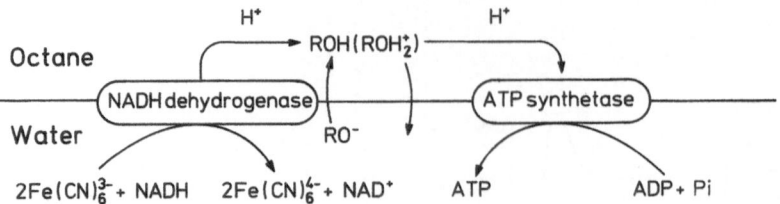

Fig. 37. Scheme of ATP synthesis at the octane/water interface due to the redox reaction [11]

The dielectric constant of a biomembrane is about 4. According to the proposed calculation scheme, the Gibbs energy of transfer of H^+, Li^+, Mg^{2+}, Ca^{2+}, and K^+ exceeds the Gibbs energy required for the synthesis of ATP and the work of charge transporting chain. Nature may utilize this energy to couple the synthesis of ATP and the work of the charge-transfer chain using catalysts such as H^+-ATPase of chloroplasts, mitochondria, and bacteria, or Na^+–K^+-ATPase and Ca^{2+}-ATPase from the sarcoplasmic reticulum. Unfortunately, we do not know of enzymes providing the electrochemical coupling at the cost of Li^+ free energy of transfer, but these systems may presumably be discovered and extracted from the cell.

Halide anions (Cl^-, Br^-, and I^-) are at first sight also capable of providing the coupling of biochemical reactions due to a sufficiently large transfer energy gain. However, the surface of a biomembrane is, as a rule, charged negatively due to the lipids. This distribution of charges in the electrical double layer provides an additional barrier affecting the reaction kinetics and making the anion mechanism of coupling impossible or ineffective [88, 89].

Experimental synthesis of ATP at the octane/water interface was carried out [11, 12, 20]. The proton flow through the ATP-synthetase complex from octane to water was provided by creating an excess (relative to equilibrium) concentration of undissociated (or Lewis) acid in the octane phase (Fig. 37). This was achieved in three ways: by direct addition of acid-pentachlorphenol to octane, by the action of NADH-ferricyanidreductase of the respiratory chain of submitochondrial particles, and also through the action of the H^+-pump of bacteriorhodopsin sheets from *Halobacterium halobium*.

References

1. Boguslavsky, L.I.: In: Progress in Surface Science, (Davison, S.G., ed.), New York, *19*, 1 (1985)
2. Dehmlow, R.V., Dehmlow, S.S.: Phase Transfer Catalysis, Second Revised Edition, Verlag Chemie, Weinheim 1983
3. Grätzel, M.: Biochim. Biophys. Acta *683*, 221 (1982)
4. Racker, E.: A New Look at Mechanisms in Bioenergetics, Acad. Press, New York, San Francisco, London 1976
5. Läuger, P., Richter, J., Lesalauer, W.: Ber. Buns. Ges. Phys. Chem., *71*, 906 (1967)
6. Tien, H. Ti.: Nature *219* (1968)
7. Hong, F., Manzerall, D.: Biochim. Biophys. Acta *266*, 584 (1972)
8. Volkov, A.G., Lozhkin, B.T., Boguslavsky, L.I.: Dokl. Akad. Nauk SSSR *220*, 1207 (1975)
9. Boguslavsky, L.I., Volkov, A.G., Kozlov, I.A., Metelsky, S.T., Skulachev, V.P.: ibid. *218*, 963 (1974)

10. Boguslavsky, L.I., Kondrashin, A.A., Kozlov, I.A., Metelsky, S.T., Skulachev, V.P., Volkov, A.G.: FEBS Letters 50, 223 (1975)
11. Yaguzhinsky, L.S., Boguslavsky, L.I., Volkov, A.G., Rachnaninova, A.B.: Dokl. Akad. Nauk SSSR 221, 1465 (1975)
12. Yaguzhinsky, L.S., Boguslavsky, L.I., Volkov, A.G., Rachmaninova, A.B.: Nature 259, 494 (1976)
13. Martinek, K., Semenov, A.N.: Uspechi Khimii 50, 1376 (1981)
14. Ayala, G., Nascimento, A., Gómez-Puyou, A., Darscon, A.: Biochim. Biophys. Acta 810, 115 (1985)
15. Skulachev, V.P.: Transformation of Energy in Biomembranes, Nauka, Moscow, p. 203, 1972
16. Yaguzhinsky, L.S., Ismailov, A.D., Boguslavsky, L.I.: Biochim. Biophys. Acta 368, 22 (1974)
17. Masters, B.D., Mauzerall, D.: J. Membrane Biology 41, 377 (1978)
18. Boguslavsky, L.I., Volkov, A.G., Kondrashin, A.A., Skulachev, V.P., Yasaitis, A.A.: Bio-organich. Khimiya 1, 1783 (1975)
19. Boguslavsky, L.I., Boizov, V.G., Volkov, A.G., Kozlov, I.A., Kondrashin, A.A., Metelsky, S.T., Yasaitis, A.A.: Biokhimiya 41, 1047 (1976)
20. Yaguzhinsky, L.S., Volkov, A.G., Boguslavsky, L.I.: ibid. 41, 1203 (1976)
21. Samec, Z., Marecek, V., Weber, J.: J.Electroanal.Chem. 103, 11 (1979)
22. Brody, S.S., Owens, H.F.: Z. Naturforsch. 31 c, 567 (1976)
23. Kharkats, Yu.I., Volkov, A.G., Boguslavsky, L.I.: J.Theor.Biol. 65, 379 (1977)
24. Kharkats, Yu.I., Volkov, A.G., Boguslavsky, L.I.: Dokl. Akad. Nauk SSSR 220, 1441 (1976)
25. Kharkats, Yu.I., Volkov, A.G., Boguslavsky, L.I.: Biofizika 21, 634 (1976)
26. Boguslavsky, L.I., Bogolepova, F.I., Lebedov, A.V.: Chem.Phys. Lipids 6, 296 (1971)
27. Tien, H. Ti.: In: Topics in Photosynthesis (Barber, J., ed.) Elsevier, Amsterdam, 3, 115 (1979)
28. Boguslavsky, L.I.: Electron Transfer Effects and the Mechanism of the Membrane Potential, in: Modern Aspects of Electrochemistry (White, R.E., ed.), Plenum Press, 18, 1986 (in Press)
29. Bell, R.P.: J.Phys.Chem. 32, 882 (1928)
30. Guinazzi, M., Silvestri, G., Serravable, G.: J.Chem.Soc.Chem.Com. 200 (1975)
31. Volkov, A.G., Gugeshashvili, M.I., Mironov, A.F., Nizhnik, A.N., Boguslavsky, L.I.: Elektro-khimiya 18, 1628 (1982)
32. Volkov, A.G., Gugeshashvili, M.I., Mironov, A.F., Boguslavsky, L.I.: Bioelectrochem. Bioenergetic 9, 551 (1982)
33. Davies, J.T., Rideal, E.K.: Interfacial Phenomena, Acad. Press 1963
34. Hajkova, P., Homolka, D., Mareček, V., Volkov, A.G., Samec, Z.: Elektrokhimiya 21, 209 (1985)
35. Rabinowitch, E.I.: Photosynthesis and Related Processes, Interscience Publishers Inc., New York, V. 1, 1945
36. Rabinowitch, E., Weiss, J.: Proc.Roy.Soc. (London) A-162, 251 (1937)
37. Krasnovsky, A.A.: Dokl. Akad. Nauk SSSR 60, 421 (1948)
38. Oparin, A.I., Serebrovskaya, K.V., Lozovaya, T.I.: In: Functional Biochemistry of Cell Structure, Nauka, Moscow, p. 7, 1976
39. Belamy, W.D., Gaines, G.L., Tweet, A.G.: J.Chem.Phys. 38, 2528 (1963)
40. Chin, P., Brody, S.S.: Z. Naturforsch. 31 c, 44 (1976)
41. Brody, S.S., Owens, N.F.: ibid. 31 c, 567 (1976)
42. Kandelaki, M.D., Volkov, A.G., Levin, A.L., Boguslavsky, L.I.: Dokl.Akad.Nauk SSSR 271, 462 (1983)
43. Kandelaki, M.D., Volkov, A.G., Levin, A.L., Boguslavsky, L.I.: Bioelectrochem. Bioenerg. 11, 167 (1983)
44. Krawczuk, S.: Studia Biophysica 100, 119 (1984)
45. Volkov, A.G., Lozhkin, B.T., Boguslavsky, L.I.: Dokl.Akad.Nauk SSSR 220, 1207 (1975)
46. Boguslavsky, L.I., Volkov, A.G., Kandelaki, M.D.: FEBS Letters 65, 155 (1976)
47. Boguslavsky, L.I., Lozhkin, B.T., Kiselev, B.A.: Dokl.Akad.Nauk SSSR 222, 228 (1975)
48. Boguslavsky, L.I., Volkov, A.G., Kandelaki, M.D.: Bioelectrochem. Bioenerg. 4, 68 (1977)
49. Boguslavsky, L.I., Volkov, A.G., Kandelaki, M.D.: Biofizika 21, 808 (1976)
50. Volkov, A.G., Gugeshashvili, M.I., Mironov, A.F., Boguslavsky, L.I.: Elektrokhimiya 19, 1194 (1983)

51. Volkov, A.G., Bibikova, M.A., Mironov, A.F., Boguslavsky, L.I.: Bioelectrochem. Bioenerg. *10*, 477 (1983)
52. Volkov, A.G., Bibikova, M.A., Mironov, A.F., Boguslavsky, L.I.: Elektrokhimiya *19*, 1398 (1983)
53. Lelental, M.: J. Electrochem. Soc. *120*, 1650 (1973)
54. Volkov, A.G., Gugeshashvili, M.I., Mironov, A.F., Boguslavsky, L.I.: Bioelectrochem. Bioenerg. *10*, 485 (1983)
55. Gugeshashvili, M.I., Volkov, A.G., Mironov, A.F., Boguslavsky, L.I.: ibid. *10*, 493 (1983)
56. Gugeshashvili, M.I., Volkov, A.G., Yaguzhinsky, L.S., Mironov, A.F., Boguslavsky, L.I.: Elektrokhimiya *19*, 1629 (1983)
57. Volkov, A.G., Mironov, A.F., Boguslavsky, L.I.: ibid. *12*, 1326 (1976)
58. Boguslavsky, L.I., Volkov, A.G.: In: Structure and Function of Active Enzymes Centres, Nauka, Moscow, p. 30, 1976
59. Volkov, A.G., Boguslavsky, L.I.: In: Biochemistry of Mitochondria (Severin, S.E., ed.) Nauka, Moscow, p. 14, 1976
60. Boguslavsky, L.I., Volkov, A.G.: In: IFIAS Workshop on Physicochemical Aspects of Electron Transfer Processes in Enzyme Systems, Stockholm, IFIAS, p. 22, 1977
61. Mirsky, V.M., Sokolov, V.S., Markin, V.S., Chekulaeva, L.N.: Biol. Membrany *1*, 1143 (1984)
62. Boguslavsky, L.I.: In: Current Topics in Membrane and Transport (Bronner, F., Kleizeller, A., eds.) Acad. Press, New York *4*, 1, 1980
63. Wikström, M., Kraab, K., Saraste, M.: Cytochrome Oxidase, A Synthesis, New York, Acad. Press, p. 198, 1981
64. Post, A., Young, S.E., Robertson, R.N.: Photobiochem. Photobiophys. *8*, 137 (1984)
65. Boguslavsky, L.I., Volkov, A.G.: Dokl.Akad.Nauk SSSR *224*, 1201 (1975)
66. Boguslavsky, L.I., Volkov, A.G., Kandelaki, M.D., Nizhnikovsky, E.A., Bibikova, M.A.: Biofizika *22*, 223 (1977)
67. Boguslavsky, L.I., Volkov, A.G., Kandelaki, M.D., Nizhnikovsky, E.A.: Dokl. Akad. Nauk SSSR *227*, 727 (1976)
68. Boguslavsky, L.I., Zhuravlev, L.T., Kandelaki, M.D., Shengeliya, V.Ya.: ibid. *240*, 1453 (1978)
69. Myasaka, T., Fujishima, A., Honda, A.: Bull.Chem.Soc. Japan *54*, 957 (1981)
70. Showell, M.S., Fong, F.K.: J.Amer.Chem.Soc. *104*, 2773 (1982)
71. Fong, F.: In: Light Reaction Path of Photosynthesis (Fong, F., ed.) Springer, Berlin, p. 1, 1982
72. Toyoshima, Y., Marino, M., Motoki, H., Sukigara, M.: Nature *265*, 187 (1977)
73. Kachan, A.A., Nechievich, L.A.: Dokl.Akad.Nauk SSSR *241*, 1204 (1978)
74. Renger, G., Eckert, H.-J.: Bioelectrochem.Bioenezg. *7*, 167 (1980)
75. Volkov, A.G., Kolev, V.D., Levin, A.L., Boguslavsky, L.I.: Photobiochem.Photobiophys. *10*, 105 (1985)
76. Kolev, V.D.: J.Mol.Struct. *114*, 257 (1984)
77. Fragata, M.: J.Colloid Interface Sci. *66*, 470 (1978)
78. Volkov, A.G., Kolev, V.D., Levin, A.L., Boguslavsky, L.I.: Elektrokhimiya *22*, 1303 (1986)
79. Kolev, V.D., Volkov, A.G.: Intern. Youth Simposium Plant Metabolism Regulation, Abstract, Varna, 24–29 September, p. 24, 1983
80. Volkov, A.G.: Photobiochem.Photobiophys. *11*, 1 (1986)
81. Volkov, A.G.: J. Electroanal. Chem. *173*, 15 (1984)
82. Volkov, A.G.: Elektrokhimiya *21*, 81 (1985)
83. Volkov, A.G.: Biofizika *30*, 491 (1985)
84. Kutjurin, V.M.: Uspekhi sovremennoi biologii *59*, 205 (1965)
85. Kutjurin, V.M., Artamkina, I.Yu., Anisimova, N.N.: Dokl.Akad.Nauk SSSR *180*, 1002 (1968)
86. Volkov, A.G.: In: Bioconversion of Solar Energy (Berezin, I.V., ed.) Pushino, ONTI NZBI Akad.Nauk SSSR, p. 149, 1984
87. Semenov, N.N., Shilov, A.E., Lichtenstein, G.I.: Dokl.Akad.Nauk SSSR *221*, 1374 (1975)
88. Kornyshev, A.A., Volkov, A.G.: J.Electroanal.Chem. *180*, 363 (1984)
89. Volkov, A.G., Kornyshev, A.A.: Elektrokhimiya *21*, 814 (1985)

Counterions and Adsorption of Ion-Exchange Extractants at the Water/Oil Interface

A. N. Popov

Abstract

This review deals with the relation between the extractant surface activity upon adsorption from the non-aqueous phase at the water-oil interface and the extraction characteristics of the systems. Alkylammonium salts, alkylphosphoric and sulfonic acids and macrocyclic ethers were taken as extractants. The properties of adsorption monolayers were compared for the above extractants at water-air and water-oil interfaces. The method of determining the extractant activity coefficients in low permittivity media from the surface pressure isotherms is discussed.

At present, liquid extraction is widely used to extract and concentrate substances [1]. Other high-efficiency processes utilizing double emulsion membranes, membranes on a porous hydrophobic support [2], and electrodialysis systems with liquid membranes [3] gradually find application in this field. In practice, many inorganic and organic ions are determined by ion-selective electrodes with a liquid membrane [4]. Interface catalysis has also been widely accepted in organic chemistry [5]. The water-oil interface is common to all the above systems. Ion transfer across the interface is of primary importance in describing processes that take place in these systems. Almost all extractants and membrane-active substances of ion-selective electrodes are insoluble in water and display surface activity at the water-oil interface upon adsorption from the non-aqueous phase. The adsorption process at the interface may be viewed as a part of a general process of ion transfer across the interface. Thus, it is only natural to expect a correlation between the extraction and adsorption characteristics in water-oil systems. The concepts developed by Rehbinder [6] serve as a basis for studying the relation between bulk and surface properties of surfactant solutions. In many important cases extraction processes are selected that are based on ion-exchange mechanisms. Liquid anion exchangers are represented by solutions of onium salts (mainly alkylammonium salts [7]) in immiscible diluents, while liquid cation exchangers are mainly represented by alkylphosphoric, carboxylic and some other acids [1]. In interface catalysis, quaternary alkylammonium salts, crown ethers and cryptands [5] are used to transfer inorganic ions to the organic phase. All these compounds are used in ion-selective electrodes [4] as membrane-active substances and in membrane systems for the ion extraction and concentration [2, 3]. The ion exchange extraction may be viewed as an exchange reaction between counterions and the monolayer of the extractant adsorbed at the interface with the subsequent transfer of the resultant compound into the organic phase. According to the modern concepts of ion exchange extraction, the counterion

The Interface Structure and Electrochemical Processes
at the Boundary Between Two Immiscible Liquids
Editor: V. E. Kazarinov
© Springer-Verlag Berlin, Heidelberg 1987

exchange occurs either at the interface or in the adjoining layers of water or the organic phase [8, 9]. Hence the interaction of the extractant and counterions in the electrical double layer at the interface, the nature of the counterion and its effect on monolayer properties are essential in interpreting the mechanism of ion transfer across the water-oil interface.

The present review is devoted to the relation between bulk characteristics (extraction constants, extractant association in the non-aqueous medium) of ion-exchange extraction systems and properties of the monolayers adsorbed at the water-oil interface.

The Water/Oil Interface and Extraction Processes

The field of liquid extraction is characterized by rapid advances in the interfacial electrochemistry of two immiscible electrolytes [10, 11] and an ever-growing attention to the water-oil interface. Interfacial phenomena and their influence on extraction kinetics of inorganic compounds are being continuously studied by Yagodin, Tarasov et al. [8, 12]. They proved that, together with the bulk reactions producing extractable compounds, an important role is also played by surface reactions. In certain cases the contribution of the latter to the total extraction rate may be prevailing. The adsorption layers in extractants are sometimes significantly catalytically active and the reaction rate at the interface appears to be 5–6 orders higher than in the bulk of the aqueous phase [13]. This is usually accounted for by the concentration and favourable orientation of the extractant molecules in the adsorption monolayer. Experiments with ionogenic surfactants established their influence on the ion extraction rate, thus supporting the above statement. For instance, in a number of cases anionic surfactants accelerate cation transfer across the interface, while cationic surfactants decelerate the same process. On the whole, problems pertaining to the effect of surfactants on the extraction rate are more complicated and related to the hydrodynamics of the two-phase systems [8, 12]. The mechanism of micellar catalysis is also closely related to the specific structure of the adsorption layer at the water-oil interface [14].

The interface in extraction systems is usually studied by measuring the interfacial tension, viscosity and potential. To study the adsorption kinetics, one usually plots the isotherms of the interfacial tension and uses the Gibbs adsorption equation to calculate the surface concentration of the extractant [15–22]. In practice, the concentration of the extractant is selected so as to saturate the monolayer at the interface. In such systems a rise in extractant concentration does not affect the extraction rate if the limiting stage is the surface reaction or a reaction in the adjoining layers [17–20, 23].

The water-oil interfacial potential is one of the most important characteristics of the electrical double layer. In the polar oil-water systems the Volta potential is determined by the salt distribution coefficient, while in the nonpolar oil-water systems it is related to the orientation of the surfactant and solvent molecules on the adsorption layer [11, 24, 25]. The measurements of the Volta potential and interfacial tension enable one to determine the zero charge potential, the value of the dipole part of the potential drop and to consider the possibility of the specific adsorption of surface-active ions [26]. However, data on the influence of various factors on the potential drop in the widely used extraction systems are scarce [27, 28]. A rigorous stirring during the extraction leads to considerable boundary deformations that generate an interfacial potential

[29, 30]. Longitudinal waves at the interface cause the appearance of an alternating electrical current [31]. The electrical fields generated in the course of rigorous stirring may significantly affect the extraction kinetics.

The interfacial viscosity and elasticity data are contradictory due to the experimental problems. The study of the rheological properties of interfacial layers gives valuable information even for the well-known systems. For instance, the study of the interfacial viscosity made it possible to detect surface films that disappear under equilibrium conditions due to their dynamic nature. The formation of high viscosity films slows down the extraction rate, changing the mechanism of the process [32].

Raman and infrared spectroscopy yield valuable data on the orientation and interaction of molecules in monolayers [33, 34]. Raman spectroscopy was used to study the adsorption of dyes at the water-carbon tetrachloride interface [35]. The adsorption isotherm plotted from spectral data almost coincided with the surface pressure isotherm.

Until recently, insufficient attention was paid to the relation of the interface structure and the liquid extraction mechanism. Further investigations utilizing modern methods developed in electrochemistry, micellar catalysis, and other disciplines should contribute to solving many practical and theoretical problems.

Adsorption and Extraction Constants

The free energy of adsorption is one of the vital characteristics of surfactants. It equals the work of transfer of one mole of surfactant from the bulk of a solution to the interface taken with the opposite sign. This surfactant solution, which corresponds to an infinite dilution, is taken as a standard state of the bulk phases. The selection of the standard state at the interface may vary, but in any case it is related to the selection of the state equation for the adsorption monolayer (or of the adsorption isotherm) [36–38]. Frequently the standard state is assumed to be a state of molecules in the monolayer, saturated to the degree of 0.5. Sometimes, comparing the behaviour of a number of surfactants at a certain interface, the standard state is referred to as a state of the monolayer with the same surface pressure (20 mN/M) [39]. Quite often it is difficult to select the equation of the monolayer state or adsorption isotherm. Besides, when the fractional surface coverage equals 0.5, the activity coefficient of molecules in the monolayer frequently differs from unity [37]. Therefore, at present, a method of determining the free energy of adsorption proposed by Betts and Pethica has been widely accepted [40]. They assumed that the standard state represents the state of molecules governed by the equation of state for the ideal two-dimensional gas (corresponding to Henry's adsorption isoterm) with a surface pressure of 1 mN/M. The adsorption free energy ($\Delta\mu^0$) and work of adsorption (W_{ad}) are written as follows:

$$W_{ad} = -\Delta\mu^0 = RT \ln \left(\frac{\pi}{C} \right)_{C \to 0}. \tag{1}$$

This expression of the work of adsorption is close to the expression of surface activity proposed by Rehbinder [6, p. 158]:

$$G = \left(\frac{d\pi}{dC} \right)_{C \to 0}. \tag{2}$$

The work of adsorption is determined from Eq. (1) by using the linear section of the surface pressure curve as a function of the bulk concentration of the surfactant (the surface pressure does not exceed 3 mN/M). The values obtained for the work of adsorption are usually compared with the values obtained for certain fractional surface coverages. Thus, it becomes possible to estimate the change in the free energy as a function of the molecular interaction in the monolayer. (The relation between the values of the work of adsorption determined from different isotherms is discussed in Refs. [37–39]). The adsorption entropy and enthalpy are determined from the temperature dependence of the work of adsorption [36–39].

To determine the work of adsorption, Abramson used the equation:

$$W = RT \ln \frac{\Gamma}{\delta C} + \frac{\pi_m}{\Gamma_m}, \tag{3}$$

where δ is the thickness of the adsorption layer, π_m is the surface pressure corresponding to a complete monolayer coverage, and I_m is the maximum adsorption [41]. To select the thickness of the adsorption layer, it is necessary to design and analyse models and make certain assumptions. The thickness of the adsorption layer affecting the surface tension at the air-water interface amounts to 6–7 Å, while at the water-oil interface it is 12 Å [42]. However, this approach seems to be too approximate [43].

The Gibbs equation serves as a basis for studying the adsorption thermodynamics [11, 24, 44]. For the strongly adsorbed surfactants with a small molar fraction in the solution compared to the molar fraction of the solvent (the solvents are mutually insoluble) the Gibbs equation is written as [45]:

$$d\sigma = - \sum_{i=3}^{n} \Gamma_i d\mu_i. \tag{4}$$

At low concentrations of a high-activity surfactant in the bulk phase, the surface excess is usually identified by the surface concentration.

A specific application of the Gibbs equation to the adsorption of ionogenic surfactants dissociating in the solvent volume should be pointed out. Haydon et al. apply the Gibbs equation in the form:

$$\Gamma = - \frac{1}{RT} \frac{d\sigma}{d \ln(Cf)} \bigg/ \left(1 + \frac{C}{C_i + C} \right), \tag{5}$$

where Γ is the surface excess of the extractant, C its bulk concentration, C_i the concentration of the supporting electrolyte, and f the surfactant activity coefficient [46]. Depending on the concentration of the supporting electrolyte, the $\left(1 + \dfrac{C}{C_i + C} \right)$ values range from 1 (the excess of supporting electrolyte) to 2 (the absence of supporting electrolyte). The direct radio-tracer measurements of the adsorption of sodium decylsulphate at the water-toluene interface in the presence and in the absence of the supporting electrolyte confirmed the Gibbs equation in the form of Eq. (5) [47]. The introduction of the co-ion adsorption in the Gibbs equation unreasonably complicated its application [46] since in the majority of cases the error introduced by neglecting the co-ion adsorption at the interface does not exceed the error of the experimental determination of Γ values. Γ is calculated by numerical differentiation which gives an error of up to 8% [46]. By special methods the error may be reduced to 3–5%, but it excessively complicates the data processing [48]. Therefore, to plot the adsorption isotherms and to conduct further calculations it is convenient to use splain-functions [49].

The adsorption at the water-oil interface was thoroughly studied for the case when the surfactant is insoluble in the organic phase and the phases are mutually insoluble. If the organic solvent possesses adequate polarity and is consequently well soluble in water, the surface excess of the organic solvent should not be considered as equal to zero. In this case, to calculate the surface excess of the surfactant, additional assumptions are required [47, 50]. When studying the adsorption from the organic phase at the water-oil interface two cases may be singled out: (a) the surfactant is distributed between both phases, (b) the surfactant is mainly located in the organic phase.

In the first case, to calculate adsorption, the values of the equilibrium surfactant concentrations in any phase may be taken, since under equilibrium conditions the chemical potentials of all components in the water and organic phases are equal and are interrelated through the distribution constant [44, 51]. Investigating aniline adsorption in the water-toluene system, Rehbinder was the first to experimentally show that at equilibrium the adsorption may be calculated from any phase [52]. He also proved the validity of the following equality:

$$C_{aq} \frac{d\sigma}{d \ln C_{aq}} = C_{org} \frac{d\sigma}{d \ln C_{org}}. \tag{6}$$

It follows that:

$$K_{extr} = \frac{C_{org}}{C_{aq}} = \frac{d\sigma}{d \ln C_{aq}} \bigg/ \frac{d\sigma}{d \ln C_{org}}. \tag{7}$$

The Rehbinder relation (7) was used for the thermodynamic interpretation of the hydrophilic-oleophilic relation in the surfactant molecules [53].

The surface excess in the water-oil system was also calculated for the case of the surfactant distribution between the phases [26, 54].

When measuring the surface pressure isotherms, it is desirable that the values of the interfacial tension are not time-dependent. In this case, in the interfacial region a state is reached close to the equilibrium for the surfactant distribution between the phases [55].* If the surfactant is soluble in both phases, one should be careful in calculating the surface excess in such systems, and the surfactant distribution coefficient should be determined independently. For instance, trioctylmethylammonium chloride (Oct$_3$MeNCl) in the benzene-water system has a distribution coefficient of the order of 10^2 [57]. The surface pressure isotherms at the benzene-water interface are almost independent of the phase in which Oct$_3$MeNCl is dissolved. It means that in both cases Oct$_3$MeNCl is almost completely located in the benzene phase, i.e. the surfactant distribution equilibrium is reached at the interface. Apparently, the anomalies in the

---

* Crotov showed that in the case of adsorption of several surfactants in the system of two immiscible liquids there may exist stationary conditions (differing from the equilibrium one) of the surface monolayer [56].

surface pressure isotherms for the aqueous solutions of tetra-heptylammonium chloride (Hep$_4$NCl) at the boundary with n-heptane arise for the same reasons: during the experiment the larger portion of salt is in the organic phase [58].

The work of transfer of one mole of surfactant (W_{extr}) from the aqueous phase to the non-aqueous phase may be written as:

$$W_{extr} = {}_{ad}W_{aq} - {}_{ad}W_{org},$$
(8)

where $_{ad}W_{aq}$ and $_{ad}W_{org}$ are the works of surfactant adsorption from the corresponding phases. Substituting $W_{extr} = RT\ln K_{extr}$ into Eq. (8), we have:

$$RT\ln K_{extr} = {}_{ad}W_{aq} - {}_{ad}W_{org}.$$
(9)

This equation may be derived from the Rehbinder equation (7) and the expression for the work of adsorption of Eq. (1). Equation (9) was used in several works [11, p. 158, 53, 59–62]; it serves as a basis for studying the relation between extraction and adsorption characteristics in the water-oil systems.

Counterions and Adsorption of Extractants

1. Alkylammonium Salts

Alkylammonium salts are typical cationic surfactants. The adsorption and the properties of the monolayers formed by these salts at water-air and water-oil interfaces have frequently been investigated [11, 24, 36, 37, 46, 63]. Mostly the adsorption of alkylammonium cations from the aqueous phase was studied. Special attention was paid to the relation between the structure of alkylammonium cations, their surface-active properties and the monolayer structure. Less attention was given to the influence of the anion nature of alkylammonium salt adsorption from the non-aqueous phase and competitive adsorption of counterions in cationic monolayers at the water-oil interface, although these problems are vital for understanding the mechanism of liquid ion-exchange extraction.

The lyotropic Hofmeister series have been widely used in colloid chemistry for a long time. They are arranged in the order of increasing or diminishing ion influence on the surface tension, viscosity of solutions, coagulation stability of colloids and other properties. The position of ions in these series is mainly determined by the solvation (hydration) energy [64]. In particular, the less hydrated they are the higher is the ion's ability to be adsorbed at the hydrophobic interfaces. This regularity was discovered by Frumkin for the surface activity of tetra-propylammonium salts at the water-air interface [65]. Later the series was continued [66, 67]. The interest in the interaction of counterions with monolayers at the interface has arisen again in connection with investigations on bilayer lipid membranes [68].

The most detailed study of the interaction between inorganic ions and insoluble cationic monolayers was conducted by Goddard et al. [69–72]. A specific interaction was observed between anions and the monolayer of alkylammonium cations. Its value depends on the nature of the anion with the exception of the ion of fluorine. Only in the case when F^- is taken as a counterion, the dependence of the surface pressure on the area per molecule in the monolayer obeys the Davis equation of state for the charged monolayers. As the counterion radius decreases, the surface pressure on a given area per molecule increases. A large potential drop corresponds to such a monolayer, i.e. the specific interaction results in the condensation of the monolayer. On the basis of the surface potential drop data it was concluded that the specific adsorption leads to the formation of ionic pairs between the monolayers cations and counterions [69].

The authors determined the energy of the specific adsorption of anions on the monolayer formed by cations of docosyltrimethylammonium at the water-air interface with the help of a specifically nonadsorbing electrolyte (NaF). The ionic strength of the solution was maintained constant throughout the experiments. Therefore the energy of the unspecific Coulomb interaction of counterions with the charged monolayer was considered equal in all cases. Then the salt of the anion under investigation was introduced into the aqueous phase. The amount of salt was calculated so as to reduce the surface pressure by 1 mN/m. The concentration of the added counterions in the monolayer remains the same. Taking into account these assumptions, the specific interaction energy was calculated from the Boltzmann equation written in the form:

$$C_s = C_v \exp\left(\frac{e\varphi}{kT}\right) \exp\left(\frac{\Phi}{kT}\right), \tag{10}$$

where C_s and C_v are the surface and bulk concentrations of counterions, Φ is the specific interaction energy, and $e\varphi$ the electrostatic energy. By taking the interaction energy of Cl^- ions with the monolayer as unity, the values 4.5, 5, 6.5, and 8 (in kT units) were obtained for Br^-, NO_3^-, J^-, and NCS^-, respectively. Through viscosity measurements, Davis [24] obtained values for the specific adsorption in cationic monolayers somewhat different from those reported by Goddard, although the anion series maintained the sequence.

Ter-Minasian-Saraga and Plaisance investigated the specific adsorption on the monolayers with quaternary nitrogen atoms as cations [73, 74]. They studied the selective adsorption of thiocyanate ions from solutions with constant ionic strength where KX salt ($x = F^-, Cl^-, Br^-, J^-, NO_3^-, ClO_4^-$) was present apart from NCS^-. The adsorption was studied with the help of a thiocyanate ion tracer (C^{14}). The results were interpreted on the basis of Stern theory with regard to the discrete structure of the electrical double layer [75]. The specific adsorption was considered to be adequate for the ionic pair formation on the inner Helmholtz plane. This concept with due regard to the ion-exchange reaction and several assumptions on the structure of the polymeric layer at the interface enabled the authors to connect the change in the potential drop in the monolayer with the specific adsorption. The excellent experimental results within the above assumptions made it possible to evaluate the fraction of counterions bonded in ionic pairs. This fraction appeared to be close to unity for NCS^- and ClO_4^-, close to zero for F^- and OH^-, and had intermediate values for the other anions under consideration. The specific adsorption capacity of anions forms the sequence:

$$ClO_4^- > NCS^- > J^- > NO_3^- > Br^- > F^- \simeq OH^-. \tag{11}$$

The surface activity of the soluble cationic surfactants at the water-air and water-nonpolar oil interfaces is mainly determined by the structure of the hydrocarbon chains. The work of adsorption in the homologous series of alkylammonium salts obeys the Traube rule [24, 46, 63, 76]. Shipunov studied adsorption in the homologous series of tetra-alkylammonium chlorides at the water-heptane interface. He established that the work of adsorption is proportional not to the number of methyl groups in the hydrocarbon chains but to their surface area. The increase in the cationic surface by 1 Å² cause the work of adsorption from the aqueous phase to grow by 126 J/mol [58].

The nature of the anion also affects the surface activity of the water-soluble cationic surfactants. The surface activity of tetra-butylammonium salts at the water-air and water-hexane interfaces reduces in the order iodiode > bromide > chloride [77, 78]. Boguslavsky et al. reported that the adsorption equilibrium constant and the adsorption at the water-octane interface are higher for tetra-alkylammonium bromides as compared to chlorides [76, 79]. The influence of the counterions on the adsorption of alkylpyridinium salts at the water-air interface was studied. It was established that the salts lowered the surface tension in the sequence: perchlorate > thiocyanate > iodide > bromide > chloride [80–83]. The anion influence on the work of adsorption was found to increase with the growth of the concentration of the surface-active salt [83].

The anion series in the order of their influence on the surface activity of cationic surfactants and on the properties of insoluble monolayers coincide with the lyotropic series. The anion nature is manifested in the specific interaction with cationic surfactants with the formation of ionic pairs or in the anion penetration into the adsorption monolayers. Apparently, charge transfer complexes are frequently formed at the water-air and water-oil interfaces between the adsorbed cations and anions. Mukerjee and Ray established that these complexes exist in ionic pairs, between Br^- or J^- anions and dodecylpyridine ions in chloroform, and also on the micelle surface formed by the same cations in water [84–86]. The data on the ionic pair structure at the interface can be obtained by spectral study of monolayers and solutions of ionogenic surfactants in nonpolar solvents. According to Goddard, the investigation on the mixed monolayers formed by the surface-active anions and quaternary ammonium compounds will enable us to better understand the specific properties of the interaction between the adsorbed ions [72].

The adsorption at the water-air interface is used in separation and concentration processes (foam separation and flotation extraction) [87]. To remove the traces of metals that are present in the anionic form, the method of foam fractionation is used. (The foam is formed during air bubbling through the water solution of hexadecyltrimethylammonium bromide) [88]. In this process the selectivity coefficients are determined by the anion hydration energies.

The anion position in the series (II) determines its influence on the surface activity of alkylammonium salts upon adsorption from the aqueous phase. The value of the effective permittivity of the electrical double layer is lower than in the bulk phase due to dielectric saturation caused by the strong electric field and the presence of the adsorbed monolayer [89]. Therefore, the less hydrated anions permeate more easily to the plane formed by the adsorbed alkylammonium ions and form ionic pairs due to the interaction.

In the extraction systems the surfactant is distributed between the phases. Boguslavsky et al. systematically studied the adsorption of tetra-alkylammonium salts in such systems at the water-nitrobenzene interface [11, 76, 90, 91]. The values of the Volta potential at this interface are consistent with the thermodynamic distribution theory. The adsorption isotherms are formally described by the Frumkin equation: with the increasing size of tetra-alkylammonium cation, the repulsion between the adsorbate particles increases. It testifies to the fact that the cations of alkylammonium salts (starting with tetra-butylammonium) are located in the nitrobenzene part of the electrical double layer. The adsorption theory of the surface-active ions developed by Krylov confirms this conclusion [89]. Boguslavsky used Eq. (7) to describe the distribution of salt between the bulk phase and the surface layer. He established that the change in the distribution potential may be determined from the adsorption isotherms [11, p. 158]. Undoubtedly, this result is a valuable contribution to the study of extraction kinetics in the water-polar oil system.

The adsorption of alkylammonium salts from the organic phase at the water interface was extensively studied in connection with the investigation of the mechanism of the anion-exchange extraction [16, 21, 62, 92–98]. The following alkylammonium cations were studied:* $Hept_4N^+$, Hex_4N^+, Oct_3NH^+, $Oct_2Me_2N^+$, Oct_3MeN^+, Oct_4N^+, Dec_3NH^+, Dec_3MeN^+, $DecMe_3N^+$, Dec_4N^+, Lau_3NH^+, Lau_4N^+, $LauMe_3N^+$, Cet_3NH^+ – in various solvents (n-octane, benzene, toluene, o-xylene, methylbutylketone, n-amyl alcohol, nitrobenzene). The influence of anions on the surface activity of the above alkylammonium cations forms the sequence:

$$PO_4^{3-} > SO_4^{2-} > F^- \simeq OH^- > Cl^- > Br^- \simeq [Fe(CN)_6]^{4-} > NO_3^-$$
$$\simeq [Fe(CN)_6]^{3-} > J^- > PdCl_4^{2-} > NCS^- > ClO_4^-$$
$$\simeq ReO_4^- > AuCl_4^- > Ag(CN)_2^- > Au(CN)_2^- . \qquad (12)$$

This series as opposed to the extraction series [7] coincides with the anion series of decreasing hydration energy [99]. The alkylammonium cation being the same, the salts of anions (better extractable into the organic phase) are worse adsorbed from the non-aqueous phase at the water-oil interface. The anion sequence (12) acquires a quite clear physical meaning: the higher the anion hydration energy, the greater the gain in the adsorption energy. This is due to the fact that during adsorption of alkylammonium salts at the water-oil interface the anions are located on the aqueous side of the electrical double layer. (The gain in the adsorption energy depends on the re-solvation energy of the anion.) The surface pressure isotherms demonstrating the effect of the anion of the alkylammonium salt on the surface activity from a non-aqueous phase are given in Figs. 1 and 2.

The transfer of hydrocarbon chains from the aqueous to the non-aqueous phase mainly contributes to the energy of alkylammonium salt adsorption from water at the boundary with organic solvents. The anion nature does not significantly influence the work of adsorption. For instance, the work of adsorption of Am_4N^+ halides from the aqueous phase in the presence of the corresponding 0.1 M potassium salts at the interface with n-heptane is 32.85, 33.15, 33.55 kJ/mol for chloride, bromide and iodide, respectively. The data reported in Refs. [77, 78, 83] estimate the influence of the anion nature on the work of adsorption of the alkylammonium salt from the aqueous phase at the boundary with an organic solvent as not exceeding 10%. Therefore, in the first approximation for a given alkylammonium cation the work of adsorption from the aqueous phase is independent of the anion nature. It follows from Eq. (9) that in this case a linear dependence should exist between the work of adsorption from the non-aqueous phase and the logarithm of the extraction constant. The data in Fig. 3 confirm

---

* The notation of hydrocarbon radicals was taken as follows: Me – methyl, Bu – butyl, Am – amyl, Hex – hexyl, Hept – heptyl, Oct – octyl, Dec – decyl, Lau – lauryl, Cet – cetyl.

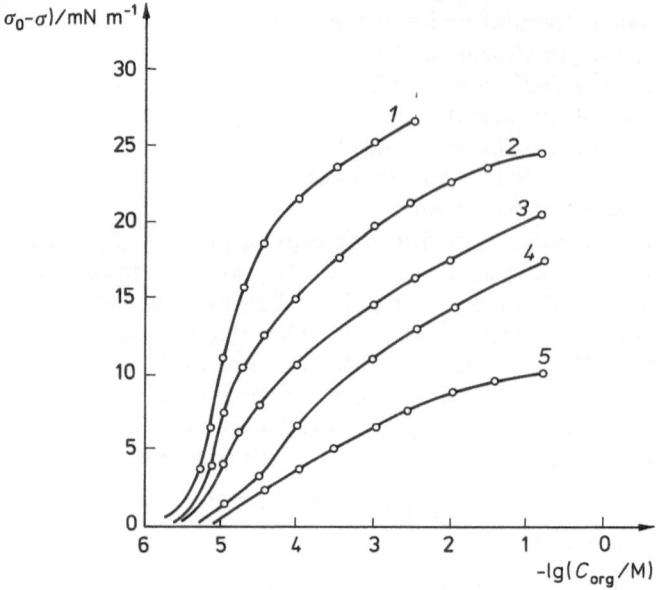

Fig. 1. Dependence of surface pressure isotherms in benzene-water solutions on the salt concentration: – in benzene: 1 – Oct_3MeNCl; 2 – Oct_3MeNBr; 3 – Oct_3MeNNO_3; 4 – Oct_3MeNJ; 5 – $Oct_3MeNClO_4$; – in water: 1 – $1\,M$ KCl; $0.1\,M$ $LaCl_3$; 2 – $1\,M$ KBr, $1\,M$ HBr, $1\,M$ $BaBr_2$; 3 – $1\,M$ KNO_3; 4 – $1\,M$ KJ; 5 – $1\,M$ $NaClO_4$

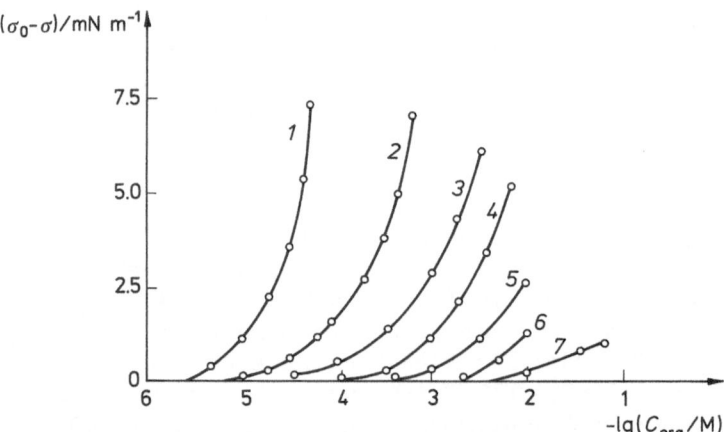

Fig. 2. Dependence of surface pressure isotherms in methylbutylketone-water solutions on the salt concentration: – in MBK: $1, 2$ – $CetMe_3NCl$; 3 – $CetMe_3NBr$; 4 – $CetMe_3NNO_3$; 5 – $CetMe_3NJ$; 6 – $CetMe_3NNCS$; 7 – $CetMe_3NClO_4$; – in water: related K^+ and Na^+ salts ($0.1\,M$)

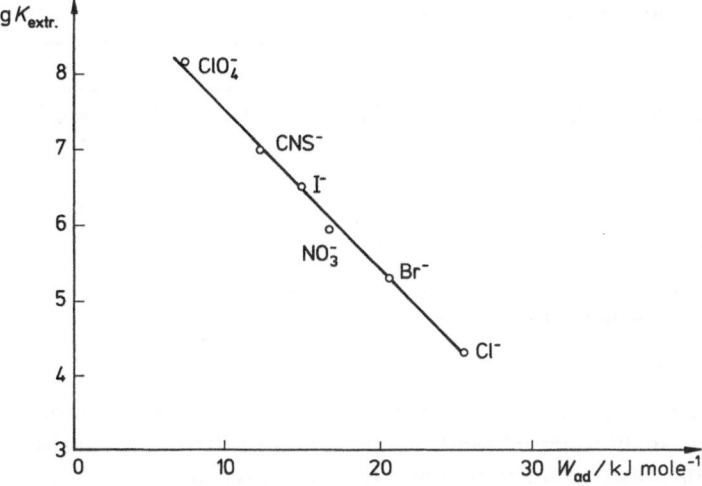

Fig. 3. Logarithm of the extraction constant as a function of the work of adsorption of $CetMe_3N^+$ salts from methylbutylketone at the interface with 0.1 M aqueous solutions of the related potassium and sodium salts

Table 1. Limiting adsorption, work of adsorption and specific energy of interaction of $CetMe_3N^+$ cations with counterions at the interface with 0.1 M aqueous solutions of the corresponding potassium and sodium salts upon adsorption from methylbutylketone

Anion	Adsorption, $\Gamma \times 10^{10}$ mol/sm^{-2}	Work of adsorption, kJ/mol^{-1}	Specific interaction energy	
			Φ/kT	Φ/kT^{a}
Cl^-	1.55	25.5	1	1
Br^-	1.45	20.6	4.5	4.5
NO_3^-	1.10	17.0	5.5	5.0
J^-	0.65	15.0	6.5	6.5
NCS^-	0.40	12.3	7.0	–
ClO_4^-	0.35	7.5	7.5	8.0

[a] Energy of the specific interaction of anions with the monolayer of docosyltrimethylammonium cations at the water-air interface [69].

this conclusion. (Extraction constants were calculated on the basis of correlations advanced by Schmidt [7, 99].)

A linear dependence is observed between the logarithm of the extraction constant for alkylammonium salts and the anion hydration energy [100]. Consequently, Eq. (9) may be rewritten:

$$_{ad}W_{org} = A\Delta G_{hydr} + B, \tag{13}$$

where A and B are constants. The experimental data prove the validity of Eq. (13) [61, 98]. It may be used for quantitative estimation of the anion hydration energy.

Table 1 lists the energies of the specific interaction of cations $CetMe_3N^+$ with the counterions at the water-methylbutylketone interface. This energy was determined by

Table 2. Values of limiting adsorption, attraction constant, and work of adsorption for tetra-alkylammonium chlorides at the water-benzene interface upon adsorption from the non-aqueous phase

Salt	$Hept_4N^+$	Oct_4N^+	Dec_4N^+	Oct_3MeN^+	$Oct_2Me_2N^+$	Hep_4N^{+} [a]
Limiting adsorption, 10^{10} mol/sm^2	4.8	4.8	4.8	4.7	7.6	1.8
Attraction constant	1.5	1.7	1.7	1.6	1.4	0.7
Work of adsorption, kJ/mol	33.3	36.0	36.3	32.7	30.5	–

[a] Data for the water-octane interface [76].

the technique developed by Goddard for the monolayers at the water-air interface. Extending this technique to the water-oil interface, one should assume that the growing specific interaction at the water-air interface causes the monolayer compression, while due to the specific adsorption at the water-methylbutylketone interface the ionic pair forms and is at least partially desorbed into the volume of the organic phase.

As is seen from the series (12) the surface activity of alkylammonium salts from the non-aqueous phase is determined by the specific interaction of anions with the monolayers of alkylammonium cations and not by the anion charge.

The adsorption isotherms of tetra-alkylammonium salts (beginning with $Hept_4N^*$) from benzene at the water interface formally follow the Frumkin equation:

$$BC = \frac{\theta}{1+\theta} \exp(-2a\theta),\tag{14}$$

where B is the constant of the adsorption equilibrium, and a is the attraction constant accounting for the interaction between the adsorbed particles [101, p. 68–80].

An attraction is observed between the adsorbed cations of tetra-alkylammonium at the water-benzene and water-octane interfaces (Table 2).

It can therefore be concluded that the electrical double layer is shifted to the aqueous phase at the above interfaces [11, p. 162]. In the adsorption theory of surface-active ions at the water-oil interface the nature of the organic phase is taken into account by the permittivity of the solvent in the electrical double layer [89]. According to this theory and the ideas of Boguslavsky [11], at the water-nitrobenzene interface ($D = 36$), the sites of adsorbed ions are located in the non-aqueous phase. The repulsion between these sites grows with increasing radius of the alkylammonium cation. At the water-octane and water-benzene interfaces ($D = 2$) the cation sites are in the aqueous phase and the attraction constant grows with the radius of adsorbed ions.

Such a dependence of the attraction constant makes it possible to assume that there should be no essential interaction between the adsorbed cations at the interface of water with an organic solvent with an intermediate permittivity value ($D \cong 17$). Actually, the adsorption of $CetMe_3NCl$ at the water-methylbutylketone interface ($D = 14.6$) is described by the Langmuir isotherm. Apparently, the values of the effective permittivity are equal in the Helmholtz layer of both phases and no definite conclusion can be drawn what so ever concerning the location of the adsorbed cation sites. The adsorption isotherms of the above systems are given in Fig. 4.

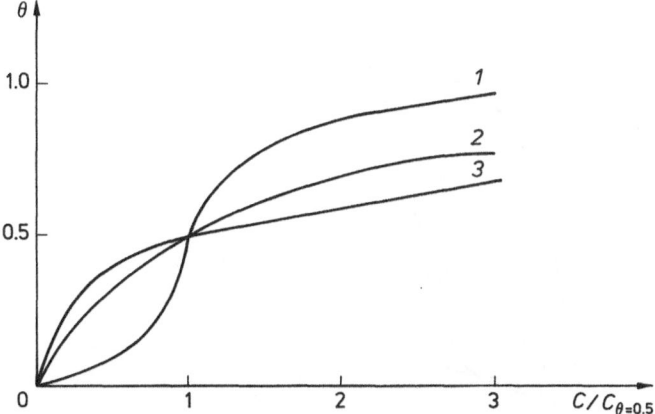

Fig. 4. Adsorption isotherms in the dimensionless coordinates at the water-oil interface: *1* – Oct$_3$MeNCl in benzene ($a = +1.6$); *2* – CetMe$_3$NCl in MBK ($a = 0$); *3* – Hept$_4$NCl in water at the interface with nitrobenzene [76] ($a = -1.2$)

The analysis of molecular models shows that hydration chains of tetra-alkylammonium ions (starting with Hex$_4$N$^+$) are quite flexible, so that the four hydrocarbon chains of the cation adsorbed at the water-oil interface are oriented towards the non-aqueous phase. At the limiting coverage, the area per hydrocarbon chain is 18.5 Å^2 [102]. This corresponds to the adsorption of 2.25 10^{-10} mol/cm^2. The values of the limiting adsorption of tetra-alkylammonium salts at the water-benzene interface considerably exceed this value (Table 2). The adsorption should not be calculated from Eq. (4) since the condition of the constant potential drop at the interface was not experimentally verified. Further studies on the subject will require highly accurate experiments with alkylammonium cations containing hydrocarbon chains of different structure.

If different anions are located in the aqueous and organic phases, the ion-exchange reaction will occur in the region of the interface, and the composition of this region will differ from that of bulk phases. Since different salts of a certain alkylammonium cation have different surface activity from the non-aqueous phase, the composition of the region adjoining the interface may be determined from the surface pressure isotherms.

The following conclusions can be drawn from the surface activity data for alkylammonium salts at the interface with aqueous solutions of various inorganic salts and acids [61, 98]. If the alkylammonium salt with the least extractable ion (Cl$^-$) is dissolved in the organic phase, while the acid or salt with the better extractable anion (ClO$_4^-$) is dissolved in the aqueous phase, the better extractable anions serve as counterions of the adsorbed monolayer. The starting part of isotherm 2–4 (Fig. 5) coincides with the surface pressure isotherm for the system containing a water solution of 1 M NaClO$_4$ and benzene solution of Oct$_3$MeNClO$_4$. When the Oct$_3$MeNCl concentration in benzene exceeds that of ClO$_4^-$ ions in the aqueous phase, the latter are replaced by Cl$^-$ ions in the adsorbed layer which results in a sharp growth of surface pressure. When a certain alkylammonium salt concentration is reached, the surface pressure attains the value close to the pressure of the system devoid of ClO$_4^-$ ions.

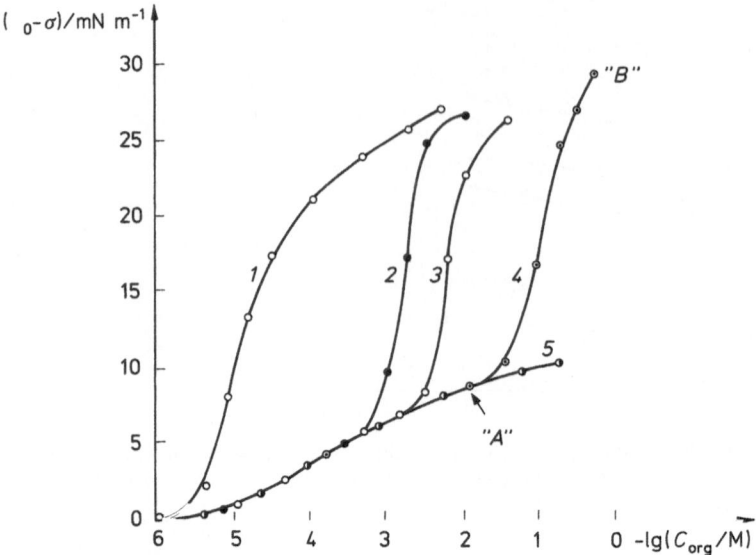

Fig. 5. Dependence of surface pressure isotherms on salt concentration in benzene-water systems: – in benzene: *1–4* – Oct_3MeNCl; *5* – $Oct_3MeNClO_4$; – in water: *1* – $1\,M$ KCl; *2* – $1\,M$ $KCl + 10^{-3}\,M$ $NaClO_4$; *3* – $1\,M$ $KCl + 10^{-2}\,M$ $NaClO_4$; *4* – $1\,M$ $KCl + 10^{-1}\,M$ $NaClO_4$; *5* – $1\,M$ $NaClO_4$

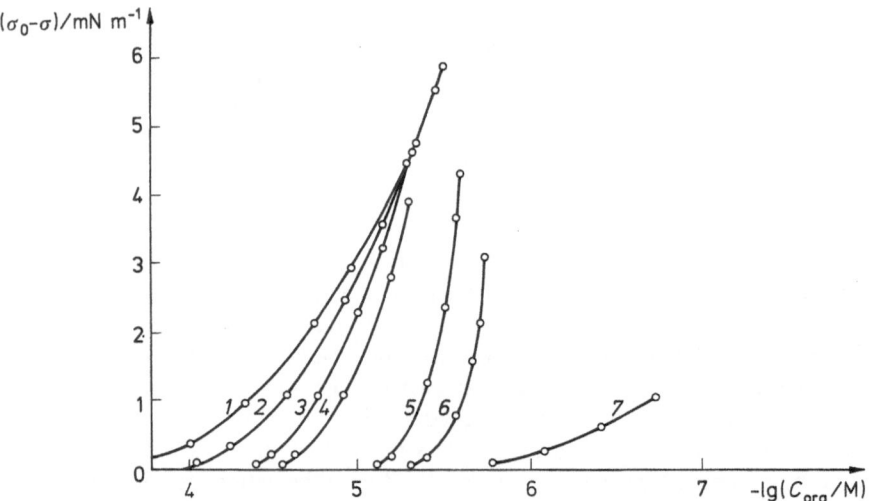

Fig. 6. Dependence of surface pressure isotherms for $CetMe_3NBr$ solutions in methylbutylketon-water solutions on the salt concentration: *1* – $0.1\,M$ KBr; *2* – $0.1\,M$ $KBr + 10^{-4}\,M$ $NaClO_4$; *3* – $0.1\,M$ $KBr + 5 \cdot 10^{-4}\,M$ $NaClO_4$; *4* – $0.1\,M$ $KBr + 10^{-3}\,M$ $NaClO_4$; *5* – $0.1\,M$ $KBr + 5 \cdot 10^{-3}\,M$ $NaClO_4$; *6* – $0.1\,M$ $KBr + 10^{-2}\,M$ $NaClO_4$; *7* – $0.1\,M$ $NaClO_4$

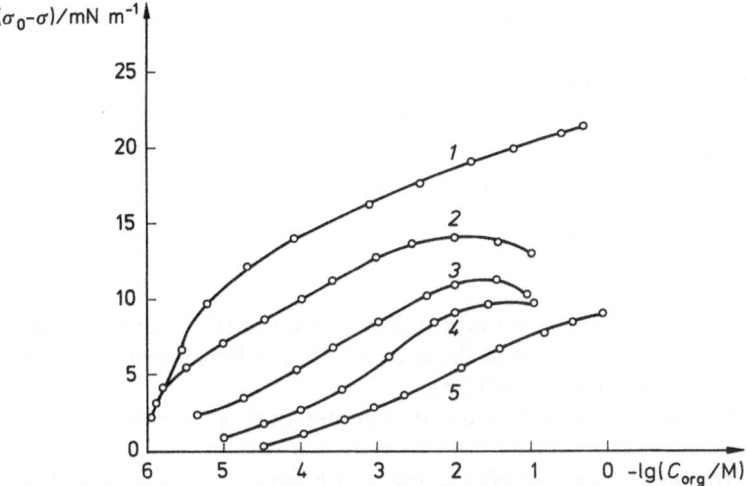

Fig. 7. Surface pressure isotherms for Oct_3NHClO_4 solutions in benzene: – curves 2–5, and $(Oct_3NH)_2SO_4$ in benzene – curve 1 at the interface with aqueous solutions: *1, 2* –0.5 *M* H_2SO_4; *3* – 1 *M* HCl; *4* – 1 *M* HBr; *5* – 1 *M* $HClO_4$

However, the isotherms do not fully coincide, which implies the presence of a small amount of perchlorate ions together with Cl^- ions in the monolayer.

Even a thousand-fold excess of the chloride ions does not significantly affect the surface pressure. A competitive anion replacement is also observed at other interfaces. Similar isotherms for the water-methylbutylketone interface are given in Fig. 6.

If the anions of the alkylammonium salt in the organic phase are better extractable than the anions of the salt in the aqueous phase, the concentration of the anion species at the interface depends on the salt concentrations and the value of the extraction constant (Fig. 7).

Thus, the surface pressure isotherms can help to determine the composition of the interfacial region, at least for the case when the non-aqueous phase contains the anion exchanger – alkylammonium salt, and the aqueous phase contains inorganic acids and corresponding salts. Apparently, this approach may be extended to the systems where the ion exchange between phases takes place and where the difference in hydration energies in anions or cations varies with the surface activity of ion exchangers from the non-aqueous phase.

The monolayers of cationic polysoaps were studied at the air-water interface [73, 74]. A direct relation between the anion adsorption and the monolayer selectivity towards a given counterion was established. In particular, such a monolayer exhibits an ideal selectivity towards rhodanide ions in the concentration regions where they were adsorbed at the interface. When other counterions in the monolayer replace rhodanide ions, the dependence of the monolayer potential on the ion concentration no longer obeys the Nernst equation.

According to Ishibashi et al. [103, 104] the composition of the region adjoining the liquid membrane-electrolyte interface differs considerably from that of a bulk solution. At the same time the composition of solutions at the interface determines the selectivity of the membrane towards a certain ion species. Hence the surface pressure isotherms can be used in predicting the selectivity of ion-selective electrodes towards various ions and in choosing the concentration of the liquid ion-exchanger [61, 98].

For example, the liquid membrane – Oct_3MeNCl solution in benzene in the electrochemical chain:

$$Hg, Hg_2Cl_2 \left| \begin{array}{c} |KCl| \\ sat \end{array} \right\| \begin{array}{c} 0.1\ M \\ NaClO_4 \end{array} \left| \begin{array}{c} |Oct_3MeNCl| \\ benzene \end{array} \right| \begin{array}{c} 1.0\ M\ KCl \\ NaClO_4 \\ 10^{-3}-10^{-1}\ M \end{array} \left\| \begin{array}{c} |KCl| \\ sat \end{array} \right| Hg_2Cl_2, Hg \tag{15}$$

– behaves as an ideal anion exchanger towards perchlorate ions in those regions of the anion exchanger and ion concentration in the aqueous phase where perchlorate ions are adsorbed at the water-oil interface (Fig. 5).

In liquid extraction the competitive anion adsorption influences the ion-exchange kinetics.

In ClO_4^--ion extraction by Oct_3MeNCl solution in benzene there are almost no Cl^- ions at point "A" and no ClO_4^- anions at point "B" of the surface pressure isotherm (Fig. 5) at the interface. The rate constant for the ion exchange in the system with the composition corresponding to point "A" almost doubles the constant for the system with the composition corresponding to point "B" [105].

Thus, the study of adsorption at the water-oil interface enables to evaluate the anion distribution constants, to judge on some specific kinetic features of ion-exchange processes, and on concentration regions and phase compositions at which the liquid anion-exchange electrode is selective towards a certain ion species.

2. Cation-Exchange Extractants

Aliphatic and alkylphosphoric acids, sulfonic acids and some other compounds are considered to be cation-exchange extractants [106]. Fatty and sulfonic acids are the classical objects of the physical chemistry of surface phenomena. Many studies were devoted to their adsorption from the aqueous phase at the interface with air and organic solvents [24, 36, 37, 39, 50, 75]. Some works are concerned with the influence of the cation nature on the monolayer properties of organic acids. Goddard et al. established the following series of cations arranged according to their interaction with the carboxylic group of arachic acid in the monolayer at the air-water interface: $Li^+ > Na^+ > K^+ > Me_4N^+ > Et_4N^+$ [70, 107]. The more condensed monolayer and smaller potential drop correspond to the stronger interaction of cations with smaller radius.

In sulfonic acids (monolayers of dodecyl- and docosylsulfonic acids) this series reverses: the less hydrated cations form more condensed monolayers. The cation series coincides with the lyotropic series [70–72, 108, 109]:

$$Cs^+ > Rb^+ > K^+ > Na^+ > Li^+. \tag{16}$$

Calcium ions interact with sulfonic acids more strongly than univalent cations. For instance, the ion-exchange constants in the monolayer of dodecylsulfate are equal to 300, 90, and 60 for the pairs: Ca^{2+}/Li^+, Ca^{2+}/Na^+, Ca^{2+}/Cs^+ [109].

The extraction of alkali metals by sulfonic acids grows with the decrease in the hydration energy of cations and corresponds to the series (16). When di-2-ethylhexylphosphoric acid (D2EHPA) is used as an extractant, the extraction constant is large for cations with a smaller radius [1, p. 131], and the extraction series coincides with the series of cation interaction with monolayers of carboxylic acids. (There is much in common in the structure of the carboxylic group and the polar group of alkylphosphoric acids. Due to this the regularities in ionic pair formation by anions of acids and ions of alkali metals should be observed in both cases.) The specific interaction of extractant-extractable ion significantly affects the selectivity of extraction. Hence the specific adsorption of counterions in the anionic monolayer changes in accordance with the cation extraction series. For example, the better extractable cations promote the adsorption of D2EHPA at the interface between air and water solutions of inorganic salts ($Ni^{2+} \simeq Co^{2+} > K^+$) [110].

A linear dependence of the Am^{3+} extraction constant on the surface activity of a number of dialkylphosphoric acids was established [20]. The minimum acid concentration in dodecane at which the limiting adsorption at the interface with 1 M nitric acid is attained served as a measure of the surface activity.

Rumanian workers used the position of minima on the curves of interfacial tension against concentration of $Ca(NO_3)_2$, $Sr(NO_3)_2$, HNO_3 to determine the composition of the extractable complex. The concentration of the extractant (solution of tri-n-butylphosphate in benzene) was kept constant [111, 112]. However, the determination of the composition of the complex from the ratio of bulk concentrations and not the surface ones seems to be inadequately justified.

Some data on the interfacial tension of extractants used for extracting the cations of metals is adduced in a review [28].

3. Macrocyclic Ionophores

With the use of valinomycin as a K^+ carrier across the bilayer lipid membranes, a number of works appeared devoted to its adsorption and complex formation with potassium salts at water-air and water-oil interfaces. In spite of the considerable efforts of scientists, these problems are far from being solved. Boguslavsky et al. studied the contact phenomena at the water-heptane interface in the presence of valinomycin and its complexes with K^+. They found the potential values to be abnormally high [11, 113]. The physical nature of these potentials is not clear. They are either related to the bulk charge in the heptane phase or to the formation of submolecular structures at the interface.

Valinomycin possesses high surface activity at water-air and water-oil interfaces [114]. However, the value of the limiting adsorption at the water-heptane interface ($5.7 \cdot 10^{-10}$ mol/cm^2) does not agree well with the specific features of the molecular structure of valinomycin.

Valinomycin dissolved in heptane interacts with aqueous solutions of potassium and sodium picrates. This does not change the interfacial tension as compared to the systems without salt [115]. On the other hand, the study of valinomycin monolayers at the water-air interface proved KCl and NaCl to similarly interact with monolayers only up to a concentration of 0.5 M. At a KCl concentration of 0.7 M the surface potential amounts to 1050 mV, while it is only 550 mV at the same concentration of

NaCl (area per molecule in the monolayer is $100\,\text{Å}^2$) [116]. It was also found that the anion of the potassium salt influences the surface properties of the monolayer. The anions form the following series arranged according to their influence on the value of the surface potential: $Br^- > Cl^- > NCS^- \gg J^- \simeq F^-$, SO_4^{2-}, at the potassium salt concentration of 0.7 M. Therefore, the behaviour of valinomycin and its complexes at the water-oil interface should be further investigated by new (for example, spectral) methods.

Temkin's isotherm describes the adsorption of crown ethers with aliphatic radicals (C_8H_{17}, $C_{10}H_{21}$, $C_{12}H_{25}$) at the water-air interface. For crown ethers containing an oxygen atom, the increase of the aliphatic chain by a CH_2 group changes the free energy of adsorption by $\sim 3.5\,\text{kJ/mol}$. It means that the Traube rule is applicable to the homologue adsorption. As opposed to 18-crown-6 and dicyclohexyl-18-crown-6, alkylcrown-ethers form micelles in aqueous solutions. The critical micellar concentration depends on the crown structure and is within a range of 10^{-4} M. The effect of the composition of the aqueous phase on the surface properties of the above polyethers was not investigated [117].

The isomers A and B of dicyclohexyl-18-crown-6 have an approximately similar surface activity at the water-air interface. Starting with 10^{-4} M, the limiting adsorption is attained, the area per molecule being around $120\,\text{Å}^2$. An aqueous solution of 0.1 M $Ba(NO_3)_2$ slightly reduces the surface activity of the isomer A of dicyclohexyl-18-crown-6 but leads to an increase in its adsorption, the limiting area per molecule in the monolayer being $87\,\text{Å}^2$ [22].

Dibenzo-18-crown-6 (DBC) has no clearly manifested surface activity at either the water-air or water-oil interface [118–120]. The acylation of DBC by carboxylic acids produces derivatives that form stable insoluble monolayers at the water-air interface. The chlorides of alkali metals in the aqueous phase increase the area per diacryloyldibenzo-18-crown-6 molecule in the monolayer. The area per molecule depends on the nature of the cation and increases in the sequence: $Cs^+ < K^+ < Li^+ < Na$. The complex formation constant of the crown ethers with a chloride of an alkali metal increases in the same sequence [118].

If DBC is dissolved in benzene, and the inorganic salt is present in the aqueous phase, the resultant complex $DBC \cdot MX$ (where MX is the metal salt) possesses a higher surface activity (Fig. 8). The surface pressure depends also on the salt concentration in the aqueous phase. Consequently, the surface concentrations of DBC and inorganic salt may be determined from surface pressure isotherms. The surface concentrations calculated for DBC and KNCS under the same conditions coincided to within 15%. That means that the complex formed at the interface has a stoichiometry of close to 1 : 1 [120].

DBC complexes with metal salts in the water-benzene system are the products of the interfacial reaction [119]. The measured work of adsorption (15.9 kJ/mol) can be taken as a sum of the free energy of complex formation and work of adsorption of DBC at the water-benzene interface. The dissolution of complexes in bulk phases was neglected. The interfacial constants of complex formation (Table 3) calculated from the work of adsorption are close to the constants determined in the mixed solvent – water-tetrahydrofuran [114]. The only exceptions are the complexes of DBC with Ba^{2+} and La^{3+} salts. Apparently this is due to stronger Coulomb repulsion of ions in DBC-salt complexes at the interface as compared to the bulk phases. (At the interface, the anions

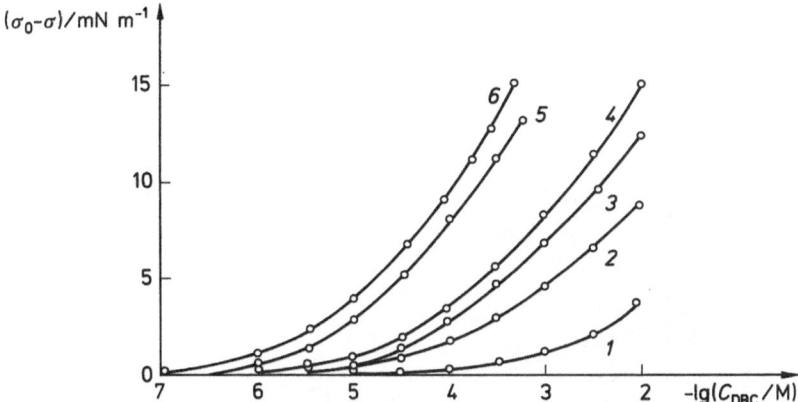

Fig. 8. Surface pressure isotherms for DBK solutions in benzene at the interface with aqueous solutions: $1 - H_2O$; $2 - 0.5\ M\ K_2SO_4$; $3 - 1\ M\ KCl$; $4 - 1\ M\ KNO_3$; $5 - 1\ M\ KNCS$; $6 - 1\ M$ KAg(CN)$_2$

Table 3. Work of adsorption (W_{ad}), limiting adsorption (Γ_{max}), and logarithm of the complex-formation constant ($\lg K$) of DBK-metal salt complexes at the interface of aqueous solutions of inorganic salts and benzene solutions of DBK

Salt	Salt concen. M	Γ_{max} 10^{-10} mol/cm^2	W_{ad}, kJ/mol	$\lg K$
KCl	1	1.15	27.2	2.00
KNO$_3$	1	1.35	28.0	2.15
KNCS	1	1.35	32.1	2.88
KAg(CN)$_2$	1	1.95	33.0	3.04
K$_2$SO$_4$	0.5	0.85	23.9	1.41
K$_2$PdCl	0.125	0.90	27.1	1.99
K$_3$[Fe(CN)$_6$]	0.33	1.05	27.0	1.97
K$_4$[Fe(CN)$_6$]	0.25	0.95	21.9	1.06
KNCS	0.33	1.15	30.0	2.50
NaF	0.8	0.60	16.0	0
NaNCS	1	1.00	22.0	1.08
NaClO$_4$	1	0.80	29.2	2.36
Ba(NCS)$_2$	0.33	1.30	21.4	0.97
RbCl	1	1.00	25.1	1.63
CsCl	1	0.85	20.8	0.86
NaCl	1	0.75	19.9	0.71
NH$_4$Cl	1	0.75	18.3	0.42
BaCl$_2$	1	0.55	16.6	0.13
LaCl$_3$	1	0.70	16.4	0.08
LiCl	1	0.60	15.8	0
HCl	1	0.65	14.9	0
KNO$_3$	0.33	1.15	24.9	1.60
AgNO$_3$	0.33	1.10	24.0	1.44
TlNO$_3$	0.33	0.95	21.9	1.06
CsNO$_3$	0.33	0.75	19.6	0.65
Ba(NO$_3$)$_2$	0.33	0.60	17.1	0.20

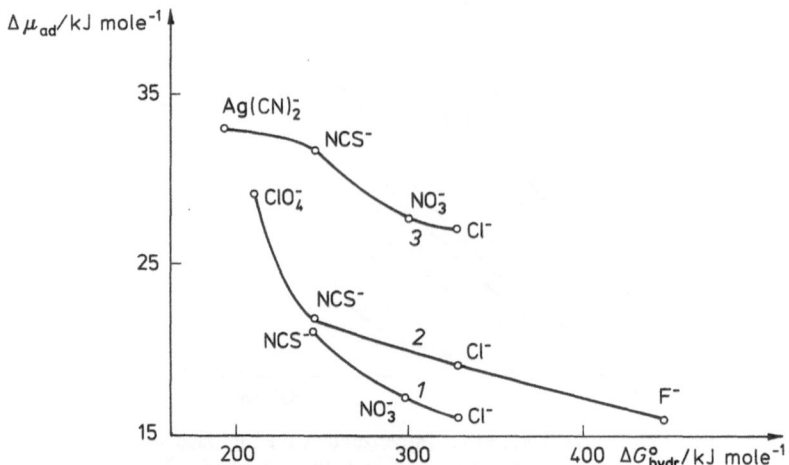

Fig. 9. Work of adsorption of DBC · MX complex as a function of hydration energy of the anion:
1 – barium salt; *2* – sodium salt; *3* – potassium salt

are located on one side of the plane formed by the ring of the macrocyclic polyether in
the aqueous part of the electrical double layer).

The surface activity of DBC · MX complexes at the water-benzene interface would
be in the sequence:

chlorides

$$K^+ > Rb^+ > Cs^+ > Na^+ > NH_4^+ > Ba^{2+} > La^{3+} > Li^+ > H^+$$

nitrates   (17)

$$K^+ > Ag^+ > Tl^+ > Cs^+ > Ba^{2+}.$$

With some exceptions these series coincide with metal series exhibiting the lowering of
the complex formation constants [114].

The surface activity is considerably affected by the anion of the metal salt: the work
of adsorption increases with the decrease in hydration energy of anions (Fig. 9). Similar
effects were observed when the transfer kinetics of alkali salts across the liquid
membranes containing DBC was studied [121–123]. Evidently, this deviation from the
linear dependence is due to the specific interaction of the macrocyclic polyether with
certain anions in the DBC · MX complexes [124], which co-exists with the decreasing
repulsion of low-hydration-energy anions from the boundaries between the low-
permittivity media [66, 67].

The adsorption isotherms of DBC complexes with metal salts are formally
described by Frumkin's isotherm [14]. The attraction constant for all complexes is
within the range of -8 to -2.5, suggesting a strong repulsion of adsorbed molecules in
the monolayer. The limiting areas per molecule in DBC – 1 M potassium salt
complexes are 125–150 Å^2, and only in the DBC-KAg(CN)$_2$ complex is the area
reduced to 85 Å^2 (the attraction constant for this complex being -0.3 ± 0.2).

Determination of Activity Coefficients for Extractants in Low-Permittivity Media from Surface Pressure Isotherms

The aggregation of surfactant molecules in nonpolar solvents is due to dipole-dipole interaction or to the the the formation of intermolecular hydrogen bonds. The micelle formation in water solutions is mainly due to hydrophobic interactions between hydrocarbon chains in molecules. The driving forces of aggregation, and the physicochemical properties of surfactants, in water and nonpolar solvents, differ considerably. The following assumption was corroborated for aqueous solutions. Below certain critical concentrations of micelle formation (CCM) the micelles are absent. Furthermore, above CCM the monomer concentration remains constant with the monomer units in micelles being in the order of 10^2. The physicochemical properties of a solution drastically change due to such sharp variations of its composition [125]. Solutions of nonpolar solvents contain monomers together with various small aggregates (dimers, trimers, etc.) in a dynamic equilibrium. The monomer concentration grows with an increase in the total concentration of surfactant [126–128]. Gradual changes in the composition of the solution and total concentration of the surfactant gradually changes its physicochemical properties. Hence, Kertes suggests that CCM as it is used to describe the state of aqueous solutions of surfactants is inapplicable as an aggregation characteristic for nonpolar media [127]. According to Eicke, the CCM concept should not be used for such binary systems as surfactant-nonpolar oil ones. Therefore, the CCM should be determined from the solubilization of a third component [128].

The surfactant aggregation in the non-aqueous phase usually begins at $10^{-7} - 10^{-6}$ M and is several orders lower than the CCM in the aqueous phase [126]. Therefore, the study of the aggregation and the determination of activity coefficients at such concentrations turn out to be a complicated experimental task. The surface pressure isotherms at the boundary between water and surfactant solutions in nonpolar solvents and other methods can be employed for this purpose. (Surfactants dissolved in the organic phase significantly affect the water-air interfacial tension [129]).

The surface pressure isotherms of surfactant solutions in nonpolar solvents at the water-oil interface are specifically S-shaped (Fig. 1). Unlike the surface pressure isotherms of aqueous solutions of surfactants they do not have sharp bends of curves that correspond to the beginning of micelle formation. Beyond certain concentration the $\partial \sigma / \partial \ln C$ value decreases with the increase in the surfactant concentration. After the beginning of micelle formation in the aqueous solutions, this value tends to zero.

Italian workers [93] studied the adsorption of tetraheptylammonium salts at the benzene-water interface. They claimed that the reason for the S-shape of the surface pressure isotherms is the surfactant aggregation in the hydrocarbon. It was assumed that only monomers (ionic pairs) are adsorbed at the water-oil interface. This assumption was confirmed by the dependence of the surface pressure on the logarithm of the monomer concentration having a shape typical of surface pressure isotherms of surfactant solutions without aggregation (the monomer concentration was determined independently). The limiting adsorption values determined from both dependences $(\pi - \lg C, \pi - \lg C_{mon})$ coincided. Thus, it was concluded that aggregation is absent when the surfactant concentration corresponds to the beginning of the limiting monolayer coverage. Later, this conclusion was corroborated by the adsorption data for trilaurylamine salts and dinonylnaphthosulfonic acid at the boundary between toluene and water electrolyte solutions [97, 130]. The monomer concentration and the average aggregation degree were calculated from the surface pressure isotherms of dinonyl-naphthosulfonic acid solutions in the toluene-water system [139]. The data agree with independently obtained results. The average aggregation for alkylammonium salts in benzene was determined by the same method [131].

In fact, the determination of the degree of aggregation is reduced to the determination of surfactant activity coefficients if the association is the only reason why the solution properties deviate from the ideal ones (model of the ideal associated solution [132]). The interaction between ionic pairs and solvent and between solvent and aggregates is neglected, the activity coefficients of aggregates and monomers taken to be unity. It is also assumed that only monomers are adsorbed at the interface, while associates are surface-inactive. This is valid for aqueous solutions of surfactants [125] and is corroborated by the experimental adsorption data for nonpolar solvents at the boundary with aqueous solutions [93, 97, 130]. The surfactant aggregation starts at a concentration where the limiting coverage of the interface is attained. The change in free energy of adsorption from the aqueous phase at the interface is always greater than in the micelle formation process [133]. Similarly, the change in free energy of the surfactant association in non-aqueous solvents is less than that of the adsorption at the water-oil interface. It follows that there is no association in the solution, and the activity coefficients are equal to unity for surfactant concentrations smaller or equal to the concentration at which the maximum value of $\partial\sigma/\partial\ln C$ is reached.

The activity coefficients of tetraoctylammonium salts in toluene were also determined [95] on the assumption that adsorption isotherms obey the Langmuir equation. This method, however, is less generalized since it is based on the selection of a concrete adsorption isotherm.

Shipunov investigated the adsorption of associated compounds from non-aqueous solutions at the water-oil interface [134]. It was observed that the adsorption isotherms of such surfactants follow the Freindlich equation, and the exponent in the equation is equal to the inverse value of the number of monomers in the associate.* However, this conclusion is not convincing since it is drawn on the assumption that the association in the bulk of the organic phase starts long before the limiting monolayer coverage is attained at the water-oil interface. It may be assumed that the limiting adsorption has been reached, and the monomer concentration in the solution is small as compared to the associate concentration. Then, imposing the condition that only monomers are adsorbed at the interface, the Gibbs equation can be written as [134]:

$$d\sigma = - \frac{RT\Gamma_{max}}{m} d\ln C, \tag{18}$$

where m is an average association. This equation is valid for the surfactant concentration regions where the association is already in progress. Thus, Eq. (18) yields the ratio of the maximum derivative to the derivative in a given region, and

---

* In 1932 Rehbinder pointed out the analogy between Freindlich's adsorption isotherm and the equilibrium constant of the substance distribution between the phases provided the association takes place in one of the phases [6, p. 94].

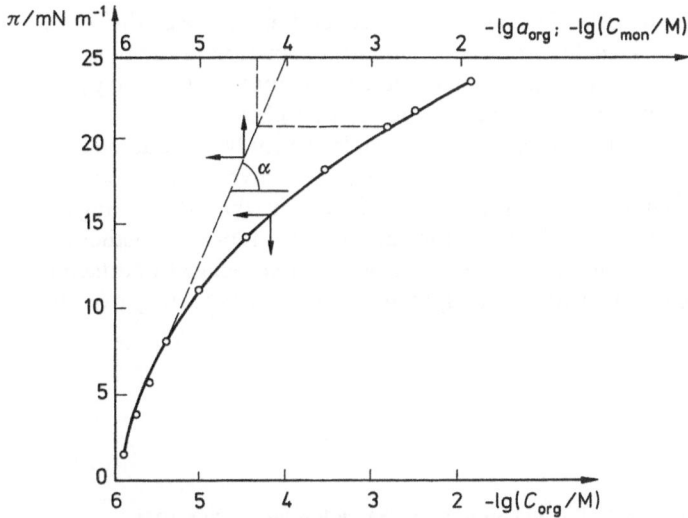

Fig. 10. Determination of the monomer activity and concentration from the surface pressure isotherms for Hept$_4$NCl solutions in benzene at the interface with water

Table 4. Monomer and associate concentration in Hept$_4$NCl solutions in benzene calculated from the surface pressure isotherms at the interface with water

Total concentration, M	Monomer concentration, M	Associate concentration, M
$2 \cdot 10^{-6}$	$2 \cdot 10^{-6}$	–
$5 \cdot 10^{-6}$	$4.0 \cdot 10^{-6}$	$2 \cdot 10^{-7}$
10^{-5}	$6.2 \cdot 10^{-6}$	$7.6 \cdot 10^{-7}$
$3.3 \cdot 10^{-5}$	$1.0 \cdot 10^{-5}$	$4.5 \cdot 10^{-6}$
10^{-4}	$1.4 \cdot 10^{-5}$	$1.7 \cdot 10^{-5}$
$3.3 \cdot 10^{-4}$	$1.8 \cdot 10^{-5}$	$6 \cdot 10^{-5}$
10^{-3}	$2.3 \cdot 10^{-5}$	$2 \cdot 10^{-4}$
$3.3 \cdot 10^{-3}$	$3 \cdot 10^{-5}$	$7 \cdot 10^{-4}$
10^{-2}	$3.5 \cdot 10^{-5}$	$2 \cdot 10^{-3}$

consequently, the average association. To determine the association, Eq. (18) is only applicable for the surfactant concentration regions where the monomer concentration is small as compared to that of associates. For instance, it follows from Fig. 10 that the average association equal to 5 is maintained within Hept$_4$NCl concentrations of 10^{-5} to 10^{-2} M. When calculated from Eq. (17), this value is only maintained within the $3.3 \cdot 10^{-4} - 10^{-2}$ M range. It means that within this range the monomer concentration is small as compared to the associate concentration (Table 4).

A number of problems arise when determining surfactant activity coefficients in nonpolar solvents from the surface pressure isotherms at the water-oil interface. Some of them pertaining to the association of alkylammonium salts have been already discussed [131]. In some cases the surface pressure isotherms are considerably affected by the composition and the concentration of electrolytes in the aqueous solution. (As for example, the adsorption of salts of tertiary amines depending on the acidity of the aqueous phase.) In these cases the level of the surfactant association in the nonpolar phase should be cautiously determined, because the composition of the compound adsorbed at the interface can differ from that in the organic phase.

To evaluate the method of surface pressure isotherms in the study of the association in nonpolar solvents further experiments are required. Moreover, the association processes should be studied by independent methods within the whole range of surfactant concentrations. Evidently, the most topical questions remain: whether any noticeable association can start unless the limiting monolayer coverage by surfactant molecules is attained at the water-oil interface, and how the associates participate in the adsorption processes.

References

1. Osnovy zhidkostnoj ekstraktsii. (Ed. Yagodin, G.A.) M., Khimiya, p. 399, 1981
2. Trouve, G.: Inf. Chim. 239, 87–90 (1983)
3. Purin, B.A.: Izv. AN Latvia SSR 5, 31–36 (1971)
4. Lakshminarayanayah, N.: Membrannije elektrody, L., Khimiya, p. 358, 1979
5. Veber, V., Gokel, G.: Mezfaznij kataliz v organicheskom sinteze, M., Mir, p. 327, 1980
6. Rehbinder, P.A.: Izbrannije trudy, M., Nauka, p. 368, 1978
7. Schmidt, V.S.: Ekstraktsiya aminami, M., Atomizdat, p. 262, 1980
8. Yagodin, G.A., Tarasov, V.V.: in: Khimiya ekstraktsii, Novosibirsk, Nauka, p. 35–53, 1984
9. Danessi, P.R., Vandergrift, G.F., Horwitz, E.P.: J. Phys. Chem. 84 (26), 3582–3587 (1980)
10. Koryta, I.: Ion-selective Electrode Rev. 5, 131–164 (1983)
11. Boguslavsky, L.I.: Bioelektrokhimiya i granitsa razdela faz, M., Nauka, p. 360, 1978
12. Yagodin, G.A., Tarasov, V.V., Pichugin, A.A.: Kinetika ekstraktsii neorganicheskih veschestv, Itogy Nauky. Inorganic Chemistry. M., VINITI 11, 158 (1984)
13. Kletenik, Yu.B., Sedova, S.A., Skvortsova, L.I.: Izv. SO AN SSSR, Chem. ser. 9 (4), 3–7 (1977)
14. Yatsimirsky, A.K., Levashov, A.V., Beresin, I.V.: in: Mitselloobrazovanije, solubilizatsiya i mikroemulsii. (Ed. Mittel, K.M.) M., Mir, p. 224–246, 1980
15. McDowell, W.J., Coleman, C.F.: J. Inorg. and Nucl. Chem. 29 (5), 1325–1343 (1967)
16. Yagodin, G.A., Tarasov, V.V., Yurtov, E.V.: Dokl. Acad. Nauk SSSR 218 (3), 647–649 (1974)
17. Danessi, P.R., Chiarizia, R.: J. Appl. Chem. Biotechnol. 28 (3), 581–598 (1978)
18. Al-Diwan, T.A., Hughes, M.A., Whewell, R.J.: J. Inorg. and Nucl. Chem. 39 (8), 1419–1424 (1977)
19. Vandegrift, G.P., Horwitz, E.P.: ibid. 39, 1425–1432 (1977)
20. Vandegrift, G.F., Horwitz, E.P.: ibid. 42 (1), 119–125 (1980)
21. Vandegrift, G.F., Horwitz, E.P.: ibid. 42 (1), 127–130 (1980)
22. Vandegrift, G.F., Delphin, W.H.: ibid. 42, 1359–1361 (1980)
23. Oh, J.S., Freiser, H.: Anal. Chem. 39 (3), 295–298 (1967)
24. Davies, J.T., Rideal, E.K.: Interfacial Phenomena, N.Y.-L., Acad. Press, p. 480, 1963
25. Scibona, G., Danessi, P.R., Fabiani, C.: Ion Exch. and Solv. Extraction 8, 95–227 (1981)
26. Gros, M., Gromb, S., Cavach, C.: J. Electroanal. Chem. 89 (1), 29–36 (1978)
27. Ortiz, S.: J. Appl. Chem. Biotechnol. 1, 149–156 (1978)
28. Cox, M., Flett, D.S.: ISEC-77: Proceeding of the International Solvent Extraction Conf. 1, 63–72 (1979)
29. Guastalla, J., Bertrand, C.: Comp. rend. serC 277 (3), 279–282 (1973)
30. Kotowski, J., Kalinska, A., Koczorowski, Z.: J. Colloid Interface Sci. 86 (2), 442–448 (1982)
31. Joos, P., Vanden Bogaert, R.: ibid. 56 (2), 213–217 (1976)
32. Yagodin, G.A., Tarasov, V.V., Kruchinina, I.E., Nickolaeva, G.D., Novicov, A.P.: Dokl. Acad. Nauk SSSR 249 (3), 662–665 (1979)
33. Takenaka, T.: Adv. in Colloid Interface Sci. 11, 291–313 (1979)

34. Möbius, D.: Ber. Bunsenges. Phys. Chem. *82*, 848–858 (1978)
35. Takahashi, H., Umemura, J., Takenaka, T.: J. Phys. Chem. *87* (5), 739–741 (1983)
36. Lucassen-Reyndess, E.: Progress in Surface and Membrane Sci. *10*, 253–360 (1978)
37. Aveyard, R., Haydon, D.: An Introduction to the Principles of Surface Chemistry, Cambridge Univ. Press, p. 235, 1973
38. Aveyard, R., Vincrnt, B.: Progress in Surface Science *8*, 59–102 (1977)
39. Rosen, M.J.: Surfactants and Interfacial Phenomena, N.Y., Wiley Interscience, p. 72, 1981
40. Betts, I., Pethica, B.: Trans. Faraday Soc. *56* (8), 1515–1528 (1960)
41. Abramson, A.A.: Poverchnostno-aktivnie veshestva, L., Khimiya, p. 304, 1981
42. Abramson, A.A., Abramova, N.V., Malakhova, E.E.: Kolloid. Zh. *33* (4), 475–479 (1971)
43. Kruglyakov, P.M., Rovin, Yu.G.: Physiko-khimiya tchernikh uglevodorodnikh plenok, M., Nauka, p. 183, 1978
44. Ohno, S., Kondo, S.: Moleculyarnaya teoriya poverkhnostnogo natyazeniya v zhidkostyakh, M., IL, p. 285, 1963
45. Cook, G., Redwood, W., Taylor, A., Haydon, D., Kolloid, Z.: Polymere *227* (1–2), 28–37 (1968)
46. Carrol, B., Haydon, D.: J.C.S. Faraday Diss., Part 1 *71* (3), 361–377 (1975)
47. Tajima, K., Murata, H., Tsutsui, T.: J. Colloid Interface Sci. *85* (2), 534–539 (1982)
48. Levie, R., Sarangapani, S., Czekaj, P., Benke, G.: Anal. Chem. *50* (1), 110–115 (1978)
49. Forsyth, T.J., Malcolm, M., Mouler, K.: Maschinnije metody matematicheskih vychislenij, M., Mir, p. 269, 1980
50. Hutchinson, E., Randall, D.: J. Colloid Sci. *7*, 151–165 (1952)
51. Rusanov, A.I.: Fazovye ravnovesiya i poverkhnostnije yavleniya, L., Khimiya, p. 388, 1967
52. Rehbinder, P.: Biochem. Z. *187* (1), 19–33 (1927)
53. Kruglyakov, P.M., Koretskiy, A.F.: Dokl. Acad. Nauk SSSR *197* (5), 1106–1109 (1971)
54. Avaeyrd, R., Saleem, S.: Can. J. Chem. *55* (23), 4018–4027 (1977)
55. Pierson, F.W., Whitaker, S.: J. Colloid Interface Sci. *54* (2), 203–248 (1976)
56. Crotov, V.V.: Kolloid. Zh. *43* (3), 475–486 (1981)
57. Jnoue, Y., Tochiyama, O.: Bull. Chem. Soc. Jpn. *53* (6), 1618–1624 (1980)
58. Shipunov, Yu.A.: Kolloid. Zh. *43* (2), 394–396 (1981)
59. Abramson, A.A., Slavina, Z.I.: ibid. *35* (1), 118–121 (1973)
60. Starobinetz, G.L., Egorov, V.V.: ibid. *41* (2), 377–379 (1979)
61. Popov, A.N., Purin, B.A.: Dokl. Acad. Nauk SSSR *246* (3), 659–661 (1979)
62. Nikitin, S.D., Schmidt, V.S.: Kolloid. Zh. *45* (5), 1011–1013 (1983)
63. Uryu, S., Yamanaka, M., Aratono, M., Motomura, K., Matuura, R.: Memoirs of the Faculty of Sci., Kyushu Univ. ser. C *14* (1), 11–18 (1983)
64. Jong, H.: in: Colloid Science. (Ed. Kruyt, H.), Amsterdam-L.-Brussel: Elsevier Publ. Comp., p. 259–330, 1949
65. Frumkin, A.N., Reichstein, N., Kulwarskaya, R.: Kolloid Z. *40* (1), 9–17 (1926)
66. Randles, J.E.: in: Advances in Electrochemistry and Electrochemical Engineering. N.Y.-L., Interscience Publ., p. 1–30, 1963
67. Llopis, J.: in: Modern Aspects of Electrochemistry., N.Y., Plenum Press *6*, 91–158 (1971)
68. Ter-Minnassian-Saraga, L.: in: Progress in Surface and Membrane Science, N.Y.-L., Academic Press *9*, 223–256 (1975)
69. Goddard, E.D., Kao, O., Kung, H.: J. Colloid Interface Sci. *27* (4), 616–624 (1968)
70. Goddard, E.D.: Croat Chem. Acta *40* (3–4), 134–150 (1970)
71. Goddard, E.D., Kung, H.: J. Colloid Interface Sci. *37*, 585–594 (1971)
72. Goddard, E.D., Matteson, G.H., Totten, G.E.: ibid. *85* (1), 19–27 (1982)
73. Plaisance, M., Ter-Minnassian-Saraga, L.: ibid. *56* (1), 33–41 (1976)
74. Plaisance, M., Ter-Minnassian-Saraga, L.: ibid. *59*, 113–122 (1977)
75. Feat, G., Levine, S.: in: Monolayers (Advances in Chemistry Series *144*). Washington, Amer. Chem. Soc., p. 98–122, 1975
76. Boguslavski, L.I., Frumkin, A.N., Manvelyan, M.A.: Dokl. Acad. Nauk SSSR *233* (1), 144–147 (1977)
77. Tamaki, K.: Bull. Chem. Soc. Jpn. *40* (1), 38–41 (1967)
78. Tamaki, K.: ibid. *47* (10), 2764–2767 (1974)
79. Manvelyan, M.A., Boguslavski, L.I.: Electrokhimiya *12* (12), 1805–1808 (1976)

80. Waligora, B., Steczko, K., Czarnecki, J.: Bull. Acad. Polon. Sci., ser. Sci. Chim. *18* (8), 481–487 (1970)
81. Waligora, B., Goralczuk, D.: ibid. *19* (8), 465–470 (1971)
82. Waligora, B., Rodakiewicz-Nowak, J.: Tenside Detergents *19* (5), 228–231 (1982)
83. Goralczyk, D.: ibid. *20* (5), 228–231 (1983)
84. Ray, A., Mukerjee, P.: J. Phys. Chem. *70* (12), 2138–2144 (1966)
85. Mukerjee, P., Ray, A.: ibid. *70* (12), 2144–2150 (1966)
86. Mukerjee, P., Ray, A.: ibid. *70* (12), 2150–2155 (1966)
87. Rusanov, A.I., Levichev, S.A., Zarov, V.T.: Poverhnostnoje razdelenije veshestv, L., Khimiya, p. 184, 1981
88. Grieves, R.B., Charevicz, W.: The P. J. Separation Sci. *10* (1), 77–92 (1975)
89. Krylov, V.S., Myamlin, V.A., Boguslavski, L.I., Manvelyan, M.A.: Electrokhimiya *13* (6), 834–840 (1977)
90. Manvelyan, M.A., Neugodova, G.L., Boguslavski, L.I.: ibid. *12* (2), 300–313 (1976)
91. Manvelyan, M.A., Heugodova, G.L., Boguslavski, L.I.: ibid. *12* (8), 1250–1254 (1976)
92. Gavach, C., Bellet, E., Davion, N., Seta, P.: J. Chim. Phys. *68* (6), 1005–1007 (1971)
93. Scibona, G., Danessi, P., Conte, A., Scuppa, B.: J. Colloid Interface Sci. *35* (4), 631–635 (1971)
94. Popov, A.N.: in: All-Union Conference on Extraction. Abstracts, Riga, Zinatne *3*, 23–27 (1977)
95. Heifetz, V.L., Shneerson, A.A.: Zh. P. Chem. *50* (3), 534–538 (1977)
96. Popov, A.N.: Izv. AN Latvia SSR, Chem. ser. *2*, 175–179 (1978)
97. Pizzichini, M., Chiarizia, R., Danessi, P.: J. Inorg. and Nucl. Chem. *40* (4), 669–771 (1978)
98. Popov, A.N.: Elektrokhimiya *20* (12), 1571–1577 (1984)
99. Schmidt, V.S., Rybakov, K.A., Rubisov, V.N.: Zh. Anal. Chem. *38* (7), 1182–1187 (1983)
100. Ivanov, I.M.: Izv. SO AN SSSR, Chem. ser. *6*, 13–21 (1980)
101. Damaskin, B.B., Petriy, O.A., Batrakov, V.V.: Adsorbtsiya organicheskikh soedininij na elektrodakh, M., Nauka, p. 334, 1968
102. Kitajgorodsky, A.I.: Molekuljarnije krystally, M., Nauka, p. 79, 1971
103. Ishibashi, N., Mihara, H., Iyo, A.: Denki Kagaku *44* (4), 268–272 (1976)
104. Yoshida, N., Ishibashi, N.: Bull. Chem. Soc. Jpn. *50* (12), 3189–3193 (1977)
105. Popov, A.N., Timofeeva, S.K., Kulikova, L.D.: Izv. AN Latvia SSR, Chem. ser. *1*, 100–101 (1982)
106. Martynov, V.V.: Ekstraktsiya organicheskimi kislotami i ikh solyami. Reference book, M., Atomizdat, p. 366, 1978
107. Goddard, E.D., Kao, O., Kung, H.C.: J. Colloid Interface Sci. *24* (2), 297–303 (1967)
108. Weil, J.: J. Phys. Chem. *70* (1), 133–136 (1966)
109. Hendrikx, Y., Mari, D.: J. Colloid Interface Sci. *78* (1), 74–86 (1980)
110. Charewicz, W.A., Radzicka, W., Strzelbicki, J.: ibid. *76* (2), 290–297 (1980)
111. Tomoaja, M., Andrei, Z., Chifu, E.: Rev. Roum. Chim. *18*, 1547–1553 (1973)
112. Chifu, E., Andrei, Z., Tomoaja, M.: Anal. Chim. (Rome) *64* (11–12), 869–871 (1974)
113. Gugeshashvili, M.I., Boguslavsky, L.I., Frumkin, A.I.: Dokl. Acad. Nauk SSSR *206* (4), 585–587 (1972)
114. Ovtchinikov, Yu.A., Ivanov, T.A., Schkrob, A.M.: Membranno-selektivnije kompleksony, M., Nauka, p. 463, 1974
115. Schljahter, T.A., Tzagina, L.V., Lev, A.A.: in: Biophysika membran. Kaunas, p. 662–667, 1973
116. Colaccico, G., Gordon, E.E.: J. Colloid Interface Sci. *63* (1), 76–88 (1978)
117. Kuo, P.Z., Ikeda, I., Okahara, M.: Tenside Detergents *19* (4), 204–206 (1982)
118. Zaitzev, S.Yu., Lutzenko, V.V., Zubov, V.P.: Bioorganitcheskaya khimiya *9* (4), 567–568 (1983)
119. Danessi, P., Chiarizia, R., Pizzichini, M., Saltelli, A.: J. Inorg. and Nucl. Chem. *40* (9), 1119–1123 (1978)
120. Popov, A.N., Serga, V.E., Purin, B.A.: Dokl. Acad. Nauk SSSR *281* (1), 109–112 (1985)
121. Lamb, J., Chistensen, J., Izatt, R. et al.: J. Amer. Chem. Soc. *102* (10), 3399–3403 (1980)
122. Popov, A.N., Tinofeeva, S.K.: in: All-union conference on extraction. Abstracts, Riga, Zinatne *3*, 9–12 (1982)
123. Yatzimirsky, K.B., Talanova, G.G.: Dokl. Acad. Nauk SSSR *273* (4), 903–905 (1983)

124. Host Guest Complex Chemistry II, (Ed.) F. Vögtle, Berlin-Heidelberg-N.Y., Springer-Verlag, p. 43–45, 1982
125. Mittel, K.L., Mukerjee, P.: in: Mitselloobrazovanije, solubilizatziya i mikroemulsii. (Ed. Mittel, K.M.), Mir, p. 11–31, 1980
126. Kertes, A.S., Gutmann, H.: Surface and Colloid Science 8, 193–295 (1976)
127. Kertes, A.S.: in: Mitselloobrazovanije, solubilizatziya i mikroemulsii. (Ed. Mittel, K.M.), Mir, p. 214–223, 1980
128. Eicke, H.F.: Topics in Current Chemistry 87, 85–145 (1980)
129. David-Auslaender, J., Gutmann, H., Kertes, A.S., Zangen, M.: J. Solution Chem. 3 (4), 251–260 (1973)
130. Chiarizia, R., Danessi, P., D'Alessandro, G., Scuppa, B.: J. Inorg. and Nucl. Chem. 38 (8), 1367–1369 (1976)
131. Popov, A.N.: Zh. Phys. Khim. 55 (2), 466–469 (1981)
132. Komarov, E.V., Kopyrin, A.A., Proyaev, V.V.: Teoreticheskije osnovy ekstraktsii assotsiirovannymi reagentami, M., Energoatomizdat, p. 126, 1984
133. Mukerjee, P.: Ber. Bunsenges. Phys. Chem. 82 (10), 931–937 (1978)
134. Shipunov, Yu.A.: Zh. Phys. Khim. 56 (11), 2783–2788 (1982)

Kinetics of the Photochemical Charge Separation in Micellar Solutions

M. G. Kuzmin and N. K. Zaitsev

The photochemical charge separation in micellar and other organized assemblies has been extensively investigated within recent years as of their features in mimicking biological processes, their use in the storage of solar energy and in molecular electronics. Our review aims to derive some correlations to improve the understanding of these processes. Photochemical reactions also provide valuable information about the structure and dynamic properties of organized assemblies [1–4].

Section 1 deals with the formal kinetics of the photochemical reactions in micellar solutions and its application to determine the rate constants of photoprocesses, the critical concentration of micellization (CMC) and the aggregation number of micelles from experimental data. In Section 2 the correlations of the rate constants of the photochemical charge separation and mass transfer processes with its thermodynamics, and also the microviscosity and effective polarity of organized assemblies are considered.

The interface can make the charge separation more effective and prevent their recombination. For this purpose an effective generation of charge carriers should take place, the diffusion of the carriers to the interface and their separation should compete with their geminate recombination, and the interface should also provide a kinetic barrier for the bulk recombination of the carriers.

1 The Formal Kinetics of Reactions in Micellar Systems

1.1 Solubilization of Molecules by Surfactant Micelles and the Distribution of Reactant Molecules among Micelles

1.1.1 The Pseudophase and Microscopic Models of Solubilization

The formal kinetics of reactions in micelles requires one to take into account the distribution of the reactant molecules between the micelles and the bulk solution, and also among the micelles. There are two approaches that take into account the distribution of the solubilizates between the micelles and the bulk phase. The first one is called "the pseudophase model", the second one is referred to as the "microscopic approach".

The pseudophase model distinguishes the solubilized and nonsolubilized molecules and never takes into account the number of molecules solubilized in any particular micelle. The transfer of molecules from the bulk phase into the micellar pseudophase is

The Interface Structure and Electrochemical Processes at the Boundary Between Two Immiscible Liquids
Editor: V. E. Kazarinov
© Springer-Verlag Berlin, Heidelberg 1987

considered in terms of an equilibrium:

$$A_v \underset{k_-}{\overset{k_+}{\rightleftarrows}} A_m \tag{1}$$

characterized by an equilibrium distribution coefficient $\varrho = k_+/k_-$. The indices "v" and "m" denote the bulk and the micellar phases, respectively. The part α of the solubilized molecules A is given by the expression:

$$\alpha = \varrho X/(1 + \varrho X) \tag{2}$$

where X is the volume concentration of the micellar phase.

In the microscopic approach the transfer of A molecules from the bulk phase to micelles is considered as a multistep chemical equilibrium:

$$M + A \underset{k_{out}}{\overset{k_{in}}{\rightleftarrows}} MA + A \underset{k_{out2}}{\overset{k_{in2}}{\rightleftarrows}} MA_2 + A \underset{k_{out3}}{\overset{k_{in3}}{\rightleftarrows}} \dots . \tag{3}$$

The usual approximation is that the solubilization of one or few molecules does not change the properties of a micelle:

$$k_{in} = k_{in2} = k_{in3} = \dots \quad \text{and} \quad k_{out} = k_{out2} = k_{out3} = \dots .$$

Such an assumption leads to the Poisson distribution of molecules A among the micelles [5] and scheme (3) can be replaced by the following one:

$$A + M \overset{K}{\rightleftarrows} AM \tag{4}$$

which resembles the mass action law, but here at the solubilization of A molecules only the number of A molecules is changed and the micelle remains unchanged. The degree of the solubilization α is given by the following expression:

$$\alpha = K[M]/(1 + K[M]) \tag{5}$$

where $K = k_{in}/k_{out}$. The results of the pseudophase and microscopic approaches coincide for the description of the equilibrium state and $K = \varrho m V$, where V is the molar volume of the surfactant and m is the aggregation number.

The relation between K and ϱ depends on the different standard conditions for the two models: in the microscopic model the standard concentrations of micelles, the solubilized A molecules and unsolubilized A molecules in overall volume of the solution are equal to $1 \text{ mole} \cdot \text{dm}^{-3}$; in the pseudophase model the standard concentration of A molecules in each phase is equal to $1 \text{ mole} \cdot \text{dm}^{-3}$. The monomolecular rate constants for exit from the micellar phase are independent of the model, but the rate constants for entrance into the micellar phase depend on the units in which the surfactant concentration is expressed as:

$$k_{in} = mVk_+ .$$

1.1.2 Solubilization as a Kind of Interphase Equilibrium

At the interphase equilibrium the electrochemical potentials of each substance in both phases are equal:

$$\bar{\mu}_{im} = \bar{\mu}_{iv} \quad \text{or} \quad \mu_{im}^0 + RT \ln a_{im} + z_i F \varphi_m = \mu_{iv}^0 + RT \ln a_{iv} \tag{6}$$

where $\bar{\mu}_{im}$ and $\bar{\mu}_{iv}$, μ_{im}^0 and μ_{iv}^0, a_{im}, and a_{iv} are the electrochemical potentials, the standard chemical potentials and activities of a substance (or an ion) with the number i in the micellar and bulk phases, respectively; z_i is the electric charge of this ion and φ_m is the electrostatic potential of the micelles (or the micellar phase) with respect of the bulk phase. One can see, that the α value of any given $[M]$ is determined by the difference of μ_{im}^0 and μ_{iv}^0 (that is, mainly the hydrophobic balance of the ion or the substance with the number i) the electric charge of the substance, the electrostatic potential of the micelles with respect of the bulk phase, and also by the ratio of the activity coefficients of A molecules in the micellar and bulk phases.

The solubilization changes the equilibrium constants of chemical reactions. For example, the equilibrium constant of the protolytic dissociation reaction:

$$AH^{(n)} \rightleftarrows A^{(n-1)} + H^+ \tag{7}$$

for the micellar phase reads as follows:

$$K_a^m = \frac{[H^+]_m[A^{(n-1)}]_m}{[HA^{(n)}]_m} = K_a \frac{\varrho(H^+)\varrho(A^{(n-1)})}{\varrho(HA^{(n)})} \tag{8}$$

where K_a is the equilibrium constant in an aqueous phase.

If we determine the value of pK_a^m for the standard concentration of H^+ in aqueous phase, then:

$$pK_a^m = -\log\frac{[H^+]_v[A^{(n-1)}]_m}{[HA^{(n)}]_m} = pK_a - \log\frac{\varrho(A^{(n-1)})}{\varrho(HA^{(n)})} \tag{9}$$

and the so-called pK_a shift depends on the ratio of distribution coefficients of the acid and the conjugated base. For the electron transfer reaction:

$$A + D \rightarrow A^{\bar{\cdot}} + D^{\dagger} \tag{10}$$

the ΔG value is determined by the difference of the redox potentials of A and D. In the micellar phase the redox potentials are also shifted:

$$E^0(A/A^{\bar{\cdot}})_m = E^0(A/A^{\bar{\cdot}}) - \frac{2.3RT}{\mathscr{F}}\log\frac{\varrho(A)}{\varrho(A^{\bar{\cdot}})}. \tag{11}$$

The electrostatic potential difference between the microphase and the bulk phase strongly affects the charge separation. Calvin et al. [6] have shown that the electrostatic potential is even more important than the chemical nature of microheterogenous systems. They compared the quantum yields of hydrogen formation in the catalytic photoreaction of $[Ru(bpy)_3]^{+2}$ with dialkylviologensulfonate and EDTA as a sacrificial electron donor in the presence of colloidal SiO_2 or sodium dodecylsulfate (SDS) micelles, which had close microphase potentials. In homogeneous aqueous solutions the photoreduction is actually absent, but in both microheterogeneous systems the values of the quantum yields and their dependence on the electrostatic potential are the same. The origin of this phenomenon is discussed in Sect. 2.1.2.

The electrostatic potential of micelles can be determined from equilibrium or kinetic data. Fernandez and Fromherz [7] studied the acid-base equilibrium in the ground state of two similar but differently charged coumarins in different micelles:

$$\text{C}_{11}\text{H}_{23} \quad \text{ROH} \qquad\qquad \text{C}_{17}\text{H}_{35} \quad \text{RNMe}_2\text{H}^+ \tag{12}$$

$$\text{ROH} \overset{K_{a1}}{\underset{}{\rightleftharpoons}} \text{RO}^- + \text{H}^+, \tag{13}$$

$$\text{RNH}_3^+ \overset{K_{a2}}{\underset{}{\rightleftharpoons}} \text{RNH}_2 + \text{H}^+. \tag{14}$$

The values of K_{a1} and K_{a2} change in micellar solutions as compared with aqueous solutions due to different changes of the standard electrochemical potentials of forms RO^- and ROH and, respectively, RNH_3^+ and RNH_2 when the system is transferred to the micellar phase. (For compounds I and II in all forms, complete solubilization is implied). The changes of the electrostatic components of the electrochemical potential due to solubilization of the forms ROH and RO^- and, respectively, RNH_3^+ and RNH_2 differ by the same value $\mathscr{F}\varphi$. Assuming that the transfer from the aqueous phase to the micellar one results in the same difference of the μ^0 value for RNH_3^+ and RO^- as compared with RNH_2 and ROH, respectively; that is, the solvation free energy of an ion, the charge of which equals unity. An expression for the micellar potential has been obtained from the experimental ΔpK_a values [7]:

$$\varphi_m = \frac{1.15RT}{\mathscr{F}}(\Delta pK_{a1} + \Delta pK_{a2}). \tag{15}$$

The micellar potential affects the rate constants if the reactants are located in different phases. Such an influence is illustrated by the rate constants of acridinium cation fluorescence quenching by the same anions at different ionic strengths of solution, which was changed by addition of NaCl (Fig. 1) [8]. The quenching is controlled by penetration of the anions into micelles of SDS and the relative values of the micellar potential can be calculated from the values of the quenching rate constants. The ratio of the quenching rate constants at two different values of the ionic strength can be written as follows:

$$\ln(k_{q1}/k_{q2}) = \ln\gamma_1 - \ln\gamma_2 + z\mathscr{F}(\varphi_1 - \varphi_2)/RT. \tag{16}$$

Here γ_i and φ_i are the activity coefficients for the quencher ion in the aqueous phase and the micellar potential at any given ionic strength I_i. The value of the ionic strength can be calculated from the following assumptions about the concentration of ions in the bulk phase: the surfactant concentration equals CMC, all added sodium cations and inorganic anions are located in the bulk phase. The activity coefficient γ_i can be calculated from the expression of the second approximation of the Debye-Hückel theory [9]:

$$\log\gamma_i = 0.51\sqrt{I_i}/(1 + 0.337a\sqrt{I_i}) \tag{17}$$

where a is the radius of the quencher ion. The plot of the micellar potential vs. the concentration of NaCl, calculated by the authors of the present review from the data of Ref. [8] using Eqs. (16)–(17) coincides with data obtained with acid-base indicators [7] as shown in Fig. 1.

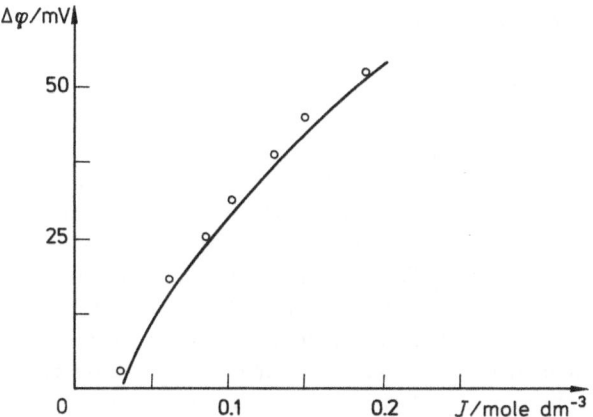

Fig. 1. The dependence of potential drop at SDS micelle/water interface on the ionic strength of the aqueous solution. Circles are calculated from the data on fluorescence quenching [44], solid line is calculated from the data for prototropic equilibria for special probes [7]

The pseudophase model fails if the concentration of the solubilized reactant influences the reaction rate constant. This was observed for the interaction of p-nitrophenyldiphenyl phosphate with fluoride ions in a micellar solution of cetyltrimethylammonium fluoride [10]. The bromide ions inhibit the reaction because they compete with fluoride ions in the occupation of micelles. However, the use of the microscopic model has no advantage as compared with the pseudophase model, because in both models one must take into account the dependence of the solubilization constants upon the presence of foreign ions and upon the ionic strength of the solution.

The specific interaction of counterions with the micelles can be important for the process of photochemical charge separation in microheterogeneous systems. Whitten et al. [11] studied the reaction of water-soluble porphyrins with different solubilized alkylviologens. The presence of tetraalkylammonium cations influences the yield of the primary charge separation products with the marked dependence on the length of the alkyl chain in the tetraalkylammonium cation. At present, a theoretical model which could describe such a phenomenon quantatively does not exist.

The use of the pseudophase model is convenient if the distribution of the reactants among micelles is not taken into account, i.e. for the first order reactions and also for the second order reactions if the intermicellar exchange by the reactant molecules occurs faster than the reaction in question. When the reaction is faster than or has a comparable rate to the intermicellar exchange by the reactants molecules, one should take into account the statistics of the intermicellar distribution of molecules.

1.1.3 The Effect of the Intermicellar Distribution of Reactant Molecules

Because of the small size of micelles, the use of the "mean" concentrations of the reactants in the micellar phase would not fit in all cases. The effect of the intermicellar distribution of the reactant molecules on the reaction rate was considered by Moroi [12] for three types of distributions (Poisson, binomial and Gauss) and the equations for the resulting effective rate constants were derived. If the multistep equilibrium (1–3) can be prolonged infinitely with equilibrium constants $k_{\mathrm{in}\,j}/k_{\mathrm{out}\,j}=k_{\mathrm{in}}/jk_{\mathrm{out}}$, the reactants obey the Poisson intermicellar distribution. The probability P_s of finding in the given micelle the number s of the quencher Q molecules is expressed as follows:

$$P_s = \frac{\langle s \rangle^s e^{-\langle s \rangle}}{s!} \tag{18}$$

where the mean number of Q molecules per micelle $\langle s \rangle = K[Q]/(1 + K[M])$. If the equilibrium (3) breaks at the step number n, the quencher molecules are distributed according to binomial laws with the parameter n, where n is the limiting number of solute molecules in a micelle. The Poisson distribution is the limiting case of the binomial one at $n \to \infty$. We have not found in the literature any physical model which would lead to the Gauss (normal) distribution of solubilized molecules among micelles.

The most direct evidence that a micelle can bind a limited number of solute molecules only was given by Whitten et al. [13, 13a] with the help of a non-solubilized fluorophore (4-cyano-N-benzylpyridinium or 4-cyano-N-methylpyridinium) in the solution of anionic micelles. The limiting number of the quencher cation (dimethylviologen) was found to be one cation per two to three surfactant molecules in the micelle. The Stern-Volmer plot had two linear regions. At first, the quencher ions are effectively bound by the micelles and then the binding completely ends. This is also correct for other quenchers, such as Cu^{+2} and 4-cyano-N-methylpyridinium but two linear regions are less noticeable and the binding of the cations continues at their high concentrations. Burrows et al. [14] have shown that the fluorescence quenching of fluoranthene by iodide ions in CTAB micelles does not obey the Stern-Folmer law and the fluorescence intensity reaches a limiting value at high concentrations of quencher, and they concluded that CTAB micelles can include a limited number of iodide anions only. One should pay attention to the fact, that in both cases considered the micelle and the solubilizate have opposite charges so that an interaction similar to salt-formation between the end groups of the micelles and the quencher is possible. Such an interaction implies some kind of stoichometry in solubilization.

However, the majority of authors process the kinetic data for micellar solutions in terms of the Poisson distribution, assuming that the mean number of solute molecules per micelle $\langle s \rangle$ is small as compared with the limiting value n.

1.2 The Interfacial Exchange of Molecules in Microheterogeneous Solutions

1.2.1 The Exchange of Molecules at the Interface Between Micelles and Bulk Phase

Micelles are dynamic aggregates and exchange surfactant molecules with the surrounding bulk solution. The lifetime of a surfactant molecule in a micelle is in the order of 10^{-5}–10^{-6} s and depends upon various factors. The solubilized substances can also transfer from the micelles to the bulk phase and vice versa. The rate of such a process can be comparable with the rate of the excited molecule's deactivation [15–17]. The usual implication is that the photoexcitation does not lead to any considerable change of the binding constant. The binding constant for the products of the photoreaction may differ markedly from the ones of the reactants. So the reaction can result in some photoinduced flows of substance from the micellar phase to the bulk phase [18, 19]. Such photoinduced transport processes can also be noticed where the flow of substance is caused by the photochemical change of the micellar media, leading to changes in the standard potential of the substance or a potential drop at the interface of the micelle and the bulk solution. For example, Grätzel et al. [20] studied the photoionization of a surfactant phenothiazine derivative, forming micelles by itself, and found that the micelles were dissolved when the number of photoionized phenothiazine moieties exceeded 22 per single micelle. Here the photoreaction leads to a change of the potential drop at the interface. In the case of the photoisomerization of azobenzene surfactants [21], which also form micelles by themselves, a change of the non-Couloumb part of the electrochemical potential takes place. The CMC value depends upon the contents of the thermodynamically unstable *cis*-form of the azobenzene. Thus one can drive the system from the micellar state to the premicellar one by photochemical means, i.e. by causing a flow of substance from the micellar to the bulk phase. But in most cases the substance transfer is caused by the change of its properties rather than by a change of the micellar medium.

1.2.2 First-Order Reaction in Micellar Solutions

The mode of intermicellar distribution of the reacants is not important for first-order reactions. The probability of finding more than one photoexcited molecule in a single micelle is negligible at the usual intensity of the excitation. The dependence of the yield of the photoproducts upon the surfactant concentration can be processed well in terms of the pseudophase model.

The yields of the products change upon solubilization of the reactants because of the change of the effective polarity and microviscosity of the micellar phase as compared with the bulk phase. There are reactions for which the solubilization itself changes the yields of the photoproducts due to the compartmentalization of the reactants and intermediates of the reaction. For example, the photolysis of the asymmetric ketones in homogeneous solutions results in different radicals, which form a number of products as a result of a bulk recombination [22]:

$$R-\underset{\underset{O}{\overset{\|}{}}}{C}-R' \overset{h\nu}{\longrightarrow} R^\cdot + CO + R'^\cdot, \tag{19}$$

$$R^\cdot + R'^\cdot \rightarrow RR', \tag{20}$$

$$2R^\cdot \rightarrow RR; \quad 2R' \rightarrow R'R'. \tag{21}$$

In micellar solutions the cage escape of the radicals formed due to the photolysis of the ketone does not take place and the radicals $R^\cdot$ and $R'^\cdot$ recombine with each other exclusively. It results in the product of Eq. (20) to become the only product of the reactions.

The type of reaction as in Eq. (19) can be isotopically selective. The photolysis of dibenzylketone is subjected to a magnetic isotope effect. The rate of the geminate recombination of radicals formed by the photolysis of molecules containing C^{13} atoms is more than for those ones without C^{13} atoms because of the increased superfine interaction in the molecules containing a magnetic isotope of C^{13}, which results in the increased probability of the intersystem crossing of the radical pair into the singlet state, where the recombination is spin-allowed. These effects are not important for the bulk recombination. In micellar solutions the photolysis of the ketones is accompanied by the geminate recombination, rather than the bulk recombination, because a micelle is a "macrocage" for the radicals [23, 24]. Thus, the solubilization changes the yield of the products and the isotope effect of the photoreactions.

The pseudophase model provides a means to take into account the reaction yield dependence upon the surfactant concentration for the first-order reactions. If ϕ_m is the reaction (or luminescence) yield of the solubilized reactants and ϕ_v is the one of the non-solubilized reactants, the total quantum yield of the first-order reaction ϕ can be written as follows:

$$\phi = \alpha\phi_m + (1-\alpha)\phi_v \tag{22}$$

where the part of the solubilized reactant molecules α can be calculated from Eqs. (2) or (5) [15].

1.2.3 Second-Order Photoreactions in Micellar Solutions. CMC Measurements and Aggregation Numbers

In the general case second-order photoreactions in micellar solutions depend upon the distribution of the reactants among the micelles. Aikawa, Turro, and Yekta [17] derived four limiting cases to be met which are given below together with their appearance in the fluorescence kinetics and stationary spectra. Consider a fluorescence quenching reaction of the following scheme:

$$\begin{array}{c} \xrightarrow[hv]{\text{excitation}} A^* + Q \longrightarrow \text{quenching} \\ \Big\downarrow \substack{\text{unimolecular} \\ \text{decay}} \\ \text{product} \end{array} \tag{23}$$

Here A^* denotes an excited molecule of the reactant and Q is another, non-excited reactant, called the fluorescence quencher.

Case 1. The quencher is completely solubilized ($\varrho X \gg 1$), and the fluorescence is quenched in a static way, i.e. the fluorophore A, solubilized in the micelles, containing one or more molecules of the quencher Q, does not emit light at all. The emission occurs from the part of the A molecules, solubilized in the micelles, containing no quenching Q. That part of the micelles is given by Eq. (18): $P_0 = \exp(-\langle s \rangle)$, which leads at stationary illumination to:

$$\ln(\phi_0/\phi) = [Q]/[M] \tag{24}$$

where ϕ_0 and ϕ are the fluorescence quantum yields in the absence and in the presence of quencher, respectively. At pulse excitation, the fluorescence decays exponentially with the lifetime equal to one in the absence of the quencher:

$$\ln I(t) = \ln I(0) - t/\tau_0 \tag{25}$$

where $I(t)$ is the fluorescence intensity at the time "t" and τ_0 is the decay lifetime.

Such a fluorescence dependence upon the concentration of quencher was observed for the pyrene fluorescence quenching by dimethylaniline in SDS micelles [25, 26]. Rodgers and Bexendale [27] studied the $[Ru(bpy)_3]^{+2}$ luminescence quenching by 9-methylanthracene and found that Eq. (24) fitted well in the case of sufficiently fast dynamic emission quenching, when the lifetime of an excited fluorophore molecule in a micelle, containing at least one quencher molecule, is considerably reduced. Equation (24) makes it possible to determine the concentration of micelles $[M]$ from the slope of the linear plot of $\ln(I_0/I)$ vs. $[Q]$, that is with the surfactant concentration known, and to determine the CCM and the aggregation number of the surfactant. The concentration of micelles is zero if the surfactant concentration is below CMC, and above CMC it can be found from:

$$[M] = \frac{[\text{Surfactant}] - \text{CMC}}{\text{aggregation number}}. \tag{26}$$

Thus, the plot of $[M]$ vs. the surfactant concentration is linear with a slope equal to the reverse aggregation number and the intercept with the abscissa axis equal to the CMC. This method was widely used to determine the aggregation numbers of micelles and vesicles (liposomes) [28–39].

Case 2. The fluorescence quenching is static; the quencher is partially solubilized.

If the exchange by the quencher between the micellar and bulk phases is much slower than the decay of the excited fluorophore molecules, their lifetime remains unchanged and the quantum yield depends on the concentration of quencher and micelles similiarly to case 1:

$$\ln(\phi_0/\phi) = K[Q]/(1 + K[M]). \tag{27}$$

But if the rate of the exchange is comparable with the one of the decay, the lifetime of the excited molecules depends on the Q concentration:

$$1/\tau = 1/\tau_0 + k_q[Q]_v. \tag{28}$$

Turro et al. [17] considered the fluorescence quenching rate constant by non-solubilized Q molecules k_q to equal the forward rate constant of binding k_{in}, since every association resulted in complete quenching. The fluorescence quantum yield reads as follows:

$$\phi_0/\phi = (1 + k_q \tau_0 [Q]_v) \exp\{K[Q]/(1 + K[M])\}. \tag{29}$$

The last expression may be considered as a product of the right-hand parts of Eq. (28) and Eq. (27) in non-logarithmic form. Mataga et al. [42, 43] used the two last equations to process the kinetic data for the system pyrene-dicyanobenzene and pyrene-dimethylaniline in micelles of different surfactants. The quenching in these systems is fast dynamic rather than static so the use of Eqs. (27–28) is an approximation similar to that described in Ref. [27] (see Case 1). Mataga et al. [25, 26] consider that the rate constant k_q does not coincide with k_{in}, because of the quenching being not static, i.e. not instant.

Case 3. The quenching is dynamic, that is the fluorophore reacts with the quencher within the lifetime of the excited state; both quencher and fluorophore are completely solubilized. The total intensity of the fluorescence at stationary excitation as well as at pulse excitation is the result of the simultaneous emission by fluorophore molecules from the micelles, containing 0, 1, 2, ... etc. molecules of quencher. One can consider the fluorophore molecules located in micelles with equal number of quencher molecules to be an ensemble and the total emission as a sum of the emission of all ensembles. The statistical weight of each ensemble depends on the distribution mode, for a Poisson distribution it equals P_s. The usual assumption [44, 45] is that the reaction in the microphase obeys the mass action law, i.e. the excited molecule's deactivation rate constant depends linearly upon the number of quencher molecules in the micelles of a given ensemble:

$$1/\tau_s = 1/\tau_0 + k_r s \tag{30}$$

where τ_s and τ_0 are the lifetimes of the excited fluorophore molecules in the ensembles of micelles containing s molecules and no molecules of the quencher, respectively; k_r is the rate constant of the reaction of the fluorophore in the ensemble of molecules, containing one quencher molecule each. The total emission intensity results from the summation of the ensembles intensities. At stationary excitation it yields:

$$\phi/\phi_0 = e^{-\langle s \rangle} \sum_{s=0}^{\infty} \frac{\langle s \rangle^s}{[1 + s(k_r \tau_0)] s!} \tag{31}$$

and fluorescence kinetics at pulse excitation:

$$I(t)/I(0) = \exp\{-[t/\tau_0 + \langle s \rangle (1 - e^{-k_r t})]\}. \tag{32}$$

In order to derive the latter expression, one should substitute the corresponding exponential function for the series. Such a case was observed for 2-(4-aminophenyl)-6-methylbenzotriazole fluorescence quenching by duroquinone in SDS micelles [40].

Case 4. The quenching is dynamic; the quencher is partially solubilized. This case is considered by Turro et al. [17] to be the most general one. They provided expressions for the luminescence decay function $I(t)$ and for the fluorescence quantum yield:

$$I(t)/I(0) = \exp\{-[t/\tau + \langle s \rangle \alpha_r^2 (1 - e^{-(k_r + k_{out})t})]\} \tag{33}$$

where, in order to calculate τ, one should use Eq. (28) with:

$$k_q = \alpha_r k_{in}; \quad \alpha_r = k_r/(k_r + k_{out}). \tag{34}$$

For continuous illumination, the luminescence quantum yield is given by:

$$\phi/\phi_0 = \frac{\tau}{\tau_0} e^{-\langle s \rangle \alpha_r^2} \sum_{s=0}^{\infty} \frac{(\langle s \rangle \alpha_r^2)^s}{[1 + s(k_{out} + k_r)\tau_0] s!}. \tag{35}$$

This is the case for numerous examples of quenching of aromatic compound fluorescence by metal ions in anionic micelles when the concentration of the micelles is small enough and the direct exchange by the metal ions among the micelles has no influence.

Turro et al. [17] also mention cases of fast exchange of quencher among micelles and of the escape of the fluorophore from the micelles during its lifetime with subsequent interaction with the quencher in the bulk solution. The latter situation was observed for the photoprotolytic dissociation of 1-naphthol in SDS and Triton X-micelles [16]. The escape of an excited molecule of 1-naphthol from the micelle leads to its instant dissociation. It gave the possibility of measuring the escape rate constant of 1-naphthol from SDS micelles which equals to 2×10^8 s^{-1} [16].

Of course the above cases are extreme, and combinations of these cases are possible. Mixed (dynamic and static) quenching of porphyrin fluorescence by dimethylviologen due to complex formation in the ground state was found by Whitten et al. [11]. Complex formation of reactants solubilized in micelles is a common phenomenon [41, 42]. Besides that, the mutual diffusion of the reactants within the micelles occurs in both stationary and non-stationary ways [43] which can lead to an apparent mixed fluorescence quenching. In order to analyse the kinetics in the micellar solutions and to take into account the complex formation in the ground state and the non-stationary diffusion, one should compare the changes of the absorbance spectra with the static component of quenching as well as in the case of homogeneous solutions. In micellar solutions the increment of the static fluorescence quenching must appear as a discrepancy of the values determined at stationary illumination, for example, from Eq. (35) and at pulse excitation, for example, from Eq. (33).

The most general model which takes into account the direct intermicellar exchange of the quencher molecules (the rate constant of this process is denoted k_e) was put forward by De Schryver et al. [44]. This model gives the following expression for the fluorescence decay after a pulse excitation:

$$I(t)/I(0) = \exp\{A_2 t - A_3[1 - \exp(-A_4 t)]\} \tag{36}$$

where

$$A_2 = 1/\tau_0 + \frac{k_r}{k_r + k_e[M] + k_{out}} \cdot \frac{k_{in} + k_e K[M]}{1 + K[M]} [Q] = 1/\tau_0 + S_2[Q], \tag{37}$$

$$A_3 = \left(\frac{k_r}{k_r + k_e[M] + k_{out}}\right)^2 \frac{[Q]}{K^{-1} + [M]} = S_3[Q], \tag{38}$$

$$A_4 = k_r + k_e[M] + k_{out}. \tag{39}$$

The fluorescence decay, in the general case, is non-exponential with the increment of the internal exponent dependent upon the term A_3, that is at a given Q concentration and binding constant upon the ratio $k_r/(k_r + k_e[M] + k_{out})$. If the fluorescence quenching within the micelles is almost static (i.e. faster, than the exchange processes)

$$k_r > k_{out} + k_e[M], \tag{40}$$

Eqs. (37–39) can be simplified to:

$$S_2 = (k_{in} + k_e K[M])/(1 + K[M]), \tag{41}$$

$$S_3 = (K^{-1} + [M])^{-1}, \tag{42}$$

$$S_2/S_3 = k_{out} + k_e[M], \tag{43}$$

$$A_4 = k_r. \tag{44}$$

De Schryver et al. [44] applied this model to analyse the pyrene fluorescence quenching by metal ions in SDS micelles. The situation described by the inequality (40) was observed for nickel, copper and lead ions. The k_e values were determined from the slope of the linear plot of S_2/S_3 vs. $[M]$ [see Eq. (43)]. For europium and chromium ions, both interfacial exchange processes in micelle-micelle and micelle-bulk solution are very slow as compared with pyrene fluorescence decay. Here, the kinetics fits well to Case 2 discussed above. For silver and thallium ions, the rates of the fluorescence quenching and the interfacial exchange were comparable, i.e.:

$$k_r \sim k_{out} + k_e[M].\tag{45}$$

For these quenchers, De Schryver et al. analysed the fluorescence-decay curves numerically. The fluorescence decay is monoexponential at quenching by cesium ion. Here, the inequality (40) has the opposite sense with the right-hand part being large as compared with the deactivation rate of excited pyrene molecules. The quenching obeys the Stern-Volmer law, but the quenching constant depends upon the concentration of micelles:

$$I(t) = I(0) \exp(-A_2 t),\tag{46}$$

$$1/\tau = 1/\tau_0 + k_r[Q]/(K^{-1} + [M]).\tag{47}$$

Celadé and De Schryver applied the same model to the fluorescence quenching in SDS micelles by neutral molecules [45, 46] and in inverted micelles by halogene anions [47]. The intermicellar exchange of the counterions (iodine ions) was studied for cetyltrimethylammonium chloride (CTAC) micelles [48] and the value $k_e = 9.4 \times 10^8$ dm^3 mole^{-1} s^{-1} was obtained.

The influence of the surfactant concentration upon the quenching process may be caused by the change of the micelles' shape, rather than intermicellar exchange of quencher molecules, as it was supposed for pyrene fluorescence quenching by copper ions in SDS micelles [49].

Table 1. Values of the rate constants of the quencher entrance into SDS micelles k_{in}, of its exit out of micelles k_{out}, and of the intramolecular reaction of the probe with the quencher k_r

The probe	The quencher	k_{in}, 10^9 dm^3/mole s	k_{out} 10^5 s^{-1}	k_r 10^7 s^{-1}	K dm^3/mole
Pyrene	Eu^{+3}	1 [44]	1 [44]	1.6 [44] 1.8 [50]	5500 [44] 130000 [50]
Pyrene	Cr^{+3}	1 [44]	1 [44]	0.98 [44] 1.4 [50]	5900 [44] 3700 [50]
Pyrene	Cu^{+2}	–	–	–	11000 [50]
Anthracene	Cu^{+2}	–	–	–	8800 [50]
1-Methylpyrene	Cu^{+2}	1.2 [44]	1.2 [44]	2.7 [44]	10000 [44]
Pyrene	Ni^{+2}	1 [44]	1 [44]	0.89 [44]	10900 [44]
Pyrene	Co^{+2}	1 [44]	1 [44]	0.54 [44] 0.39 [50]	8700 [44]
Pyrene	Pb^{+2}	7 [44]	3 [44]	0.91 [44]	27700 [44]
Pyrene	Tl$^+$	29 [44]	76 [44]	1.9 [44] 1.4 [50]	3800 [44] 47200 [50]
Pyrene	Ag$^+$	16 [44]	80 [44]	1.6 [44]	2000 [44]
Pyrene	Cs$^+$	20 [44]	20 [44]	0.027 [44]	10000 [44] 4900 [50]
Pyrene	Mn^{+2}	–	–	0.26 [50]	126000 [50]

Pyrene fluorescence quenching by metal ions in SDS micelles was studied at both constant and pulse illumination by Ziemiecki and Cherry [50, 51]. In order to analyse the kinetic data they applied the same kinetic model as De Schryver et al. The values of k_r obtained at steady-state excitation and at pulse excitation coincide. Thus, the pyrene fluorescence quenching by metal ions in SDS is an example of pure dynamic quenching within the micelles. The data of De Schryver et al. [44] and of Ziemiecki and Cherry [50, 51] are compiled in Table 1.

The counterion exchange among the particles of inverted microemulsions was investigated by Atik and Thomas [52–54], who also observed the exchange of the photoproduct ions among the particles.

The numerical simulation of the intermicellar counterion exchange proved [55] that this process is controlled by the Coulomb factor and its rate increases drastically with the growth of the surfactant concentration.

Theoretical works describe some models of mutual diffusion of reactants within the micelles and over the surface of micelles [56–62]. The general conclusion was that, but for a very short initial period of time, the mutual diffusion of the reactants obeys a single-exponential law.

1.2.4 Reversible Reactions

For reversible adiabatic photoreactions in micellar solution:

$$A* \underset{k_{-R}}{\overset{k_R}{\rightleftharpoons}} B* + C. \tag{48}$$

In the general case one obtains rather complicated and inconvenient equations for the relationships of observed fluorescence quantum yields and fluorescence kinetics and the rate constants of the reactions and exchange by reactants and products between micellar and volume phases and concentration of the product C. The degree of solubilization of the initial compound A can be taken into account similiarly as in the case of the irreversible reactions of Eqs. (2) and (22). If one can neglect some interphacial exchange processes, very simple equations similar to the case of reactions in homogeneous solutions is obtained:

$$\frac{\phi_m \phi'_{0m}}{\phi_{0m} \phi'_m} = a + b[C] = \frac{1}{k_1^m \tau_0^m} + \frac{k_{-1}^m \tau_0'^m}{k_1^m \tau_0^m}[C] \tag{49}$$

where k_1 and k_{-1} are effective rate constants and τ_0^m and $\tau_0'^m$ are lifetimes of $*A$ and $*B$ in the micellar phase in the absence of the reaction.

Case 1a. Reactions occur completely in the micellar phase (the exit rate constant for $B*$ $k'_{out} \ll k^m_{-R}[C]_m$; $k'_{out} \ll 1/\tau_0'^m$); the interphase exchange by product C is much faster than the reactions and decay of excited molecules (exit rate constant for C $k''_{out} \gg 0.1 k^m_{-R}$; $k''_{out} \gg 1/\tau_0'^m$). The stationary concentration $[C]_m$ of the product is equal to its equilibrium concentration in the micellar phase, which is determined by its overall concentration $[C]$ in the solution, its distribution coefficient ϱ'' and the volume concentration of the micellar phase X:

$$[C]_m = [C]\varrho''/(1 + \varrho''X) \tag{50}$$

then:

$$\frac{\phi_m \phi'_{0m}}{\phi_{0m} \phi'_m} = \frac{1}{k_R^m \tau_0^m} + \frac{k_{-R}^m \tau_0'^m}{k_R^m \tau_0^m} \cdot \frac{\varrho''}{1+\varrho''X}[C] \tag{51}$$

or:

$$k_1 = k_R^m, \tag{52}$$

$$k_{-1} = k_{-R}^m \varrho''/(1 + \varrho'' > X). \tag{53}$$

Case 1b. The reactions occur completely in the micellar phase ($k'_{out} \ll k_{-R}^m [C]_m$; $k'_{out} \ll 1/\tau_0'^m$); the interphacial exchange by product C is much slower than the reactions and decay of excited molecules ($k''_{out} \ll 0.1 k_{-R}^m$; $k''_{out} \ll 1/\tau_0'^m$). The stationary C concentration $[C]_m$ in the micelles containing excited product $*B$ is equal to the sum of its equilibrium concentration plus some additional concentration $[C]_m^0$, corresponding to one molecule of C per micelle ($[C]_m^0 \sim 0.1$ M). Then:

$$\frac{\phi_m \phi'_{0m}}{\phi_{0m} \phi'_m} = \frac{1+k_{-R}^m \tau_0'^m [C]_m^0}{k_R^m \tau_0^m} + \frac{k_{-R}^m \tau_0'^m}{k_R^m \tau_0^m} \cdot \frac{\varrho''}{1+\varrho''X}[C] \tag{54}$$

or:

$$k_1^m = k_R^m/(1 + k_{-R}^m \tau_0'^m [C]_m^0), \tag{55}$$

$$k_{-1}^m = k_{-R}^m \varrho''/(1 + k_{-R}^m \tau_0'^m [C]_m^0)(1+\varrho''X). \tag{56}$$

If the rate of interphase exchange is comparable with the rates of the reactions and decay of excited molecules, one can use equation similar to Eq. (54), taking into account the value of the coefficient δ, which determines the average excess of the concentration C in a micelle during the lifetime of the excited species:

$$\frac{\phi_m \phi'_{0m}}{\phi_{0m} \phi'_m} = \frac{1+\delta k_{-R}^m \tau_0'^m [C]_m^0}{k_R^m \tau_0^m} + \frac{k_{-R}^m \tau_0'^m}{k_R^m \tau_0^m} \cdot \frac{\varrho''}{1+\varrho''X}[C]. \tag{57}$$

Case 2a. Reactions occur in the micellar phase, but exited product partially leaves the micellar phase ($k'_{out} \sim 1/\tau_0'^m$); the interphacial exchange by product C is much faster than the reactions and the decay of excited molecules ($k''_{out} > 0.1 k_{-R}^m$; $k''_{out} \gg 1/\tau_0'^m$). The product fluorescence quantum yield is determined by both molecules located in the micelles and by molecules which exit from micelles to the bulk phase:

$$\frac{\phi_m \phi'_{0m}}{\phi_{0m} \phi'_m} = \frac{1}{1+k_{out}'\tau_0'^m}\left\{\frac{1}{k_R^m \tau_0^m} + \frac{k_{-R}^m \tau_0'^m}{k_R^m \tau_0^m} \cdot \frac{\varrho''}{1+\varrho''X}[C]\right\} \tag{58}$$

and:

$$k_1 = k_R^m(1 + k_{out}'\tau_0'), \tag{59}$$

$$k_{-1} = k_{-R}^m \varrho''/(1+\varrho''X). \tag{60}$$

Case 2b. Reactions occur in the micellar phase, but excited product partially leaves the micellar phase ($k'_{out} \sim 1/\tau_0'^m$); the interphacial exchange by product C is much slower than the reaction and decay of excited molecules ($k''_{out} \ll 0.1 k_{-R}^m$; $k''_{out} \ll 1/\tau_0'^m$). Then:

$$\frac{\phi_m \phi'_{0m}}{\phi_{0m} \phi'_m} = \frac{1}{1+k_{out}'\tau_0'^m}\left\{\frac{1+k_{-R}^m \tau_0'^m [C]_m^0}{k_R^m \tau_0^m} + \frac{k_{-R}^m \tau_0'^m}{k_R^m \tau_0^m} \cdot \frac{\varrho''}{1+\varrho''X}[C]\right\} \tag{61}$$

and:

$$k_1 = \frac{k_R^m (1 + k'_{out} \tau_0'^m)}{1 + k_{-R}^m \tau_0'^m [C]_m^0},$$ (62)

$$k_{-1} = \frac{k_{-R}^m \varrho''}{(1 + k_{-R}^m \tau_0'^m [C]_m^0)(1 + \varrho'' X)}.$$ (63)

2 Charge Separation in Micellar Systems

2.1 Micelles as Media for Photochemical Reactions

2.1.1 Models of Micellar Solutions Influencing the Reactivity of the Solubilizates

A wide range of chemical reactions in micellar systems depend mainly upon the difference of properties of the micellar phase and the bulk phase. One should only distinguish the solubilized and non-solubilized reactant molecules, the micelles altogether being considered as a pseudophase. For second-order reactions, the intermicellar distribution of reactant molecules should be taken into consideration as discussed above.

Both models fail to precisely locate the solubilizate molecules within the micelles which can evidently influence the rate constant of the reactions and interfacial exchange of the solubilizates.

Muckerjee et al. [63–65] proposed a so-called "three-phase" model of the micellar solutions which considers the molecules adsorbed at the micelle surfaces as a separate phase. This model was applied to allow a comprehensive analysis of the distribution of several organic nitroxides between dodecane and water, their surface activity at this system and the solubilization constants of these nitroxides by SDS and other micelles. The high values of the binding constants of nitroxides by micelles can be explained not by their solubility in the hydrophobic nuclei of the micelles, but by their adsorption at the interface of the micelles and the aqueous phase.

The existence of ion pairs of hydroxyaromatic anions with polar groups of cationic micelles was proposed by Zaitsev et al. [66] to explain the effective charge of anions close to zero, observed in acid-base photoreactions of hydroxyaromatics in CTAB solutions. Such a value for the effective charge was found by simulation of the values of the diffusion rate constants of hydrogen ions to excited anions of hydroxyaromatics to make the calculated diffusion-controlled protonation reaction of the excited anions rate constants close to experimentally observed ones. In aqueous solution, the excited anions are protonated with diffusional values of the rate constant with some non-significant steric factor [67, 68]. The "three-phase" model can help to interpret the reactivity of polar and charged substances in micellar solutions.

2.1.2 The Effective Microviscosity and Polarity of Micelles

The effective microviscosity and polarity of organized molecular assemblies have a great influence upon photoreaction including diffusion and interaction of polar and charged molecules. These characteristics of the inner parts of micelles, vesicles, and microemulsions can be determined by several means, some of which are given below.

Measurements of the fluorescence depolarization can give the lifetime of the rotational diffusion of a fluorophore implanted into micelles [69–76] or lipid vesicles [77, 78]. The measurement of the rotational diffusion of the same fluorophore in homogeneous solutions of different viscosity yields a calibration curve to determine the effective viscosity of the microphase (the microviscosity). The fluorescence depolarization kinetics for Bengal Rose in CTAB micelles in the picosecond range provides an opportunity to distinguish the rotation of the Bengal Rose molecule and the rotation of a micelle as a whole [69]. Usually the solubilization results in an enhancement of the fluorescence quantum yields and lifetimes. Such an enhancement was observed for the fluorescence of aromatic conjugated systems [83] and also the phosphorescence emission at room temperature in liquid solutions. Grätzel and Humphrey-Baker [79] have shown that it is the high microviscosity of micelles that enhances the fluorescence of the cyanine dyes.

The effect of added alcohols upon the microviscosity of micelles was studied by Turro et al. [76]. The low-molecular alcohols reduce the microviscosity of CTAB micelles. The high-molecular alcohols do not influence the microviscosity. Schinitzky et al. observed a phase transition of a bilayer membrane by the fluorescence depolarization technique [75].

The use of the picosecond technique enabled Pouligni et al. to observe the reorientation kinetics of 3,3'-diethyloxadicarbocyanine by absorption anisotropy in microemulsions of Aerosol OT (AOT) and methyl alcohol in cyclohexane [84, 85].

A further technique to study the microviscosity of micelles is based on the measurements of intramolecular exciplex formation [86–89]. In solvents possessing viscosities exceeding 2 cPs, the rate of the intramolecular exciplex formation depends upon the viscosity of the solvent and not upon its polarity [86]. In contrast to the intermolecular exciplex [87] whose formation rate is influenced by the distribution of the reactants among the micelles, the formation of the intramolecular exciplex is controlled by the relative rotation of two parts of one molecule only. The values of the microviscosity obtained with this method are in good agreement with ones obtained by the depolarization technique [87].

The rate of a photo-isomerization reaction can be used to characterize the structure of micelles [90]. The dark-reaction of the spiropyran cycle cleavage, yielding the corresponding merocyanine, is controlled by the high microviscosity of micelles.

The effective polarity of micelles was estimated from the form and position of emission spectra of various compounds with spectral characteristics sensitive to the polarity of homogeneous solvents. Such compounds include pyrene for which the ratio of the first and the third vibrational bands (I_1/I_3) depends upon the polarity of solvent [91–93] and also various compounds with the emission spectra sharply dependent upon the polarity [94–97]. To characterize the polarities of some micellar and membrane systems, Kano, Goto, and Ogawa [94] used the plot of an intramolecular exciplex maximum vs. the polarity function of the solvent:

$$f(\varepsilon, n) = (\varepsilon - 1)/(\varepsilon + 2) - (n^2 - 1)/(n^2 + 2). \tag{64}$$

The transition from micelles to microemulsions causes a decrease of the polarity measured with a hydrophobic probe. Gregoritch and Thomas determined the

Table 2. Effective values of microviscosities η and dielectric constants ε of some molecular assemblies

System	ε	η ($T°$ C), cPs
SDS	51–55 [97], 48 [92], 40 [94]	9 (20°) [87], 52 (21°) [93], 15–36 [86], 30 (25°) [79]
Sodium decylsulfate	51–55 [97]	
Sodium tetradecylsulfate	51–55 [97]	
Sodium dodecanoate	43–51 [97]	
Triton-X100	27–30 [97], 5,5–7,9 [94]	
Brij	27–29 [97]	
Octaethyleneglycole, mono-dodecyl ether	27–29 [97]	
CTAB	28–33 [97], 15 [92]	19–36 (27°) [86]
CTAC	31–41 [91], 28–31 [97], 24 [92], 40 [94]	31 (20°) [87]
Cetyltrimethylammonium fluoride	28–33 [97]	
Dodecyltrimethylammonium chloride	30–40 [97]	
Dodecyltrimethylammonium bromide	29–35 [97], 31–41 [91]	
Tetradecyltrimethylammonium fluoride	31–44 [97]	
Tetradecyltrimethylammonium chloride	29–35 [97]	
Tetradecyltrimethylammonium bromide	28–32 [97]	22 (25°) [71]
Benzyldimethylhexadecylammonium chloride	26–28 [97]	
Tetradecyltriethylammonium bromide	–	26 (25°) [71]
Tetradecyltripropylammonium bromide	–	32 (25°) [71]
Tetradecyltributhylammonium bromide	–	35 (25°) [71]
Dimiroystoylphosphatidylcholine	14–17 [97]	38 (23.9°) [89],
Dipalmitoylphosphatidylcholine	–	18 (41.4°) [89]
Mirystoyltrimethylammonium bromide	31–41 [91]	

photoionization quantum yields of pyrene, pyrenebutyric acid and sodium 1-pyrenylsulfonate along with the spectral properties of these compounds [98]. The transition from micelles to microemulsions causes a decrease of the photoionization quantum yield for pyrene, whereas those for the pyrene derivatives remain constant. The conclusion was drawn that the pyrene molecules are solubilized in hydrophobic nonpolar nuclei of microemulsion droplets, whereas derivative molecules locate in the polar surface layer. This result shows also the limitations of the probe technique to measure the properties of the micellar media. The results are evidently dependent upon the kind of probe used. One should suppose some gradients of viscosity, polarity, and electric potential within micelles and other molecular assemblies. The evidence that pyrene is located in two different sites of of polymerized microemulsions with different microviscosity and polarity was found by the fluorescence quenching technique [99]. A general recommendation is to determine different microenvironment properties of the probes which resemble molecules and their reactivity should be estimated.

Data from the literature on the microviscosity and polarity of some molecular assemblies are compiled in Table 2. One can see that micelles behave like viscous and polar solvents. For example, the dielectric constant for SDS exceeds 50.

In order to illustrate the relationship of polarity measurements and photochemical charge separation, let us estimate the non-Coulomb component of the change of the standard Gibbs energy ΔG^0 for the transfer of a singly-charged ion with a radius $r = 5$ Å from the micellar phase into the aqueous phase [9]:

$$\Delta G^0_{n-c} = \frac{e^2}{r}\left(\frac{1}{\varepsilon_m} - \frac{1}{\varepsilon_{aq}}\right). \tag{65}$$

Here, ε_m and ε_{aq} are the dielectric constants of the micellar and aqueous phases and e is the charge of an electron. One can see that for SDS micelles $|\Delta G^0_{n-c}| < 4$ kJ/mole. This value is small compared with the electrostatic component $\Delta G^0_C = \mathscr{F}\Delta\varphi$, as usual values of $\Delta\varphi$ for SDS micelles are about -150 mV [7]. This is why the $\Delta\varphi$ value at the interface can prove more important for the charge separation yield than the chemical nature of the microphase (see Sect. 1.1.2).

2.2 Initial Charge Separation and Geminate Recombination in Micellar Systems

The formal kinetics of photoreactions in micellar solutions has been studied quite well. Rather less explored however, is the problem of the dependence of rate constants on the nature of the reactants and detergents. Such specific features of the reactions in the micellar phase, particularly the correlations of reactivity with the properties of the reactants and surfactants are discussed in this section.

2.2.1 Photoionization in Micellar Solutions

The value of the quantum yield of the primary charge separation products depends upon the ratio of the carriers' separation and geminate recombination rate constants. The threshold (minimal energy) of the photoionization is determined by the energy necessary for an electron to leave any given molecule. As the photoionization takes place at the instant of the photon absorption the medium has no time to solvate the photoionization products (the electronic polarization of the medium, not the orientational, is important), so the effect of the medium upon the ionization threshold is relatively weak. For studies of the photoionization processes, the electron traps located in the bulk phase are usually used.

There are several examples described in the literature where the compounds which undergo the photoionization in micellar solutions are not photoionized by light of the same wavelength in homogeneous solutions. There can be two reasons for this phenomenon: the decrease of the photoionization threshold due to the influence of the potential drop at the interface, or a strong increase of the photoionization quantum yield due to charge separation process becoming faster. The photoionization of perylene and tetracene in micelles of SDS and of Triton X100 and also in tetramethylsilane and methyl alcohol was studied by Bernas et al. [100, 101]. The shape of the plot of the photoionization quantum yield φ_e vs. the wavelength of the incident light for the neutral micelles of Triton X100 resembles the ones for organic solvents. The shape of this plot for negatively charged SDS micelles differs markedly (Fig. 2). For SDS micelles this plot can be linearized in coordinates φ_e vs. $h\nu$

$$\phi_e = p(h\nu - I_{min}) \tag{66}$$

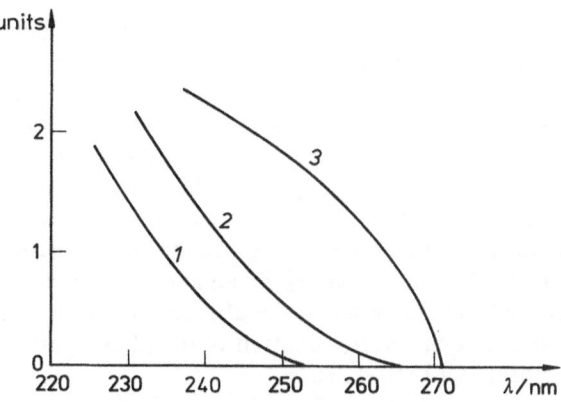

Fig. 2. The dependence of perylene photoionization relative quantum yields on the excitation wavelength in methanol (1), in Triton X100 micelles (2), in SDS micelles (3) [101]

which resembles the same plot for the gas-phase photoionization (where there is no solvent cage, and therefore no geminate recombination). At any given wavelength of the incident light, the photoionization quantum yield in the SDS micelle is more than in Triton X100 micelles. Nevertheless, the photoelectrons appear in SDS and Triton X100 at the same wavelength of the incident light. Thus, the ionization threshold of perylene in SDS micelles coincides with the one in Triton X100 micelles. Bernas et al. concluded that the anionic micelles help the charge separation so that the geminate recombination in SDS micelles does not take place but the ionization threshold does not depend on the potential drop at the interface of the micelle and the aqueous solution. Very low values of the ionization potential obtained by Bernas et al. were explained by the simultaneous presence of the two reasons: the relatively low ionization potentials of the polycyclic aromatic hydrocarbons and of highly negative values of the medium polarization by an electron energy V_0 for water. This conclusion relates to the photoionization of the Frank-Condon state of the molecules and it cannot be applied to systems where the electron is ejected into solution from a thermalized excited state.

In a study of the biphotonic ionization of pyrene with a tunable laser, Wallace et al. [102] found that the ionization potential in SDS is lower than in the gas phase and in methyl alcohol. They applied the following expression for the ionization potential of a molecule in a medium:

$$I_m = I_{vac} + P^+ + V_0 \tag{67}$$

where I_{vac} is the ionization potential of the same substance in vacuo and P^+ is the polarization energy of the medium by the radical cation formed. Wallace et al. suppose that the value of V_0 for the micellar solution coincides with the one for water and give an estimate of $V_0 = -1.3$ eV.

The photoionization of substituted aromatic hydrocarbons caused some discussion. Thomas and Piciulo [103, 104] studied the photoionization of 3-aminoperylene in SDS and CTAB micelles and also in homogeneous solutions of a wavelength of 530 mm. In SDS micelles the photoionization efficiency depends linearly upon the power of the light, in CTAB and neutral micelles the dependences are quadratic. Thomas and Piciulo supposed that because of the highly negative V_0 in SDS, a single-photoionic ionization threshold is reached. The single-photonic ionization by green light was, however, challenged [105].

The photoionization of phenothiazine and its derivatives in SDS micelles, which is not observed in homogeneous solutions and the premicellar region was studied by Grätzel et al. [106–109]. They suppose that the anionic micelles decrease the ionization potential of the phenothiazine derivatives and the photoionization includes an electron tunnelling from the micelle into the solution through the Stern layer. However, there are no direct data for these systems to give a value of the photoionization threshold. A drastic increase of the photoionization quantum yield at the transition to anionic micelles could also be caused by a decrease of the role of the geminate recombination.

The use of micellar solutions enables one to study in aqueous solutions the reactivity of substances which are insoluble in water in the absence of additives, in particular the photoionization and electrochemical oxidation of the photoxynthetic dyes – chlorophyll and bacteriochlorophyll [110–113].

2.2.2 Photoprotolytic Dissociation in Micellar Solutions

Photpprotolytic dissociation occurs from a relaxed state and the Frank-Condon states are of no importance [114, 115]. So the photoprotolytic dissociation is not affected by the energy of the incident photons but is determined by the pK_a value of the subsance in question in the excited state (pK_a^*).

The first studies of the photoprotolytic dissociation of 1-naphthol, 2-naphthol and 1-aminopyrene cation in micellar solutions were made by Klein and Hauser [116] and also by Weller and Selinger [117, 118] who have shown that the photoprotolytic dissociation slows down in anionic micelles. Fendler [119] observed the photoprotolytic dissociation of 1-hydroxypyrene-trisulfonate in CTAB micelles. The inhibition of dissociation was explained in terms of the relatively low polarity and high microviscosity of the micellar phase [116–119] and also the effect of the negative micellar potential.

Photoprotolytic reactions in micellar solutions were systematically studied by the authors of the present review with coworkers [15, 16, 66, 120–122]. Effective dissociation rate constants k_1^m for a set of hydroxyaromatic compounds in anionic, cationic and uncharged micelles were determined from the dependence of fluorescence quantum yields on the concentration of micelles [Eq. (22)] or on the concentration of acid [Eq. (49)]. The latter method was also used for determining the effective equilibrium constants k_1^m/k_{-1}^m and effective rate constants of the reverse protonation reaction k_{-1}^m [Eq. (49)]. The data obtained are presented in Table 3.

The difference of free energies of protolytic reactions in the micellar phase ΔG_m and in aqueous solution were determined in the pseudophase model or in the microscopic model by the ratio of distribution coefficients (binding constants) of the initial acid (ϱ) and its anions (ϱ') and protons (ϱ''):

$$\Delta G_m^0 = \Delta G^0 - 2.3RT \log(\varrho'\varrho''/\varrho). \tag{68}$$

It is convenient to exclude the distribution coefficient of protons ϱ'' and to define pK_m in such a way that it is related to the pH value of the aqueous phase:

$$pK_m = pK - \log(\varrho'/\varrho). \tag{69}$$

The ratio k_1^m/k_{-1}^m determined from Eq. (49) is not identical to the equilibrium constant. For the cases 1a–2b discussed in Sect. 1.2.4:

$$pK_m = -\log(k_1^m/k_{-1}^m) + \log(1 + k_{out}'\tau_0') + \log(1 + \varrho''X). \tag{70}$$

Table 3. Values of the rate constants of the photoprotolytic dissociation of hydroxyaromatics (k_1^m) and the rate constants of protonation of corresponding anions (k_{-1}^m), and the effective values of the excited state pK_a (pK_a^*) in various micelles at complete solubilization in water

Surfactant	None (in H_2O)		CTAB		
Proton donor	k_1, $10^8 s^{-1}$	k_{-1} 10^{10} dm^3 s^{-1} mol^{-1}	pK_a^*	k_1^m 10^8 s^{-1}	k_{-1}^m 10^{10} dm^3 s^{-1} mol^{-1}
2-Naphthol	0.73	4.7	1.0	1.05	0.11
2-Naphthol-8-sulfonate	30	7.9	0.7	2.86	0.15
2-Naphthol-6-sulfonate	7.7	3.5	1.2	0.71	0.11
1,7-Dihydroxynaphthalene	0.32	4.3	1.9	0.42	0.36
2,7-Dihydroxynaphthalene	1.25	3.1	1.3	0.81	0.2
2-Anthrol	1.40	2.8	1.1	3.3	0.56
1-Hydroxypyrene	0.3	3.8	2.9	0.047	0.35
2-Naphthol-3,6-disulfonate	34	2.6	1.1	3.7	0.47
8-Hydroxypyrene-1,3,6-trisulfonate	89	1.9	0.9	4.8	0.4
1-Chloro-2-naphthol	7.6	0.6	1.1	2.1	0.3
1,4-Dichloro-2-naphthol	12	0.2			

Surfactant	Brij 56	$C_{10}H_{21}SO_4Na$	$C_{12}H_{25}SO_4Na$	$C_{16}H_{33}SO_4Na$
Proton donor	k_1^m, 10^8 s^{-1}	k_1^m, 10^8 s^{-1}	k_1^m, 10^8 s^{-1}	k_1^m, 10^8 s^{-1}
2-Naphthol				
2-Naphthol-8-sulfonate				
2-Naphthol-6-sulfonate				
1,7-Dihydroxynaphthalene				
2,7-Dihydroxynaphthalene	0.1			
2-Anthrol				
1-Hydroxypyrene				
2-Naphthol-3,6-disulfonate				
8-Hydroxypyrene-1,3,6-trisulfonate				
1-Chloro-2-naphthol	0.3	0.2	0.1	
1,4-Dichloro-2-napthol	4	1	0.5	0.3

The last term in this sum reflects the decrease of H^+ concentration in the aqueous phase due to solubilization by micelles, and can often be neglected for usual experimental conditions. The second term is close to zero in cationic and neutral micelles, where $*A^-$ remains in the micellar phase and can be greater (but less than unity for investigated compounds) in anionic micelles, and where A^- leaves the micelles of a rate constant of 10^7–10^8 s^{-1}.

The values of k_1^m/k_{-1}^m for the majority of hydroxyaromatic compounds in cationic micelles are greater than in aqueous solution for the same compounds, in neutral micelles they are smaller and in anionic micelles much smaller than in aqueous solution. Such an influence of the charge of micelles can be attributed to the decrease of the distribution coefficient ϱ' for anions in this series, while the distribution coefficients ϱ for the initial acid form are approximately constant.

A noteworthy "compensation effect" is observed in cationic micelles: a smaller increase of k_1/k_{-1} is found for hydroxyaromatic compounds with smaller pK_a values and vice versa.

This effect results in contraction of the range of pK_a for the investigated compounds in micellar solutions compared to aqueous solutions. The nature of this effect can be explained by the existence of a correlation between pK_a and the solution energy of the anion in aqueous solution which assumes the opposite correlations with the distribution coefficient ϱ'. A lower polarity of the micellar phase results in smaller distribution coefficients for the anion relative to the initial acid in neutral and negatively charged micelles, increasing the value of pK_m and decreasing the value of k_R^m.

The protolytic dissociation rate constants k_1^m in cationic micelles are not significally different from the ones for aqueous solutions, in neutral micelles they are an order of magnitude smaller, and in anionic micelles they are one to two orders of magnitude smaller than in aqueous solution. The rate constants k_{-1} for the protonation reaction of excited anions in micellar solutions are not strongly dependent on the nature of the surfactant. They are in the range $1–4 \times 10^9 \, \text{M}^{-1}\text{s}^{-1}$, which is about an order of magnitude smaller than in aqueous solution. The increase of the number of carbon atoms in alkyl chains of alkylsulfates from decylsulfate to cetylsulfate results in a decrease of k_1 and k_1/k_{-1}. An opposite effect can be expected for positively charged acids, which produce uncharged bases after protolytic dissociation.

A quite good linear correlation between $\log k$ and $\log(k_1/k_{-1})$, similar to the same correlation well-known for aqueous solutions, is observed in micellar solutions with the same slope equal to unity but shifted to the left. Such correlations coincide with the assumption that a typical linear dependence of the activation free energy on the reaction free energy (for endothermic reactions) is also valid for the micellar phase:

$$\Delta G_R^{\ddagger} = a + b\Delta G_R^0. \tag{71}$$

This yields:

$$\log k_R = a^0 - a/2.3RT + b\log\varrho'' - b\log K_m$$

where a^0 is the logarithm of the preexponential factor (usually $b \sim 1$ for endothermic reactions) and:

$$\log k_1^m = a^0 - a/2.3RT + \log[\varrho''/(1 + \varrho''X)]$$
$$- \log(1 + k_{-R}^m \tau_0'^m[\text{H}^+]_m^0) + \log(k_1^m/k_{-1}^m). \tag{72}$$

It is quite surprising that the investigated reactions in different surfactant solution have the same shift relative to aqueous solutions. This means that the values of distribution coefficients for H^+ are close to each other for different detergents and also that $\log(1 + k_{-R}^m \tau_0'^m[\text{H}^+]_m^0)$ can be neglected.

The photoprotolytic reactions in micellar solutions show the usual values of the deuteration isotope effect [122] which gives evidence that the photoprotolytic dissociation is not controlled by the exit of the excited molecules from the micelles. The product of the dissociation A^-* can be formed within a micelle initially and then A^-* can leave the micelle. The increase of the number of the carbon atoms in surfactant molecule causes an increase of the A^-* exit rate constant.

2.2.3 Second-Order Intramicellar Electron Transfer Reactions

For the most common exciplex-forming systems, like pyrene-dimethylaniline, only a quenching is observed in the micellar phase, but no exciplex emission [25, 26]. The interaction of the excited pyrene with dimethylaniline is controlled by diffusion [123] with every collision resulting in quenching, which was also shown by Mataga et al. [25, 26] to yield pyrene anion and dimethylaniline radical cation. The quantum yield of radical ions is slightly greater in cationic CTAB micelles than in acetonitrile solutions, in neutral Brij 35 micelles a little less and in anionic SDS micelles considerably less than in acetonitrile.

The intermolecular electron transfer in micellar solutions can also proceed by static means. Complex formation in the ground state in micellar solutions was observed for some typical pairs of electron donors and acceptors: dimethylviologen and stilbene [25], dimethylviologen and porphyrines [11], tetracyanoethylene and dimethylaniline [26].

Costa and Macanita [43] were among the first to observe the correlation between the value of the quenching rate constant in the micellar solution and the electrochemical potentials of the quenchers for the quenching of aromatic esters by electron donors in neutral micelles of Triton X100 (Fig. 3).

The effect of the electrochemical and hydrophobic properties of the reactants on the intramicellar charge separation rate constant k_R for model systems of ruthenium-tris-bipyridile, tris-phenanthroline and tris-diphenylphenanthroline complex quenching by organic cations in SDS micelles was studied by Miyashita et al. [124, 125]. The plot of k_R vs. the free energy of the electron transfer reaction calculated from electrochemical and spectral properties of the reactant (Fig. 4) differs from the well-known Rehm-Weller and Marcus plots for homogeneous solutions.

A correlation similar to the one for homogeneous solutions can be obtained for a series of quenchers with similar geometry and charge only. The values of k_R for highly exergonic reactions are smaller than the diffusional limit for quenching within the total volume of the pseudophase. Miyashita et al. interpreted such difference in terms of high microviscosity of micelles. The k_R value for quenching by dihexylviologene is an order of magnitude less than the k_R value for dimethylviologene. Miyashita et al. explain this phenomenon by the difference in diffusion coefficients of these viologenes, the diffusion being assumed to occur in the surface layer only. This, however, is not the only possible explanation. The decrease of the k_R value with the increase of the alkyl radical length in the quencher molecule can be described in terms of a three-phase model (see Sect. 2.1.1) assuming that the alkylviologene is distributed between the micellar surface (where the fluorescent probe is localized) and the hydrophobic nucleus of the micelle. The increase of the alkyl chain shifts the equilibrium to the side of the micellar nucleus and the intramicellar rate constant is decreased.

A correlation between the rate constants k_q and free enthalpy change ΔG^0 of electron transfer was studied by Hashimoto and Thomas [127] for quenching of excited singlet states of both pyrene and N-ethylcarbazol and of the triplet state of N-methylphenothiazine by a number of metal ions and for back electron transfer reactions in micellar sodium taurocholate and sodium dodecylsulfate solutions. Quenching rate constants were determined from Stern-Volmer plots obtained for lifetimes of excited states at high concentration of micelles, where the exponential decay

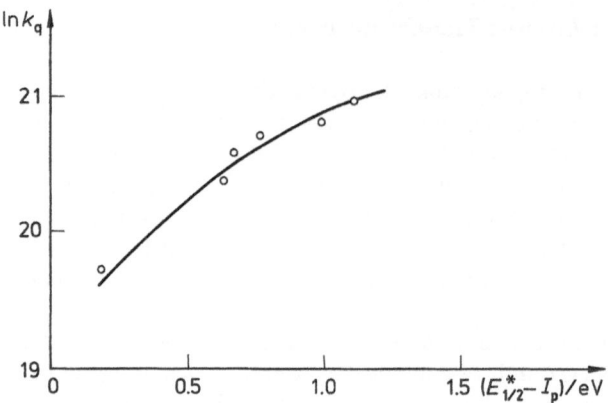

Fig. 3. The dependence of intramicellar electron transfer rate constant on the difference of the electron acceptor reduction potential and electron donor oxidation potential [104]

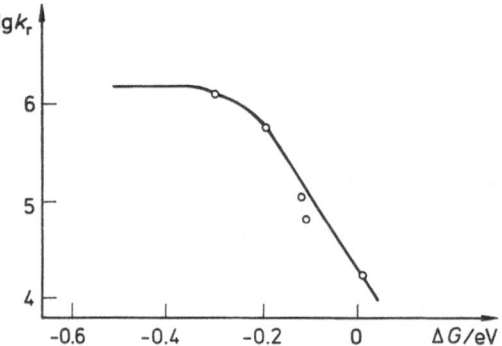

Fig. 4. The dependence of intramicellar quenching rate constant on Gibbs energy changes of electron transfer for quenching of ruthenium-tris-4,7-diphenyl-1,10-phenanthrolin by substituted pyridinium cations in SDS micellar solution [124]

is observed in the presence of quencher due to very fast exchange of the reactants among the micelles. Back electron transfer rate constants were obtained from second-order decay reactions of radical cations formed in the quenching reaction. The authors discussed the data obtained in terms of the Rehm-Weller correlation:

$$k_q = k_{12}[1 + (k_{21}/k_{23}^0)\exp(\Delta G^*/RT) + (k_{21}/k_{30})\exp(\Delta G^0/RT)]^{-1}, \tag{73}$$

$$\Delta G^* = \Delta G^0/2 + [(\Delta G^0/2)^2 + (\Delta G_0^{\ddagger})^2]^{1/2} \tag{74}$$

and obtained different values of the parameters k_{12} and $\Delta G_0^{\ddagger}$ for aqueous solutions and for each surfactant and also for forward and back reactions. But in this analysis the authors neglected the change of redox potentials in the micellar phase and the difference between the local concentration of the quencher in the micellar phase and its overall concentration in the solution used for the calculating the rate constants. This shortcoming causes a large dispersion in the Rehm-Weller correlation and shows the necessity to use different values of the parameters k_{12} and $\Delta G_0^{\ddagger}$. To overcome such difficulties we analysed the same experimental data neglecting the last term in Eq. (73) using a simpler linear approximation for $\Delta G^* = \Delta G_0^{\ddagger} + \beta\Delta G^0$ which proved much more convenient for the processes where only relative values of ΔG^0 are known. These approximations were shown to give very good results in homogeneous solutions [128].

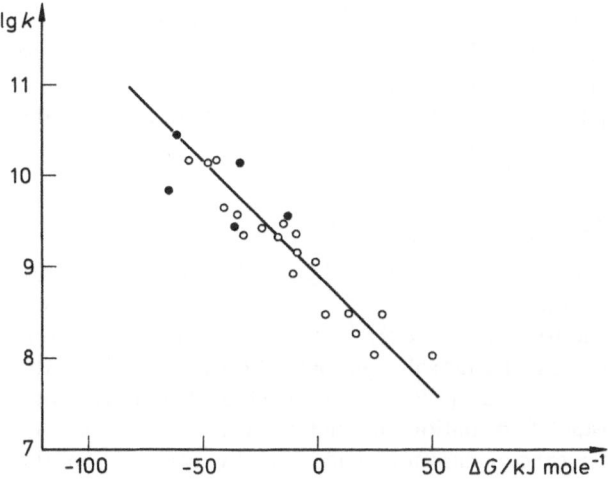

Fig. 5. The dependence of the rate constants for pyrene and N-methylphenothiazine fluorescence quenching (o) by metal ions and back electron transfer reaction (●) on Gibbs energy change [Eq. (2.12)] in sodium taurocholate micellar solution [127]

Table 4. Values of the parameter U [Eq. (2.12)] for quenching and recombination reactions in aqueous, sodium taurocholate (STC), and sodium dodecylsulfate (SDS) solutions

Ion	Quenching		Recombination	
	H_2O	SCH	SDS	SCH
Hg^{+2}	8.9	1.6	–	–
Fe^{+3}	8.9	2.4	–	8.0
Cu^{+2}	8.9	3.8	4.1	7.3
Eu^{+3}	8.9	4.5	–	–

In such a case one can obtain:

$$\log(1/k_q - 1/k_0) = U - \beta \Delta G^0/2.3RT \tag{75}$$

where ΔG^0 is the free energy of electron transfer in an aqeuous solution, k_0 is the "diffusional" rate constant ($\sim 10^{10}$ $M^{-1}s^{-1}$) and U includes terms due to the solubilization energies of quencher, excited molecules and the products of electron transfer and also $\Delta G_0^{\neq}$. The value of U is supposed to be sensitive mainly to the nature of the quencher and surfactant because the difference of solubilization energies for excited molecules and their radical ions is rather small and similar for different fluorophors. Actually for the quenching of different excited states by the same quencher in the same surfactant solution we obtained quite good linear relationships with similar slopes ($\beta = 0.14$) (Fig. 5). Table 4 shows the values of U chosen to combine all relationships for different ions and surfactants for quenching and back reaction. An important conclusion is the strong deceleration of the electron transfer reaction in micellar solutions compared to the same reaction in aqueous solutions which is determined by the nature of the ion and reaches for the systems in questions some 4–7 orders of magnitude. This gives in principal the possibility to choose such electron carriers which have sufficiently long relaxation times providing high effectively of secondary reactions consuming the energy of electron transfer photoproducts. The recombination reaction within the micellar phase is decelerated much less in the systems studied. One can hope to find ions with appropriate ratios of solubilization energies of oxidized and reduced forms which provide effective separation of primary products by predomination of the exit from the micellar phase over the recombination.

2.2.4 Bimolecular Proton Transfer Reactions

Proton transfer reactions between excited molecules and ground-state molecules are considered to be a model for the primary step of some photobiological processes such as bacterial photosynthesis and photoreception [129]. Proton transfer photoreactions of 1-hydoxypyrene with some bases in micellar solutions of CTAB were studied by the authors of the present review with co-workers [130]. Addition of bases (sodium acetate, hydroxylamine) increases the rate of formation of excited hydroxypyrene anions (Fig. 6). The ratio of the fluorescence quantum yields linearly depends on the concentration of H^+ and reverse values of the overall concentration $[B]_0$ of the base in the solution:

$$\frac{\phi_m \phi'_{0m}}{\phi_{0m} \phi'_m} = \frac{(1 + y k^m_{-R} \tau'^m_0)(1 + \varrho'' X)}{k^m_R \tau^m_0 \varrho''[B]_0} + \frac{(1 + y k^m_{-R} \tau'^m_0)(1 + \varrho''' X)[H^+]}{k^m_R \tau^m_0 \varrho'' K_a [B]_0} + \frac{k^m_{-R} \tau'^m_0 \varrho'''}{k^m_R \tau^m_0 \varrho'' K_a}[H^+]$$

$$(76)$$

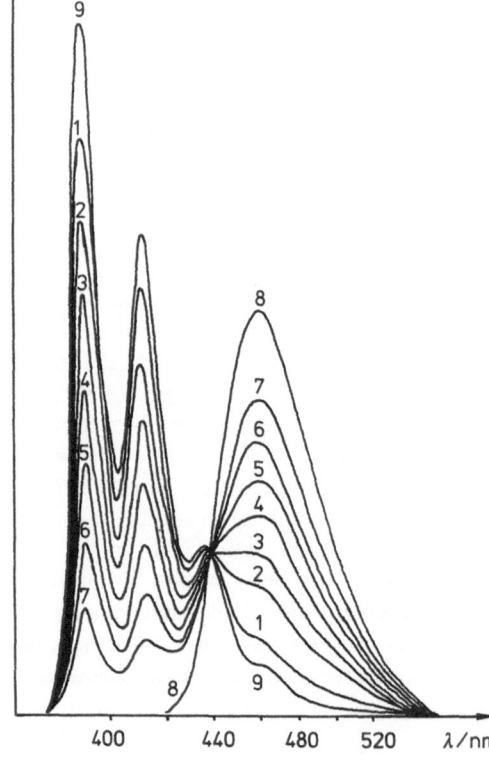

Fig. 6. Fluorescence spectra of 1-hydroxypyrene in CTAB micellar solution in the presence of sodium acetate [163]. Concentrations of acetate: 0 (1), 0.0029 (2), 0.0057 (3), 0.011 (4), 0.023 (5), 0.046 (6), 0.152 mol/dm³ (7); 1 mmol/dm³ KOH (8), and 0.33 mol/dm³ HCl (9) in the absence of acetate. CTAB concentration is 0.05 mol/dm³ [163]

where k_R^m and k_{-R}^m are the rate constants of proton transfer from excited molecules to the base B and reverse reaction, ϱ'' and ϱ''' are distribution coefficients for B and HB$^+$, K_a is a acidity constant of HB$^+$ for aqueous solutions, [H$^+$] is the concentration of protons in the aqueous phase, y is the stationary overequilibrium concetration of HB$^+$ in the micellar phase due to its formation in the proton transfer reaction, which depends on the rate of exchange by HB$^+$ between micellar and aqueous phases and the lifetime of excited species $(0 \leq y \leq 0.1$ mole dm^{-3}; the upper limit corresponds to one proton per a micelle), the other notations are the same as described in previous sections. From the dependence of fluorescence quantum yields on the concentration of B and H$^+$ one can find effective values of the rate constants and equilibrium constant for the reaction in the micellar phase:

$$*AH + B \rightleftarrows \overset{*}{A}^- + HB^+.$$

The ratio of the distribution coefficients for the base B and its protonated form HB$^+$ can be evaluated from the ratio of the first two coefficients of Eq. (76). The distribution coefficient for acetic acid is somewhat greater than for acetate anion and the acetic acid formed in proton transfer reactions is not removed from the micellar phase in spite of the very fast proton transfer reaction. For hydroxylamine the ratio of distribution coefficients is more favourable for charge separation and one can expect effective separation of charge carriers and a deceleration of their recombination after the deactivation of the excited anion.

2.3 Initial Stabilization of the Charge Separation Products

The photoreaction in micellar solutions results in a change of a molecule's location within a micelle. The photoionization of a model substance – tetramethylbenzidine (TMB) – in micellar solutions was studied in detail. Bales and Kevan [131] used electron acceptors with a different position of the acceptor site in order to determine the location of TMB within SDS micelles. The acceptors were j-doxylstearic acids with the general formula:

$$CH_3(CH_2)_{17-j} \overbrace{\qquad} (CH_2)_{j-2}COOH$$
$$NO^{\cdot}$$

(77)

The illumination of TMB in the presence of electron acceptors with different j-values yields TMB$^+$ radical cations with different quantum yields and consumption of the spin label. The highest quantum yield was obtained with $j = 10$ and especially with $j = 16$. This was explained in terms of TMB location in the hydrophobic region of the micelle asymmetrically to the electron acceptor molecule. However, another interpretation is possible: as the stearic acid molecule does not coincide in the number of carbon atoms with the micelle-forming surfactant, one can assume that the carbon atom C-13 is located near the end of the methyl groups of the surfactant, i.e. near the centre of the micelle. In this case the results of Bales and Kevan show that TMB molecules are located inside the micelle symmetrically with the photoionization probability maximum near the third carbon atom from the micelle centre. Anyway, TMB molecules are located in hydrophobic nuclei of the micelles. The important feature of TMB is that it undergoes photoionization even at the low-energy end of its absorption band [100].

The location of TMB$^+$ within the micelles was studied by Kevan et al. [132] with the spin-echo technique. In the deuterated SDS solution in the spin-echo spectrum of TMB$^+$ one can observe both the proton modulation from the hydrogen atoms of SDS and the deuterium atoms of water. The conclusion was that TMB$^+$ is located at the micelle-water interface. As discussed above, parent TMB is located in hydrophobic micelle nuclei. So the location of TMB$^+$ changes within its lifetime and it transfers from the hydrophobic region of the micelle to the interface. The evolution of TMB$^+$ in CTAB and SDS micelles was studied by Beck and Brus with pulse Raman spectroscopy [132]. They found that TMB$^+$ is formed in CTAB micelles with a delay and yields TMB and TMB^{+2} in the millisecond range. TMB$^+$ formed in SDS micelles is stable for several days [134]. The anionic micelles seem to stabilize TMB$^+$ and the cationic ones to destabilize it.

The stabilization of the radical cation in SDS micelles is consistent with the formation of the surface complex already mentioned in Sect. 2.1.1. There are a number of examples of the stabilization of radical ions in micelles of the opposite sign. The electrochemical reduction of nitrobenzene in homogeneous solutions is two-electronic, but in micellar CTAB solutions it is resolved into two separate reactions [135]. So the presence of cationic micelles inhibits the disproportionation of nitrobenzene radical anions. It may be caused by the reduced mobility of the radical ions in the surface layer of the micelles due to complex formation with the end groups of the micelles. It can account for the inhibition of the disproportionation because the latter needs a collision of two radical ions. The salt formation of the radical cation of N-methylphenothiazine with the end groups of SDS micelles was proposed in the study of electrochemical oxidation of N-methylphenothiazine in micellar solution [136]. Tetracyanoquinodimethane is solubilized in dodecylpyridinium-inverted micelles in the form of a radical anion with the oxidation of the iodide ion [137].

Similar examples can be given for acid-base reactions. For example, SDS micelles stabilize acridinium cation, leading to a shift of the acid-base equilibrium [8]. From the thermodynamic point of view, such a stabilization is the lowering of the energy levels of the charged forms with respect of their conjugated uncharged forms.

2.4 Bulk Recombination of Separated Charges and Mass Exchange at the Interface

2.4.1 Effect of the Micellar Potential upon the Bulk Recombination of Charge Carriers

The reactivity of solvated electrons generated by pulse radiolysis was studied with different micellar solutions [138–143]. A study of the solvation of electrons in inverted AOT micelles in iso-octane has shown that the electrons are rapidly solvated if the molar ratio H_2O/AOT is 60, and are not solvated at all if $H_2O/AOT \leq 5$ [138]. Electron solvation in normal and inverted micelles was also studied in femto- and picosecond ranges [139]. The solvated electrons react with halogenenaphthalenes in CTAB micelles [140], with the vesicle-forming surfactant in dioctadecyldimethylaminonium bromide vesicles [141], with naphthalene in SDS micelles [142], with

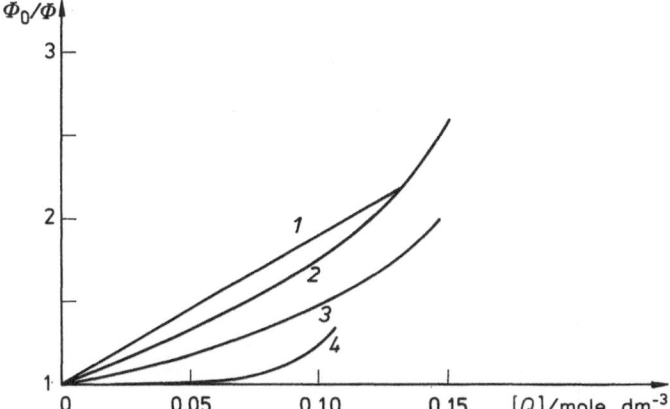

Fig. 7. Quenching of acridinium cation fluorescence in SDS micelles by iodide anion (2), by iodide anion at stabilized ionic strength (1), by bromide (3) and thiosulfate (4) anions [8]

pyrene and its derivatives [143, 144], with tris-ruthenium bipyridile [145]. The solvated electrons are concentrated by cationic micelles, the anionic micelles inhibit reactions of electrons with solubilized probes. The effect of the micellar potential upon the rate constant of the reaction of solvated electrons with solubilized probes was studied by Grätzel et al. [146]. The change of the micellar potential was achieved by the change of the ionic strength of the solution by the addition of a foreign electrolyte. The decrease of the absolute value of the negative micellar potential results in an increase of the rate constant of the reaction of electrons with a probe in accordance with the decrease of the free energy of the electron's penetration into micelles.

The penetration of other charged particles into micelles is also controlled by the micellar potential. The acridinium cation fluorescence quenching in SDS micelles by iodide, bromide and thiosulfate anions does not obey the Stern-Volmer equation, i.e. the plots of Φ_0/Φ vs. the quencher concentration are not linear [8] (Fig. 7). Stabilization of the ionic strength of the solution and thus the stabilization of the micellar potential results in linearization of the same plot. The change of the micellar potential with salt concentration calculated by the authors of the present review from the acridinium cation fluorescence quenching with account taken of the change of the activity coefficient of the quencher anion in aqueous solutions according to the second approximation of Debye-Huckel theory [9] is shown in Fig. 1. The same change determined with special probes [7] shows good agreement with these data. The key role of the microphase potential for charge separation is proven by Calvin et al. [6].

The time evolution of the optical absorption of radical ions formed by pulse laser photolysis of pyrene in the presence of dimethylaniline in different micelles was obtained by Mataga et al. [26] (Fig. 8). In CTAB micelles the recombination of the radical ions is not observed in the nanosecond time range, in neutral micelles it occurs somewhat faster than in acetonitrile solution. In SDS micelles, there is an initial period of a fast recombination, then the optical absorbance (and, hence, the radicals concentration) reach a constant value. Mataga et al. draw no conclusion about the product transfer across the micelle surface, but concluded that the micellar potential

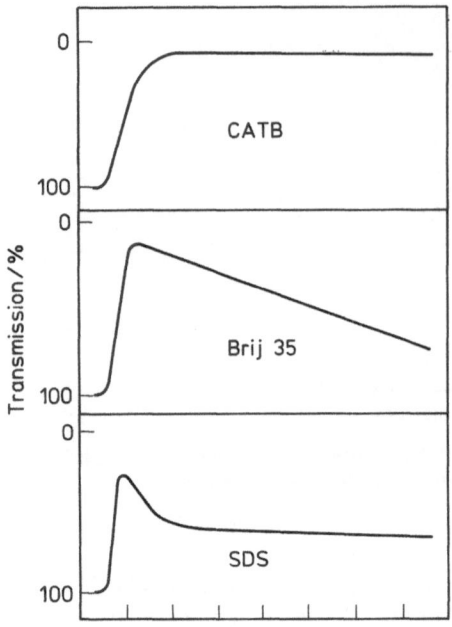

Fig. 8. The time evolution of light absorption by radical ions formed by a laser pulse due to electron transfer from N,N-dimethylaniline to excited pyrene in different micelles [26]. One division on absciss axis is equal to 100 ns

influences the radical behaviour. They assumed that dimethylaniline radical cations moved to more polar regions due to the identical charge with the CTAB micelles.

The same reaction was studied by Atik and Thomas [147]. In CTAB micelles a pyrene radical anion lives more than 1 ms and reacts with dimethylviologene, Eu^{+3}, O_2, CO_2. Thomas et al. [123] concluded that the dimethylaniline radical cation leaves the micelle within its lifetime. In the case of anionic micelles, the radical cation is trapped and fast recombination takes place. The kinetic scheme is as follows:

$$A + D \xrightarrow{h\nu} A^* + D \longrightarrow A^* \dots D \begin{array}{c} \nearrow (AD)^* \\ \searrow (A^- \dots D^+) \end{array} \rightarrow A^- + D^{+\cdot} \qquad (78)$$

The charge separation at the interface does not necessarily lead to any strong inhibition of recombination. For example, the protonation of excited anions of hydroxyaromatics in CTAB micelles is only 10–50 times slower than in aqueous solution. This is probably due to the fact that the anions exist in the form of ion pairs with the end groups of the micelles and thus are not very far from the protons in the bulk phase [66].

Initial charge separation was also obtained for vesicle (liposomal) membranes [148–150]. In some cases, the total charge separation process may be a two-quantum one [151–152].

2.4.2 Effect of the Hydrophobic Balance of Reactants upon their Solubilization

A systematic study of hydrophobic interactions upon solubilization of the typical photochemical reactants was made by Whitten et al. [13]. They measured the change of the binding constants with temperature for Cu^{++} and dimethylviologene cation and found that the most important increment in the solubilization free energy is the solubilization enthalpy for the copper ion, while it is the entropy factor for dimethylviologene; this fact is discussed by Whitten et al. in terms of hydrophobic interactions.

2.4.3 Molecule Exchange at the Interface

After generation of a pair of oppositely charged carriers within a micelle or another organized assembly, the recombination and transport of one of the carriers across the interface compete. While factors which control the recombination are well known [153–154], the kinetics of the photoproducts which exit from the micelles into the bulk phase and vice versa has not yet received a systematic study.

The rate constants of the phosphorescent probe's exit from cetyltrimethyl-ammonium chloride micelles were determined by Bolt and Turro [155]. The logarithm of the rate constant is a linear function of the number of the carbon atoms in the probe molecule. The apparent activation energy of the exit process is 9 kcal/mole.

The exit rate constants of the excited anions after the photoprotolytic dissociation of 1,4-dichloro-2-naphthol within decylsulfate, dedecylsulfate, and cetylsulfate micelles were measured with a fluorescence quencher hardly penetrating the micelles, – the nitrate ion [121]. The addition of nitrate into the solution quenched the fluorescence of those anions which escape from the micelles within the lifetime of the excited state only. The exit rate constant of the naphtholate anion increases with increasing length of the hydrocarbon radical in the micelle-forming surfactant. The exit rate is thus controlled by the lowering of the micelle polarity (i.e. by the free energy of the exit process) rather than by the micelle size or the distance that the anion must diffuse. Perhaps one can establish a kind of correlation between the rate constant of this process and its free energy as was done for photochemical electron transfer [126] and proton transfer [156, 157].

If the penetration rate of molecules into micelles is diffusion-controlled, the exit rate constants and the binding constants of these molecules should obey the following equation:

$$k_{out} = k_d/K \tag{79}$$

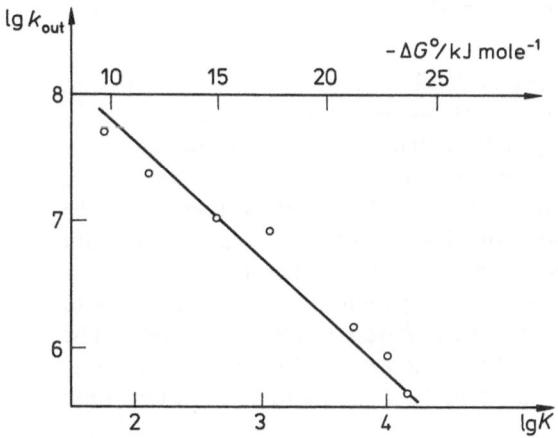

Fig. 9. The dependence of the rate constants of the molecules exit from the micelles on the solubilization Gibbs energy calculated from the data [125, 124]

where k_d is the rate constant of diffusion of the molecule to the micelle. Equation (79) is obtained from the definition of $K = k_{in}/k_{out}$ if one assumes $k_{in} = k_d$. One can derive from Eq. (79) the existence of a linear relationship between the logarithm of the exit rate constant $\log k_{out}$ and the solubilization standard free enthalpy $\Delta G_K^0 = 2.303 RT \log K$, assuming that the diffusion rate constant does not vary very much from system to system. The plot of $\log k_{out}$ vs. ΔG_K^0, calculated from published data, is shown in Fig. 9.

One can see that the linearity is quite sufficient. Thus, the rate constants of the molecules exchanged at the interface of micelle and bulk solution are controlled mainly by the solubilization free energy and the diffusion rate constants of these molecules to micelles without any additional potential barriers.

2.5 Stabilization of the Separated Charges in Systems with Mediators

Most of the work dealing with photochemical charge separation in micellar solutions aim at high quantum yields of the photoreactions and also at the long-term stabilization of initial photoproducts. Let us consider the contradictions of these two aims.

In order to achieve a high quantum yield, one must place the reactants into the same phase, either the micellar or the bulk. The micellar phase is preferable due to the shorter time for diffusion to the interface. One of the initial photoproducts should cross the interface and leave the micelle within its lifetime. Besides that, the process of the reverse transfer of this photoproduct should be slow enough. For systems of practical interest, the charge separation should occur in the nanosecond range and the recombination should be inhibited for the range of hours at least. So the binding constant of the product should be 10^{12}–10^{15} times less than that of the parent reactants.

The product binding constant depends upon one of the parent reactants. The reactants are to be well solubilized (i.e. ΔG_K^0 must be quite negative). If the charge separation process is an electron transfer from one reactant to another, a correlation can be established between the solubilization free energies of the product and the parent reactant. The reaction changes the Couloumb energy of interaction of the molecule with a micelle and the solvation free energies in the micellar and bulk phases. The full solubilization free energy of the reaction product can be written as follows:

$$\Delta G_{Kpr}^0 = \Delta G_{Kre}^0 - \mathscr{F}\Delta\varphi\left(\frac{1}{\varepsilon_m r} - \frac{1}{\varepsilon_{aq} r}\right)[z^2 - (z-1)^2] \tag{80}$$

where ΔG_{Kre} is the free energy of the reactant solubilization before it gets an extra electron, z is its initial charge, r its radius, $\Delta\varphi$ the potential drop at the interface of micelle and bulk solution (the pseudophase potential) ε_m and ε_{aq} are the dielectric constants of the micellar and bulk phases. We omit the change of the solvation free energy caused by the change of the dipole moment and higher order moments of the reactant molecule in the course of the reaction. The qualitive conclusion would not change although the solubilization free energy changes in the course of the reaction by a limited value. Let us estimate this value. The micellar potential for usual micelles does not exceed ± 150 mV [7]. Thus, the absolute value of $\mathscr{F}\Delta\varphi$ is less than 15 kJ/mole. The estimation of the change of the non-Coulombic component of the solubilization free energy in the course of the reaction is given in Sect. 2.1.2 and is not more than

4 kJ/mole. Thus, the total change of the solubilization free energy is less than 19 kJ/mole which corresponds to a change of the binding constant not more than in 10^4 times, even if the direction of the electron transfer is chosen in the most favourable way, so that the solvation and the electrostatic parts of the solubilization free energy change in the same direction. It is a considerable change of the binding constant and one could reach even better changes using less polar micelles or microemulsions and charged reactants, but it is not enough for any practical use.

A radical approach to such a problem is provided by the use of shuttle-type mediators or sacrificial systems.

The reaction of excited N-methylphenothiazine with Cu^{+2} cations was studied by Grätzel et al. [158–161, 18]. Initially electrons transfer from N-methylphenothiazine to the copper cation (+2). The copper ion (+1) thus produced is weakly bound to the micelle and transfers into the bulk phase, which contains ferricyanide ion. The latter oxidates the copper (+1) ion to the (+2) state. The highly negative charge of thus formed ferrocyanide ions inhibits their penetration into the micelles. Separated charges in this system are stabilized for several days. Later, Grätzel et al. described a system where a radical cation of N,N-dimethyl-5,11-dihydroindolo [3, 2–b] carbazole is formed. The shuttle-type mediator in this case is also the copper ion (+2).

In sacrificial systems the oxidized form of D (or the reduced form of A) produced by charge photoseparation reacts with an irreversible consumed reductant (oxidant), e.g. EDTA or triethanolamine [6, 160]. Such systems can be of some practical interest, because in principle they enable fuel production from diluted solutions of industrial and other waste.

3 Conclusion

The step of charge separation at the interface plays a key role in the use of microheterogeneous systems to produce some extraequilibrium concentrations of oxidized or reduced forms (in the case of photochemical electron transfer) or hydrogen ions (in the case of proton transfer). Consider the energetic scheme of the interfacial charge transfer for the case of electron phototransfer from donor D to acceptor A (Fig. 10).

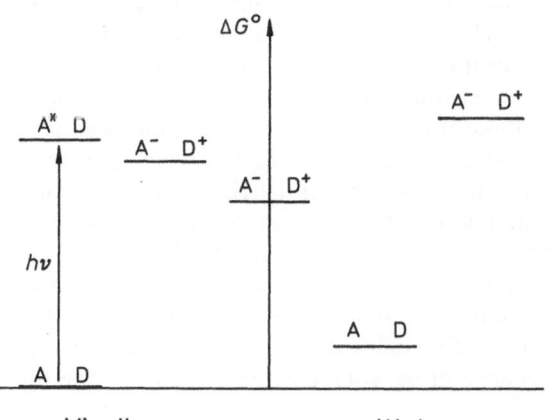

Fig. 10. The energy scheme of the interfacial charge transfer. Left side is micellar phase, right side is aqueous phase

The photochemical charge separation in micellar solutions requires, on the one hand, that the distribution coefficient of one of the product radical-ions be less than unity (ΔG_K^0 is positive). The rate of separation of charge carriers competing with their geminate recombination must depend upon the location of initially generated carriers within the micelles. If they are located in the hydrophobic nuclei of micelles, one of the carriers should diffuse to the surface. If they are located at the surface, there is no need for such diffusion. As data from Sect. 2.1.2 show, the viscosity of the micelle nuclei is about one order of magnitude higher than that of water. The time of diffusion from the micelle centre to its surface can be estimated as 10^{-8} s. As it is shown in Sect. 2.4.3, there is no specific activation barrier for the diffusion of molecules across the interface of micelle and bulk solution. So the diffusion rate constant of molecules into the bulk phase are not less than 10^9 s^{-1} for $\varrho < 1$ and decrease proportionally to $1/\varrho$ for $\varrho > 1$. These are the processes which compete with the geminate recombination, the rate of which is controlled by the spin state, the energetic and steric properties of the ion radical pairs in question.

Stabilization of separated charges can be achieved by a considerable decrease of the charge carrier recombination rate constant. This is the achievement of a rather high activation barrier upon a rather small decrease of the free energy of the separated charges which provides the energetic efficiency of the process. The stored free energy at standard conditions can be written in the following way:

$$\Delta G^0 = \mathscr{F} \Delta E^0 + RT \ln(\varrho/\varrho'). \tag{81}$$

Here, ΔE^0 is the difference of the redox potentials of donor D and acceptor A, $\mathscr{F} = 96500$ C/mol ϱ' is the distribution coefficient of the radical ion which remains in the micellar phase, and ϱ is the product of the distribution coefficients of the parent reactants A and D. If recombination can occur in the bulk phase or within the micellar phase but not at the interface, the value of recombination rate constant depends upon the minimal energy needed to bring one of the radical ions from the bulk to the micellar phase, i.e. $k = 10^{11} \varrho'$, or from the micellar to the bulk phase with $10^{10}/\varrho''$ dm^3/mole s (the highest of these values). So, the stabilization of the radical ions needs ϱ' to be small enough (less than 10^{-5}) and ϱ'' to be high enough (more than 10^5).

The excitation energy should obviously exceed the value $\mathscr{F} \Delta E^0 + RT \ln(\varrho/\varrho')$. If charge generation occurs in the nuclei of the micelles and not at their surfaces, the excitation energy should exceed $\mathscr{F} \Delta E^0 + RT \ln(\varrho/\varrho'\varrho'')$. One can expect a correlation between the photochemical charge transfer rate constant and the free energy of this charge transfer for micellar solutions as well as in homogeneous solution [124, 125, 127]. If the order of the intramicellar charge transfer reaction rate constant k_R is about 10^9 dm^3/mole s, the effective charge carrier generation can be provided by the reactant concentration in the micellar phase of about 0.1 mole/dm^3 or several molecules per micelle. If the charge generation occurs at the interface, the higher polarity of this region can decrease the necessary energy and the activation barrier of the reaction.

Thus, for the effective charge separation in micellar systems, the most convenient location of the reactants is at the interface, and after the reaction, one of the products should exit into the bulk phase. This product should have a very small solubilization coefficient. The other charge carrier produced should move towards the micelle nucleus. If one aims to obtain very long relaxation times, it is possible to overcome the difficulties of obtaining a very strong change of the solubilization coefficient due to charge transfer reactions by the use of shuttle-type mediators (Sect. 2.5). The proton transfer reactions can be considered in the same way as electron transfer, but the efficiency of the energy storage is lower due to a loss of energy in the radiative or radationless deactivation of the reaction product.

References

1. Tuzuke, S., Kitamura, N.: Pure & Appl. Chem. *56*, 1269 (1984)
2. Zamaraev, K.I., Parmon, V.N.: Usp. Khimii (USSR) *52*, 1403 (1983)
3. Vladimirov, Yu.A., Dobretzov, G.E.: Fluorescent Probes for the Study of Biological Membranes, Moscow, Nauka Publishers 1980
4. Lakowicz, J.R.: Principles of Fluorescence Spectroscopy N.Y., Plenum Press 1983
5. Snegov, M.I.: J. Phys. Chem. (USSR) *58*, 2012 (1984)
6. Calvin, M., Willner, I., Laane, C., Otvos, J.M.: Journ. Photochem. *17*, 195 (1981)
7. Fernandez, M.S., Fromherz, P.: J. Phys. Chem. *81*, 1755 (1977)
8. Zaitsev, A.K., Pavlov, A.A., Zaitsev, N.K., Kuzmin, M.G.: Chem. Physics (USSR) *4*, 182 (1985)
9. Entelys, S.G., Tiger, R.P.: The Reaction Kinetics in Liquid Phase. Moscow, Chimia Publishers, p. 61, 1973
10. Bunton, C.A., Frankson, J., Romsted, L.S.: J. Phys. Chem. *84*, 2607 (1980)
11. Schmel, R.H., Whitten, D.G.: ibid. *85*, 3473 (1981)
12. Moroi, J.: ibid. *84*, 2186 (1980)
13. Bonilha, J.B.S., Foreman, T.K., Whitten, D.G.: J.Am.Chem.Soc. *104*, 4215 (1982); Foreman, T.K., Sobol, W.M., Whitten, D.G.: ibid. *103*, 5333 (1981)
14. Burrows, H.D., Formosinho, S.J., Paira, M.F., Rasburn, E.Y.: J.Chem.Soc.Far.II *76*, 685 (1980)
15. Kuzmin, M.G., Pavlov, A.A.: Journ. Appl. Spectr. (USSR) *82*, 891 (1980)
16. Kuzmin, M.G., Pavlov, A.A., Zaitsev, A.K.: Dokl. AN SSSR *257*, 929 (1981)
17. Yekta, A., Aikawa, M., Turro, N.J.: Chem.Phys.Lett. *63*, 543 (1979)
18. Turro, N.J., Grätzel, M., Brown, A.M.: Angew.Chem. Int.Ed.Engl. *19*, 675 (1980)
19. Thomas, J.K.: Chem.Reviews *80*, 283 (1980)
20. Humphry-Baker, R., Brown, A.M.: Grätzel, M.: Helv.Chim. Acta *67*, 2036 (1981)
21. Shinkai, S., Matsuo, K., Harada, A., Hanabe, O.: J.Chem.Soc.Perkin Trans. II, 1261 (1982)
22. Turro, N.J., Cherry, W.R.: J.Am.Chem.Soc. *100*, 7431 (1978)
23. Tarasov, V.F., Buchachenko, A.L.: Isv.AN SSSR (chem. series), 86 (1983)
24. Turro, N.J., Kracutler, B.: Acc.Chem.Res. [13], 369 (1980)
25. Waka, J., Hamamoto, K., Mataga, N.: Chem.Phys.Lett. *53*, 242 (1978)
26. Waka, J., Hamamoto, K., Mataga, N.: ibid. *63*, 364 (1979)
27. Rodgers, M.A.J., Bazendale, J.H.: ibid. *81*, 347 (1981)
28. Koglin, P.K.F., Miller, D.J., Steinwandel, J., Hauser, M.: J.Phys.Chem. *85* (16), 2363 (1981)
29. Lianos, P., Zana, R.: ibid. *84* (25), 3339 (1980)
30. Infelta, P.P.: Chem.Phys.Lett. *61* (1), 88 (1979)
31. Kratohvil, J.P.: J.Colloid and Interface Sci. *75* (1), 271 (1980)
32. Miller, D.J.: Ber.Bunsenges. P.C. *85*, 337 (1981)
33. Miller, D.J., Klein, U.K.A., Hauser, M.: Ber.Bunsenges.Phys.Chem. *84*, 1135 (1980)
34. Lianos, P., Lang, J., Zana, R.: J.Colloid and Interface Sci. *91* (1), 276 (1983)
35. Lianos, P., Lang, J., Zana, R.: J.Phys.Chem. *86*, 4809 (1982)
36. Turro, N.J., Lee, P.C.C.: ibid. *86* (17), 3367 (1982)
37. Turro, N.J., Yekta, A.: J.Am.Chem.Soc. *100* (18), 5951 (1978)
38. Almgren, M., Swarup, S.: J.Phys.Chem. *87* (5), 876 (1983)
39. Almgren, M., Löfroth, J.-E.: J.Colloid and Interface Sci. *81* (2), 486 (1981)
40. Henglein, A., Scheerer, R.: Ber.Bunsenges. Phys. Chem. *82* (2), 1107 (1978)
41. Russel, J.C., Whitten, D.G.: J.Am.Chem.Soc. *103* (11), 3219 (1981)
42. Masuhara, H., Tanabe, H., Mataga, N.: Chem.Phys.Lett. *63* (2), 273 (1979)

43. Costa, S.M.B., Maçanita, A.L.: J.Phys.Chem. *84* (19), 2408 (1980)
44. Dederen, J.C., Van der Auweraer, M., De Schryver, F.C.: ibid. *85* (9), 1198 (1981)
45. Van der Auweraer, M., Dederen, C., Palmans-Windels, C., De Schryver, F.C.: Am.Chem.Soc. *104* (7), 1800 (1981)
46. Croonen, Y., Celade, E., Van der Zegel, M., Van der Auweraer, M., Vandeudricssche, M. et al.: J.Phys.Chem. *87* (8), 1426 (1983)
47. Celade, E., De Schryver, F.C.: Journ.Photochem. *18*, 223 (1982)
48. Grieser, F.: Chem.Phys.Lett. *83* (1), 59 (1981)
49. Nakamura, T., Kira, A., Imamura, M.: J.Phys.Chem. *88* (16), 3435 (1984)
50. Ziemiecki, H.W., Cherry, W.R.: J.Am.Chem.Soc. *103* (15), 4479 (1981)
51. Ziemiecki, H.W., Holland, R., Cherry, W.R.: Chem.Phys.Lett. *73* (1), 145 (1980)
52. Atik, S.S., Thomas, J.K.: ibid. *79* (2), 351 (1981)
53. Atik, S.S., Thomas, J.K.: J.Am.Chem.Soc. *103* (12), 3543 (1981)
54. Atik, S.S., Thomas, J.K.: J.Phys.Chem. *85* (25), 3921 (1981)
55. Atik, S.S., Singer, L.A.: Chem.Phys.Lett. *59* (3), 519 (1978)
56. Gösele, V., Klein, U.K.A., Hauser, M.: ibid. *68* (2), 291 (1979)
57. Hatlee, M.D., Kozak, J.J., Rottenberger, G., Infelta, P.P., Grätzel, M.: J.Phys.Chem. *84* (12), 1508 (1980)
58. Tachia, M.: Chem.Phys.Lett. *69*, 605 (1980)
59. Van der Auweraer, M., Dederen, J.C., Celadé, E., De Schryver, F.C.: J.Chem.Phys. *74*, 1140 (1981)
60. Vanderkooi, J.M., Fischhoff, S., Andrich, M., Rodo, F., Owen, C.S.: ibid. *63* (8), 36611 (1975)
61. Sano, H., Tachia, M.: J.Chem.Phys. *75* (6), 2870 (1981)
62. Hatlee, M.D., Kozak, J.J., Grätzel, M.: Ber.Bunsenges. Ph.C. *86*, 157 (1982)
63. Pyter, R.A., Raneachandran, C., Mukerjee, P.: J.Phys.Chem. *86* (16), 3206 (1982)
64. Mukerjee, P., Cardinal, J.R.: ibid. *82*, 1620 (1978)
65. Mukerjee, P.: in "Solution Chemistry of Surfactants" (Mittal, K.L., Ed.), Plenum Press: N.Y. *1*, 153, 1979
66. Zaitsev, A.K., Iljichev, Yu.V., Gorelik, O.F., Zaitsev, N.K., Kuzmin, M.G.: Chim. Physika (USSR), *4*, 1384 (1985)
67. Demjashkevitch, A.B., Zaitsev, N.K., Kuzmin, M.G.: ibid. *16*, 60 (1982)
68. Zaitsev, N.K., Demjashkevitch, A.B., Kuzmin, M.G.: Dokl. AN SSSR (USSR) *255*, 622 (1980)
69. Reed, W., Politi, M.J., Fendler, J.H.: J.Am.Chem.Soc. *103* (15), 4591 (1981)
70. Turro, N.J., Tanimoto, Y.: Photochem.Photobiol. *34*, 157 (1981)
71. Lianos, P., Zana, R.: Journ.Coll.Interface Sci. *88* (2), 594 (1982)
72. Blatt, E., Ghiggino, K.P., Sawyer, W.H.: J.Phys.Chem. *86* (22), 4461 (1982)
73. Zinsli, P.E.: ibid. *83* (25), 3223 (1979)
74. Grätzel, M., Thomas, J.K.: J.Am.Chem.Soc. *95*, 6085 (1973)
75. Shinitzky, M., Dianoux, A.C., Gitler, C., Weber, G.: Biochemistry *10*, 2106 (1971)
76. Kubota, Y., Kodama, M., Miura, M., Miura, M.: Bull.Chem.Soc.Jap. *46*, 100 (1973)
77. Heyn, M.P., Blume, A., Rehorek, M., Dencher, N.A.: Biochemistry *20* (25), 7109 (1981)
78. Wolber, P.K., Hudson, B.S.: ibid. *20* (10), 2800 (1981)
79. Humphry-Beker, R., Grätzel, M.: J.Am.Chem.Soc. *102* (2), 847 (1980)
80. Woods, R., Love, L.J.C.: Abstr. pap. presented Pittsburgh Conf. and Expos.Anal.Chem. and Appl.Spectr., Atlantic City, N.Y., March *8–13*, S.L. S.A. p. 711, 1982
81. Viktorova, E.N., Veselova, T.V., Snegov, M.I., Cherkasov, A.S.: Optica i Spectroscopia. *53* (2), 252 (1982)
82. Cherkasov, A.S., Veselova, T.V., Viktorova, E.N. et al.: Izv. AN SSSR (physics series) *46*, 311 (1982)
83. Aikawa, M., Yekta, A., Lin, J.-M., Turro, N.J.: Photochem.Photobiol. *32*, 297 (1980)
84. Pouligny, B., Lalanne, J.R., Conilland, B., Ducasse, A., Sarger, L.: Proc. 2nd Int.Symp. Ultrafast Phenomena Spectroscopy. Reinhardsbrunn Oct. 30–Nov. 5 1980, *2*, Jena S.A., 312–316, 317–318, 1980)
85. Pouligny, B., Lalanne, J.R., Conilland, B., Ducasse, A., Sarger, L.: Opt. Commun. *37* (4), 271 (1980)
86. Emert, J., Behrens, C., Goldenberg, M.: J.Am.Chem.Soc. *101* (3), 770 (1979)

87. Turro, N.J., Aikawa, M., Yekta, A.: ibid. *101* (3), 772 (1979)
88. Georgescauld, D., Desmasez, J.P., Lapouyade, R., Babeau, A., Richard, H., Winnik, M.: Photochem.Photobiol. *31* (6), 539 (1980)
89. Zacharisasse, K.A., Kühnle, W., Weller, A.: Chem.Phys.Lett. *73* (1), 6 (1980)
90. Sunamoto, J., Iwamoto, K., Akutagawa, M., Nagase, M., Kondo, H.: J.Am.Chem.Soc. *104* (28), 4904 (1982)
91. Offen, H.W., Turley, W.D.: J.Phys.Chem. *86* (18), 3501 (1982)
92. Turro, N.J., Okube, T.: ibid. *86* (2), 159 (1982)
93. Almgren, M., Grieser, F., Thomas, J.K.: J.Am.Chem.Soc. *102* (9), 3188 (1980)
94. Kano, K., Goto, H., Ogawa, T.: Chemistry Lett. 653 (1981)
95. Sudhölter, E.J.R., Van de Langkrnis, G.B., Engberts, J.B.F.N.: Recuile Journal of the Netherlands Chem.Soc. *99* (3), 73 (1980)
96. Dochi, K.F., Robert, G.P., Termai, B., Dorrick, P.J.: Aust.J.Chem. *33*, 2199–2206 (1980)
97. Zacharlasse, K.A., Phuc, N.V., Kozanklewicz, B.: J.Phys.Chem. *85* (18), 2676 (1981)
98. Gregoritch, S.J., Thomas, J.K.: ibid. *84* (12), 1491 (1980)
99. Lianos, P.: ibid. *86* (11), 1935 (1982)
100. Bernas, A., Grand, D., Hautecloque, S., Chaurbaudet, A.: ibid. *85* (24), 3684 (1981)
101. Grand, D., Hauteclorue, S., Bernas, A., Petit, A.: ibid. *87* (25), 5236 (1983)
102. Wallace, C.I., Hall, G.E., Kenney-Wallace, G.A.: Chem.Phys. *49* (2), 279 (1980)
103. Thomas, J.K., Piciulo, P.: J.Am.Chem.Soc. *100* (10), 3239 (1978)
104. Thomas, J.K., Piciulo, P.: ibid. *101* (9), 2502 (1979)
105. Hall, G.E.: ibid. *100*, 8263 (1978)
106. Alkaitis, S.A., Beck, G., Grätzel, M.: ibid. *97*, 5723 (1975)
107. Alkaitis, S.A., Grätzel, M.: ibid. *98*, 3549 (1976)
108. Grätzel, M.: in "Micellization and Microemulsions" (Ed. Mittal, K.L.) Plenum Press, N.Y., *2*, 591, 1977
109. Moroi, Y., Infelta, P.P., Grätzel, M.: J.Am.Chem.Soc. *101* (3), 573 (1979)
110. Bernas, A., Grand, D., Hautecloque, S., Myasoedova, T.: ibid. *104* (1), 105 (1984)
111. Chauvet, J.-P., Viovy, R., Santus, R., Land, E.: J.Phys.Chem. *85* (23), 3449 (1981)
112. Chauvet, J.-P., Viovy, R., Land, E.J., Santus, R., Truscott, T.G.: ibid. *87* (4), 592 (1983)
113. Chauvet, J.-P., Viovy, R., Bazin, M., Santus, R., Patterson, L.K.: Chem.Phys.Lett. *86* (2), 135 (1982)
114. Demiashkevitch, A.B., Zaitsev, N.K., Kuzmin, M.G.: ibid. *55* (1), 80 (1978)
115. Demjashkevitch, A.B., Zaitsev, N.K., Kuzmin, M.G.: Dokl AN SSSR *231*, 126 (1976)
116. Klein, U.K.A., Hauser, M.: Zeitschrift für Phys.Chem.N.F. *96* (1–3), 135–146 (1975)
117. Selinger, B.K., Weller, A.: Austral.Journ.Chem. *30*, 2377 (1977)
118. Khuanga, U., McDonald, R., Selinger, B.K.: Zeit.Phys.Chem.N.F. *101* (1–3), 209 (1976)
119. Politi, M.J., Fendler, J.H.: J.Am.Chem.Soc. *106* (2), 265 (1984)
121. Abou Al Einin, S., Zaitsev, A.K., Zaitsev, N.K., Kuzmin, M.G.: Chim. Physika (USSR), *5*, 219 (1986)
122. Abou Al Einin, S., Zaitsev, A.K., Zaitsev, N.K., Kuzmin, M.G.: Chim. Vysok. Energij (USSR), *20*, 521 (1986)
123. Katusun-Razem, B., Wong, M., Thomas, J.K.: J.Am.Chem.Soc. *100* (6), 1679 (1978)
124. Miyashita, T., Murakata, T., Jamaquchi, J., Matsuda, M.: J.Phys.Chem. *89*, 497 (1985)
125. Miyashita, T., Murakata, T., Matsuda, M.: ibid. *87*, 4529 (1983)
126. Rehm, D., Weller, A.: Ber. Bunsenges. Phys.Chem. *73* (6), 834 (1969)
127. Hashimoto, S., Thomas, J.K.: J.Phys.Chem. *89*, 2771 (1985)
128. Soboleva, I.V., Sadovski, N.A., Kuzmin, M.G.: Dokl. AN SSSR (USSR), *233*, 635 (1977)
129. Abdullaev, N.G., Kiselev, A.V., Ovchinnikov, Yu.A. et al.: Biol. Membr. (USSR) *2*, 453 (1985)
130. Zaitsev, A.K., Il'idev, Yu.V., Zaitsev, N.K., Kuzmin, M.G.: Dokl. AN SSSR *283*, 900 (1985)
131. Bales, B.L., Kevan, L.: J.Phys.Chem. *86* (19), 3836 (1982)
132. Narayana, P.A., Li, A.S.W., Kevan, L.: J.Am.Chem.Soc. *103*, 3603 (1981)
133. Beck, S.M., Brus, L.E.: ibid. *103* (5), 1106 (1983)
134. Plonka, A., Kevan, L.: J.Chem.Phys. *80* (10), 5023 (1984)
135. Peuillen, P., Martre, A.-M., Martinet, P.: Electrochim. Acta *27* (7), 853 (1982)
136. McIntire, G.L., Blount, H.N.: J.Am.Chem.Soc. *101* (26), 7720 (1979)
137. Harada, S., Schelly, Z.A.: J.Phys.Chem. *86*, 2098 (1982)

137. Harada, S., Schelly, Z.A.: ibid. *86* (11), 2098 (1982)
138. Pileni, M.P., Hickel, B., Ferradini, C., Pucheault, J.: Chem. Phys. Lett. *92* (3), 308 (1982)
139. Gaudel, Y., Migus, A., Martin, J.L., Antonetti, A.: ibid. *108* (4), 319 (1984)
140. Aikawa, M., Sumiyoshi, T., Miura, N., Katayama, M.: Bull. Chem. Soc. Japan, *55*, 2352 (1982)
141. Almgren, M., Thomas, J.K.: Photochem.Photobiol. *31*, 329 (1980)
142. Evers, E.L., Jayson, G.G., Raff, I.D., Swallow, A.J.: J.Chem. Soc. Faraday I 7, 528 (1980)
143. Frank, A.J.: in "Micellization and Microemulsions"/Ed. by K. L. Mittal. Plenum Press, N.Y., 1977
144. Proske, Th., Fischer, Ch.H., Grätzel, M., Henglein, A.: Ber.Bunsenges. *81* (9), 816 (1977)
145. Meisel, D., Matheson, M.S., Rebani, J.: J.Am.Chem.Soc. *100* (1), 117 (1978)
146. Frank, A.J., Grätzel, M., Janata, M.: Ber.Bunsenges. Phys.Chem. *80* (6), 547 (1976)
147. Atik, S.S., Thomas, J.K.: J.Am.Chem.Soc. *103* (12), 3550 (1981)
148. Grätzel, M.: Energy Resources by Photochemistry and Catalysis, N.Y., Academic Press 1983
149. Lymar, S.V., Tsvetkov, I.M., Parmon, V.N., Zamaraev, K.I.: Khimicheskaja Fizika *1*, 405 (1982)
150. Parmon, V.N., Tsvetkov, I.M., Lymar, S.V., Zamaraev, K.I.: Dokl. AN SSSR *272*, 1164 (1983)
151. Inoue, H., Hida, M.: Bull.Chem.Soc.Jap. *35* (6), 1880 (1982)
152. Parmon, V.N., Lymar, S.V., Tsvetkov, I.M., Zamaraev, K.I.: J.Mol.Cat. *21*, 353 (1981)
153. Ulrich, T., Steiner, U.: Chem.Phys.Lett. *12* (4), 365 (1984)
154. Haar, H., Klein, U.K.A., Hauser, M.: ibid. *58* (4), 525 (1978)
155. Bolt, J.D., Turro, N.Y.: J.Phys.Chem. *85*, 4024 (1981)
156. Zaitsev, N.K., Demjashkevitch, A.B., Kuzmin, M.G.: Dokl.AN SSSR *255*, 622 (1980)
157. Demjashkevitch, A.B., Zaitsev, N.K., Kuzmin, M.G.: Khimia Vysokich Energii *16*, 60 (1980)
158. Grätzel, M.: Isr.Journ.Chem. *18* (3–4), 364 (1979)
159. Moroi, Y., Braun, A.M., Grätzel, M.: J.Am.Chem.Soc. *101*, 567 (1979)
160. Moroi, Y., Infelta, P.P., Grätzel, M.: ibid. *101*, 573 (1979)
161. Frank, A.J., Grätzel, M.: Inorg.Chem. *21* (10), 3834 (1982)
162. Tunuli, M.S., Fendler, J.H.: J.Am.Chem.Soc. *103*, 2507 (1981)

Subject Index